Hugh Cottle
2 Creek Place
NASHUA NH 03063

Systems, Experts, and Computers

Dibner Institute Studies in the History of Science and Technology
Jed Buchwald, general editor, Evelyn Simha, governor

Anthony Grafton and Nancy Siraisi, editors, *Natural Particulars: Nature and the Disciplines in Renaissance Europe*

Frederic L. Holmes and Trevor H. Levere, editors, *Instruments and Experimentation in the History of Chemistry*

Agatha C. Hughes and Thomas P. Hughes, editors, *Systems, Experts, and Computers: The Systems Approach in Management and Engineering, World War II and After*

N. M. Swerdlow, editor, *Ancient Astronomy and Celestial Divination*

Systems, Experts, and Computers

The Systems Approach in Management and Engineering, World War II and After

---

edited by Agatha C. Hughes and Thomas P. Hughes

The MIT Press
Cambridge, Massachusetts
London, England

This book was set in Bembo by Asco Typesetters, Hong Kong, and printed and bound in the United States of America

Library of Congress Cataloging-in-Publication Data

Systems, experts, and computers : the systems approach in management and engineering, World War II and after / edited by Agatha C. Hughes and Thomas P. Hughes.
p. cm. — (Dibner Institute studies in the history of science and technology)
Includes bibliographical references and index.
ISBN 0-262-08285-3 (hardcover)
1. Systems engineering. 2. Expert systems (Computer science) 3. Management information systems. I. Hughes, Agatha C. II. Hughes, Thomas Parke. III. Series.
TA168.S887 2000
620′.001′171—dc21 00-021959

Contents

Introduction

Thomas P. Hughes and Agatha C. Hughes

After World War II, a systems approach to solving complex problems and managing complex systems came into vogue among engineers, scientists, and managers. In 1964, the *Engineering Index* had no entry for "systems engineering" and only two pages for "operations research," both variations upon a systems approach. By 1969, the number had jumped to eight pages of citations for "systems engineering" and to ten for "operations research." Enthusiasm for a systems approach peaked during the early Lyndon Johnson administration (1963–1969), after which the trajectory of advocacy moved downward in step with the reverses of the Vietnam War and the rise of a counterculture. The counterculture associated large systems with the military/industry/university complex and with the Vietnam quagmire. The decline in the popularity of the systems approach also resulted from the frequent failure of its practitioners to cope with complex urban problems involving political and social factors.

Before this downward turn, however, an articulated systems approach unfolded in rich and unprecedented ways during and after World War II. The approach spawned new academic fields, new "sciences of management," and new modes of engineering practice. It effloresced into a number of forms, including operations research, systems engineering, systems analysis, and system dynamics.

Operations research usually referred to a systematic analysis of operating systems, or operations, such as military bombing raids during World War II; *systems engineering* mostly designated the management of the design and development of technological systems, such as the intercontinental ballistic missile system in the United States in the 1950s; *systems analysis* often dealt with the comparison of systems that offered alternative solutions to problems, such as the use of long-range bombers versus intercontinental missiles; and *system dynamics* offered models that could be used in policy making by predicting and comparing the downstream consequences of outcomes of alternative policies.

In general it can be said that the systems approaches had their origins in the military realm in the period 1939–1960. After 1960, proponents of the systems approach increasingly emphasized its possible applications in the civil realm. While physicists, mathematicians, and engineers were its early practitioners, social scientists, including management specialists, started adopting systems techniques after World War II.

Practitioners and proponents embrace a holistic vision. They focus on the interconnections among subsystems and components, taking special note of the interfaces among the various parts. What is significant is that system builders include heterogeneous components, such as mechanical, electrical, and organizational parts, in a single system. Organizational parts might be managerial structures, such as a military command, or political entities, such as a government bureau. Organizational components not only interact with technical ones but often reflect their characteristics. For instance, a management organization for presiding over the development of an intercontinental missile system might be divided into divisions that mirror the parts of the missile being designed.[1]

The Conference

Participants in the Dibner conference on the systems approach explored the common and contrasting characteristics of the approach as it was developed in the military and civil realms by various disciplines and practitioners. They also considered the ways and the places in which the systems approach was applied, with a particular focus on applications to the social problems of urban areas in the 1960s.

The papers and the subsequent commentaries and discussions made clear the increasing dependence of the systems approach upon the digital computer, especially in the 1950s and 1960s. Conference participants stressed the fact that experts based their claims for decision-making authority in matters both military and civil upon their putative mastery of a specifically computer-based systems approach.

Historians presented papers, and engineers, scientists, and managers who played leading roles in the spread of the systems approach acted as commentators or participated in panel discussions. The conference included a roundtable presentation on the history of the International Institute for Applied Systems Analysis. Those participating in the roundtable had helped make the history they presented. Because the historians presented analytical narratives that sometimes involved their commentators, exchanges during discussion proved lively and enlightening.

Summaries of the papers as well as several of the commentaries can suggest the essential character and spirit of the conference.[2] They are organized according to several themes that emerged from the conference.

Origins of the Systems Approach

David A. Mindell, in "Automation's Finest Hour: Radar and System Integration in World War II," finds the roots of a systems approach in communities of engineers and scientists who dealt with problems of gunfire control, which has been a prime source of cutting-edge technology in this century. In World War I, Elmer Sperry, an American inventor of feedback controls, sought means to combine optical target-sensing and target-tracking devices, mechanical analogue computers, and manual operators into gunfire control assemblage for dreadnought battleships. During the interwar years, scientists and engineers at MIT, Bell Telephone Laboratories, and elsewhere used detailed analyses of practice to place control engineering on a rational theoretical basis.

Ivan Getting, a young engineer and scientist with an MIT degree, a Rhodes Scholarship, and membership in the Harvard Society of Fellows, stands out among gunfire control experts of World War II. He was strongly motivated by a holistic vision of systems integration stimulated by his work in controls. He did research and development at the MIT Radiation Laboratory, which had been given responsibility for developing microwave radar during the war. He also worked with Division 7 of the National Defense Research Committee, which presided over the development of new fire-control systems.

Getting advocated viewing fire control as a system rather than as an assemblage of separately designed radar and fire-control sensing and computing devices. Perceiving the human operators as the weak link, or "reverse salient," in gunfire control, he also pushed for increased automation. He embedded his concepts in the design of the Mark 56 Gunfire Control System.

Getting's influence helps explain why engineers and scientists experienced in radar and gunfire control became advocates of the systems approach following World War II. Their advocacy reinforced the efforts of others proposing a systems approach on the level of operations research and systems analysis.

The war provided a variety of opportunities for systems approaches to flourish. Erik P. Rau, in "The Adoption of Operations Research in the United States during World War II," discusses the activities of individuals and organizations nourishing the transfer of the British systems approach, known as operational research, to the United States, where it became known as operations research (OR). This transfer was not without bumps. Vannevar Bush, who mobilized civilian scientists and engineers for the war effort through his Office of Scientific Research and Development (OSRD), did not see operations research as a hard science. Never an advocate of social, or soft, science, he resisted diluting the research and development activities of OSRD divisions with an influx of social scientists. He also sensed that the high-ranking military officers with whom he desired to maintain good relations were ill informed about the successes of British operational research units and distrustful of taking advice from civilians on operational matters. So he tried to situate nascent operations research endeavors in the organizational framework of the nongovernment, civilian National Academy of Sciences.

What Bush failed to comprehend was the extent to which precisely quantified information from operations research personnel concerning the success and nonsuccess of operations involving innovative weapons systems designed by OSRD divisions could assist OSRD engineers and scientists in modifying existing weapons or introducing new ones. Despite his indifference, U.S. operations research groups were established. For example, a U.S. Army Air Force wing stationed in Great Britain used American operations research civilians who had learned British techniques to analyze bombing operations. Plagued by the successes of German U-boats, the U.S. Navy established an effective Antisubmarine Warfare Operation Research Group under the guidance of MIT physicist Philip M. Morse, who later became a major promoter of operations research in the civil sector. Even within Bush's OSRD, there were some advocates of operations research, including John Burchard, an MIT architect and professor who headed an OSRD division studying the effects of artillery and bombs on structures.

After a Bush-organized study group recommended that the OSRD play a major role in fostering operations research, especially if the research group could be staffed by people deeply informed about scientific method and engineering practice, Bush relented. In 1943 he established an operations research division in OSRD that was designated the Office of Field Service (OFS). It was intended to extend OSRD's reach into the front lines, especially in the Pacific theater. Rau follows the fortunes of the Office of Field Services and concludes that it failed to fulfill its mission of extending OSRD's research agenda to the military front. The OFS mistakenly chose problems, for example, that were mathematically intractable or inherently unanswerable because they involved values or computations beyond the state of the art. The reasons for this "utter failure" beg explanation.

Military organizations in the field, including that of Army General Douglas MacArthur, often failed to use—even to cooperate with—the Office of Field Service units. The Army Air Force and the navy preferred their own OR units. Nevertheless, the OSRD venture into operations research plowed ground that would be cultivated during the Cold War by civilian scientists intent upon shaping military strategy.

ORGANIZATIONS AND INDIVIDUALS

Several postwar organizations cultivated the systems approach. The RAND Corporation, founded in 1948 with U.S. Air Force funding, developed systems analysis techniques for evaluating weapons systems; the MITRE Corporation, founded in 1958 with air force funds, concentrated on a systems-engineering approach to designing command and control systems; the Ramo-Wooldridge Corporation, organized in 1953, developed advanced systems-engineering management for large military projects; and the International Institute for Applied Systems Analysis, established in 1966, took a systems approach to the study of civil problems common to advanced industrial societies. RAND, MITRE, and Ramo-Wooldridge also applied the systems approach to civil problems. The Bechtel Corporation and Parsons Brinckerhoff were among leading engineering design and consulting firms deploying systems engineering techniques for managing large construction projects in the 1960s and later.

Not only organizations but also several individuals emerge from the history of the systems approach as influential innovators and practitioners. Philip Morse, as noted, transferred operations research from a military to an industrial context. *Time* magazine cover stories featured Simon Ramo and Dean Wooldridge, engineer/scientists, and Bernard Schriever, an air force general, after they used systems engineering to manage the Atlas Intercontinental Ballistic Missile Project during the 1950s. Charles Hitch and Alain Enthoven moved from the RAND Corporation to Robert McNamara's Pentagon to introduce a systems approach to budgeting and strategic planning, which then spread to other government agencies. Jay Forrester, after having led the development of the Whirlwind computer system for the 1950s SAGE air defense project, moved to the Sloan School at MIT and published *Industrial Dynamics* (1961) and *Urban Dynamics* (1969), both of which stimulated widespread discussion of the application of computer modeling to the implementation of a systems approach for problem solving and decision making.

Systems and Projects

Developments in the 1950s prepared the ground for the spread of the systems approach in the 1960s. During the latter part of the presidential administration of Harry Truman, as well as during that of Dwight Eisenhower, the military funded three major defense projects—the Intercontinental Ballistic Missile Project, the SAGE Air Defense Project, and the Polaris Intermediate Range Missile Project—that provided rich learning experiences for thousands of engineers and scientists. The Polaris Project, managed by the U.S. Navy Special Projects Office, introduced Program Evaluation Review Technique (PERT), a computer-scheduling program later widely used in conjunction with a systems approach to project management. Academic scientists and engineers analyzed, codified, and rationalized information garnered from these projects in order to develop system sciences, which they disseminated through teaching and publications.

The U.S. Air Force and the Office of the Secretary of Defense (OSD) in the 1960s also rationalized and applied systems-management techniques developed at RAND and during Atlas and other large-scale military-funded projects. Stephen B. Johnson's paper, "From Concurrency to Phased Planning: An Episode in the History of Systems Management," explores the efforts of the Defense Department to refine and revise the systems approach by shifting from a concurrency approach to phased planning. This shift followed Defense Secretary Robert McNamara's decision to emphasize cost reduction rather than rapid deployment of weapons systems as the Cold War attenuated.

Schriever and Ramo had introduced concurrency during the Atlas Project. Straining to deploy the missiles as quickly as possible, they called for simultaneous, or concurrent, development of the missiles, the facilities to manufacture and test them, and the command and communication system for their operational control. As Johnson shows, concurrency often increased costs because design changes in a system component being developed often required costly changes in others developed concurrently and already being manufactured.

The phased-planning approach introduced by the Defense Department required a preliminary design phase that would produce enough details about a proposed weapons system to allow cost estimates to be projected for its development. This would enable the Defense Department to review the design and the cost projection before authorizing a project. Schriever, who had become head of the Air Force Systems Command, strongly objected to this sequential approach because it would lead to time delays and because with such an "overly conservative approach . . . the timid will replace the bold and we will not be able to provide the advance weapons the future of the nation demands" (p. 106). Others in the air force feared "creeping centralization" under McNamara's civilians. Nonetheless, phased planning became a cornerstone of the military research and development system.

Glenn Bugos, in "System Reshapes the Corporation: Joint Ventures in the Bay Area Rapid Transit System, 1962–1972," also explores the management of large projects. Stressing that large construction projects require the use of a systems approach to coordinate and control the activities of numerous semiautonomous contractors, he points out that firms responsible for such a multifaceted task have often resorted to joint ventures in order to pool the needed organizational and technical resources. Joint ventures also spread the risks of the heavy financial liabilities for nonperformance that companies must carry in guaranteeing and fulfilling performance specifications.

Bugos uses the history of the Bay Area Rapid Transit System (BART) to exemplify the joint-venture approach to system building. BART is a seventy-five-mile-long rapid transit system that rings the lowlands around San Francisco Bay. It includes a Bay tunnel connecting Oakland and downtown San Francisco. Begun in 1966, the system was completed in 1975 at a cost of $1.6 billion.

BART serves as an instructive example of joint venturing. Not only did a managerial joint venture (JVCO) parented by Parsons Brinckerhoff Hall & McDonald, an engineering firm specializing in design, and by the Bechtel Group, specializing in managing construction, become BART's project manager, but also a number of the project's construction contractors formed joint ventures to pool resources.

Bugos concludes that the BART Project provides an outstanding example of the need in an increasingly complex human-built world for organizational forms such as joint venturing. They provide vast networks, or systems, of diverse organizations able to cooperate for a specified period to carry out a defined project. He calls this a "postmodern" turn in management, one comparable to the spread of scientific management early in the twentieth century.

EXPERTS AND EXPERTISE

The systems approach spread in the United States in the 1950s and early 1960s concurrently with the increasing deference of the public, the military, and the civil government to experts and expertise, especially to engineers, scientists, and managers who practiced systems sciences. Even though a few prominent engineers insisted that systems engineering as a commonsense approach had a long history extending back at least to those who organized the building of the Egyptian pyramids, others involved in the application of operations research, systems analysis, and system dynamics often associated their approach with a new variation on the scientific method. This bound the systems approach ever more tightly to expertise.

Imitating the physicists who had garnered enormous prestige as designers and developers of atomic and other weapons, the expert practitioners of system sciences touted their reliance on theory, method, and mathematics. People making these claims are sometimes described as having physics envy. Like so many physicists, they looked rather askance at the soft social sciences, making an exception only for economics as an auxiliary tool for problem solving. As a result, the systems approach as taken by many practitioners became less holistic and more specialized.

Experts rose to positions of influence not only in the United States but also in France. Gabrielle Hecht, in "Planning a Technological Nation: Systems Thinking and the Politics of National Identity in Postwar France," writes about French engineers, economists, and professional administrators during the French Fourth Republic (1946–1958) and the early years of Charles de Gaulle's Fifth Republic (1958–1969). Graduates of the prestigious *grandes écoles,* they cultivated an elitist ideology of public service. They believed that they were the ones who should preside over national reconstruction and modernization.

The experts argued that France needed to cultivate technology and applied science if it hoped to establish itself once again, after the humiliating defeat of World War II, as a great power from which French political influence, cultural traditions, and style could radiate. Many public intellectuals agreed that the country needed to rely on engineering, managerial, and scientific experts, though they disagreed on the extent of such reliance. Should the experts make high-level policy by crossing the boundary presumed to exist between technology and politics, or should they only offer alternatives to the political decision makers who would take into account factors other than the rational, quantifiable ones with which the experts dealt.

Humanistic and social science intellectuals who opposed a policy-making role for the experts pejoratively labeled them "technocrats." Technocrat to these critics connoted an elite who mistakenly and destructively believed that social problems, as well as technical ones, could be solved by a technically informed, reductionist systems approach based on simplifying assumptions and the discounting of human values. Those engineers, managers, and scientists who believed that the modernization of France and restoration of French prestige rested in their hands countered that well-trained and responsible experts took an all-embracing systematic approach (*vue d'ensemble*) to solve national problems compounded of technical, political, economic, and social factors. Further, they contended that because of their scientific training, their policy recommendations would be value free.

The experts demonstrated their systems approach through the work of a Planning Commission, a state agency founded after World War II to develop multiyear plans to reconstruct and modernize the nation. Hecht characterizes these plans as producing

> a set of models that described the national economy as a heterogeneous system composed of technological and economic artifacts, people, and social relationships, and driven by the interactions of these various components. (p. 151)

She suggests that the technological experts "sought to erode" the boundary between technology and politics, and that "systems thinking provided an ideal means for describing and legitimating this goal" (p. 154).

Hecht describes negative French reactions to the extreme claims of the experts; similar criticism arose in the United States, even within the ranks of systems-approach practitioners. Russell Ackoff, previously a leading academic theorist and consultant in operations research, published critical articles in the 1970s with titles like "The Future of Operations Research Is Past" and "Resurrecting the Future of Operational Research." He asserted that American operations research "is dead even though it has yet to be buried."[3] Academics without field experience, with only "textbook" acquaintance, he argued, taught most operations research courses, and, as a result, their students had little contact with "real world" problems.[4] Because operations research had become mathematically sophisticated but contextually naïve and value free, managers pushed OR practitioners down into the bowels of their organizations. In Ackoff's opinion, most humans seek value-laden goals. For this reason, advisers to policy makers needed to incorporate in their recommendations a humanistic point of view.[5] Ackoff spoke less of problem solving and more of "managing mess."[6]

Computer Mystique

Experts not only used the systems approach; they also drew on the arcane mystique that surrounded early computers. In the 1960s, manufacturers, especially IBM, introduced mainframe computers into military, academic, and business organizations. Popular imagination envisioned room-sized computers as great brains tended by white-coated specialists. Enthusiasts painted a halcyon picture of computers masterminding society's future. Programmers developed software supposedly able to take over management functions. Graduates of management schools trained in the systems approach presented themselves as masters of computer tools.

The computer program PERT greatly enhanced the mystique of computer-empowered expertise. Booz, Allen & Hamilton, a management firm with operations research experience, took the lead in developing PERT for the U.S. Navy's Special Projects Office (SPO), which had the task of managing the development of the submarine-launched Polaris intermediate-range missile. Prepared over a period of several months in 1958, PERT made a large and lasting impression on contemporaries in engineering and management.[7]

Programmed by PERT, a mainframe computer generated flow charts that portrayed a network of actions and events, thereby improving on the nondynamic bar charts long used for scheduling activities in the construction industry. By mid-1959 the Special Projects Office had applied PERT to much of the Fleet Ballistic Missile Program. Interest in PERT spread through the industrial, military, and management communities. The peak year for the introduction of PERT-like management programs, 1962, became known among aerospace firms as the "Year of Management Systems."[8]

Harvey Sapolsky, a historian of Polaris, finds PERT, as used by the SPO, an effective management tool, but not for the reasons advanced by PERT advocates. Referring to it as "an alchemist combination of whirling computers, brightly colored charts, and fast-talking public relations officers," he sees PERT as persuading potential critics of Polaris management that its experts used an arcane, high-tech tool of near miraculous capabilities.[9]

Donald MacKenzie's essay, "A Worm in the Bud? Computers, Systems, and the Safety-Case Problem," also raises questions about the efficacy of computer software, as well as about the reliability of large computers. Such inefficiency and unreliability create serious problems because computers, he reminds us, are at the heart of "nearly all the large technical systems of the late twentieth century" (p. 161).

The SAGE system designed in the 1950s to provide a defense against bomber attack was one of the first of the large, computer-dependent systems. A few years later IBM installed a complex computerized reservation system for American Airlines. Within a short period computerized systems to manage the financial transactions of large organizations became common. Computer switches for the telephone networks and computer-based information systems followed. The list is long and is still growing.

Even though the computer can be programmed to respond to society's needs, this may, at the same time, cause considerable unease among knowledgeable people who realize that if crucial computerized systems go awry, social catastrophe may ensue. MacKenzie notes that computer-based command and control systems poised ready to launch intercontinental ballistic missiles are a special cause for concern. In 1986, C. A. R. Hoare, a world-renowned British computer scientist, remarked that "nobody trusts a computer; and the lack of faith is amply justified" (p. 163). Many of his fellow experts expressed similar concerns, thereby deflating computer mystique and eroding the prestige of computer experts.

Responding to the risks posed to large systems by computer software failures, experts developed several methods of assessing the likelihood of "bugs" causing a computer crash. Yet no "silver bullets" have been found. "Program testing," one expert acknowledges, "can be a very effective way of showing the presence of bugs, but it is hopelessly inadequate for showing their absence" (p. 175). Program testing brought the development of such abstruse techniques as reliability-growth modeling and random testing. Computer experts trained in mathematics attempted the difficult task of proving software reliability by deductive proof, as well as by inductive testing, but with less than satisfactory results.

Paradoxically, doubts about the safety of computer systems may have kept the number of highly disruptive failures low: "Beliefs that 'bugs' are pervasive ... that computer systems can be dangerous," MacKenzie concludes, "have been one of the factors that have kept the death toll in computer-related accidents modest" (pp. 182–183).

Commenting on the MacKenzie essay at the conference, Michael Mahoney stressed the continuation of the software crisis. Drawing an analogy between computer programs and industrial production systems, he observed that the essential process taking place in a programmed computer is a computation that transforms input data to output data. Like production engineers, software designers conceptualize a program as a system composed of many interacting computational units. Data being processed, however, do not move serially—and comprehensibly—from one processing unit to another, as is the case in a mechanical production process. Because the data move in feedback loops among the processing units, tracing the flow of production is not feasible. Software designers have tried various methods of analysis, including object-oriented programming, to control, localize, and fathom the flow, but the programmed computer shares the problems of the complex processes and organizations it simulates. Both the programmed computer and the human-built world defy control and analysis, Mahoney pointed out, because they push complexity to the limit.

Despite "the worm in the bud," computer systems made their way into government agencies. Atsushi Akera, in "Engineers or Managers? The Systems Analysis of Electronic Data Processing in the Federal Bureaucracy," analyzes the argument among agencies and experts about who should preside over data processing. His account begins in the 1950s, when a mass of data-processing responsibilities threatened to inundate the work force of the Patent Office, the Census Bureau, and the Old Age Bureau (social security records). These agencies, which employed hundreds of clerical workers using punch cards and other mechanical tabulation methods, were likely candidates for computer processing.

By 1950 the National Bureau of Standards, which presided over a number of governmental research and development activities, numbered among its staff engineers and scientists technically informed about computers. Overburdened government agencies turned to these experts for advice about procuring computers and using them for data processing. Mary Elizabeth Stevens of the Bureau of Standards took the lead in providing these advisory services. Her reports for various agencies, the Patent Office among them, revealed that the introduction of computers usually required changes in management structure. This raised a question that Akera explores in detail: should the technically trained (engineers and scientists) or the managerially trained (professional managers) have prime responsibility for the use of the computer?

Akera describes the struggle among committees of experts and organizations, including the Bureau of Standards and the Bureau of the Budget, to claim the contested high ground of expertise. Would-be advisers insisted that not only the data-processing function but also entire management systems needed reorganization to take advantage of the computer. They made general and vague claims for the then-arcane technique of systems analysis, which had been honed at the RAND Corporation, as the means for computer-grounded managerial reform. (David Hounshell, David Jardini, and Roger Levien discuss the RAND approach in their essays.) Akera astutely observes that the claims of the would-be expert advisers fell mostly on deaf ears and barren soil because experienced bureaucrats presiding over the agencies had no intention of surrendering their political power to advising experts.

Besides serving as data-processing centers for large systems, computers can simulate the worlds in which they function. Paul N. Edwards, in "The World in a Machine: Origins and Impacts of Early Computerized Global Systems Models," chronicles and analyzes the history of digital computer-based modeling in the thirty years following World War II. He argues that the development of computer models of worldwide phenomena has created among experts and the informed public a concept of the world as a system of interacting, dynamic forces that includes the physical, such as climate, and the social, such as population.

To develop his thesis, Edwards discusses the histories of climate models and of Jay Forrester's "system dynamics" models of global socio-technical change. In these models, Edwards finds a dynamic interaction between the computer modelers' hunger for data and the menu of data supplied to them. He shows that some of the most influential model conceptualizers, including Forrester, emphasize thinking through and designing a model that embraces numerous nonlinear feedback interactions of broad-ranging phenomena rather than the virtually impossible task of obtaining full and accurate data. As a result, their models are long on arbitrary standardized assumptions.

Edwards also explicates a paradox. In the case of climate models involving parameters and long time-series, such as the energy transfer from sun to earth and the behavior of the ocean as a heat sink, scientists using the global models depend heavily on data generated by interpolation of intermediate values from the scarce observed data. Programmed computers do the interpolation as well as the smoothing out of anomalous data, and so computers running and testing the models depend heavily on data generated by other computers.

Modelers check their representations by empirical data. For example, Forrester tested his model of urban development by checking to see if a computer model supplied with available data about the condition of a city around 1900 generates by iterating sets of nonlinear feedback equations the condition of the city a half century later.

Edwards characterizes the bold model builder:

> Forrester believed that sorting out the structure and dynamics of a system using a computer model was the key to understanding. Data could come later, in part because a systems model could help reveal *which* data might be most important. (p. 239)

Forrester's approach was not alone in being criticized. Critics argued that the RAND Corporation's quantifiable models eliminated too many variables, and they questioned RAND's use of systems analysis dependent upon these models. David A. Hounshell analyzes these reactions in "The Medium Is the Message, or How Context Matters: The RAND Corporation Builds an Economics of Innovation, 1946–1962."

In the 1950s RAND broadened its systems approach by including more economists and social scientists among its 400 or so researchers. Ironically, in view of RAND's former wholehearted commitment to a systems methodology, criticism of the systems-analysis approach began to fester among a small group of RAND economists, Armen A. Alchian and Burton Klein prominent among them. They decided that RAND's studies for the air force, which were designed by the systems analysis experts, depended on too many assumptions about the future conditions under which projected weapons systems would be deployed.

The air force studies in question also specified the ultimate characteristics of weapons systems before extensive research and development had been undertaken. Making such predictive assumptions led RAND's systems analysts and the air force to decide prematurely which characteristic of envisioned weapons systems should be optimized. In short, critics believed that an overconfident RAND and air force tended to rely upon the development and deployment of prematurely and rigidly specified weapons systems for an infinitely flexible and unpredictable future.

Alchian, Klein, Kenneth Arrow (a future Nobel laureate from Stanford who served RAND as a consultant), and several other economist critics argued that RAND studies were erroneously recommending procurement decisions before adequate research and development had been carried out for the weapons systems under consideration. The air force might well depend on systems analysis to choose among available weapons systems, but it should not depend on systems analysis to choose among systems that had not passed through a research and experimental development phase.

The essence of research, Klein and others asserted, is asking a variety of questions about as-yet-undiscovered territory, then taking a number of different routes to explore and map the territory. Without the map, decision makers could not accurately decide which research sites on the map should be mined.

Reinforced by research and development case studies done by Richard Nelson, a young RAND economist, critics of the way in which systems analysis was being used wrote papers advocating funding many small exploratory research projects instead of just a few linked closely with future procurement and deployment. The Sputnik crisis in 1957 reinforced their criticisms because a panic-prone America decided that faulty military research and development had led to a missile gap.

The RAND economists published their argument in various quarters. They stressed that private industry would not do the exploratory research and experimental development needed because the fruits of such basic, or pure, activity are not easily appropriated as profits on investment, the patent system notwithstanding. It followed, therefore, that the government, including the military, should subsidize diverse research projects in universities and nonprofit research centers, projects that would offer both the military and private industry a multitude of options for mission-oriented development and deployment.

FROM SWORDS TO PLOUGHSHARES

In the 1960s the systems approach spread from RAND and military-funded projects into the government and industrial sectors. Systems enthusiasts, especially management experts and social scientists, successfully promoted the introduction of the systems approach into the administrative hierarchies of the Defense Department and the civil government. The Great Society programs of the Johnson administration adopted the systems approach.[10] Such transfer from the military to the civil realm has been common in the United States. The early nineteenth-century "American system of production" and early twentieth-century "scientific management" are outstanding examples.[11]

Gullible enthusiasts as well as experts asserted in the 1960s that the systems approach would make possible the creation and control of technological and social systems in the vast and complex human world.[12] "Man must learn to deal with complexity, organized and disorganized, in some rational way," argued one pair of advocates.[13] So emboldened, the experts and their disciples confidently, even rashly, tackled a multitude of the complex technological and social problems plaguing society in the 1960s.[14]

Among these problems none seemed more pressing than the social and physical dilemmas of the large cities. Referring to the space program, which depended heavily on the systems approach, Vice President Hubert Humphrey eloquently captured a widespread faith when he said in 1968:

> The techniques that are going to put a man on the Moon are going to be exactly the techniques that we are going to need to clean up our cities: the management techniques that are involved, the coordination of government and business, of scientist and engineer . . . the systems analysis that we have used in our space and aeronautics program—this is the approach that the modern city of America is going to need if it's going to become a livable social institution. So maybe we've been pioneering in space only to save ourselves on Earth. As a matter of fact, maybe the nation that puts a man on the Moon is the nation that will put man on his feet first right here on Earth.[15]

The RAND Corporation and TRW Inc., a successor to the Ramo-Wooldridge Corporation, took leading roles in the transfer of systems techniques. David R. Jardini, in "Out of the Blue Yonder: The Transfer of Systems Thinking from the Pentagon to the Great Society, 1961–1965," follows RAND's move to civil projects. Davis Dyer, in "The Limits of Technology Transfer: Civil Systems at TRW, 1965–1975," traces the fortunes and misfortunes of TRW as it makes a similar move.

Jardini finds that by the 1960s the relationship between RAND and the air force had begun to turn sour. General Curtis LeMay, a blunt and domineering air force chief of staff, made clear that he expected RAND to do "narrowly purposive" studies and to follow the "Air Force party-line." This directive undermined the substantial independence and objectivity that RAND researchers highly valued. RAND responded by accepting contracts from other Defense Department agencies, NASA, and the Atomic Energy Commission, a step that further widened the rift.

Threatened by air force control and colonization, some top-level researchers at RAND advocated turning to the civil sector for contracts. This recommendation caused considerable stress among RAND researchers, RAND management, the RAND Research Council, and the RAND Board of Trustees. Reactions clustered in three categories: air force loyalists; those favoring a closer alliance with Robert McNamara's civilians at the Office of the Secretary of Defense; and others wishing to take contracts from civil government agencies whose budgets had been swollen by Johnson's Great Society Programs on education, poverty, and the plight of the city.

After Franklin R. Collbohm, an air force loyalist, was replaced in 1965 as RAND head, the organization took a number of civil contracts, including ones from the Office of Economic Opportunity, the Department of Transportation, the National Institutes of Health, the Department of Housing and Urban Development, the Ford Foundation, and the Russell Sage Foundation. In addition, it signed four contracts with New York City to do studies of the police, fire, health, and housing departments. RAND thus moved into a world far messier than that of strategic studies. Newly focused, RAND remained a leading nonprofit research center of its kind, but there were a growing number of competitors that challenged its preeminence. Nevertheless, RAND played a path-breaking role in systems sciences and nurtured a host of eminent scholars who used and spread its approach.

Dyer describes the efforts of TRW in the 1960s to transfer into the civil realm management skills honed by its predecessor, Ramo-Wooldridge, during the Atlas Project. TRW believed its expertise enabled it to solve problems associated with managing large and complex bodies of information, coordinating urban traffic flows, improving the physical environment, providing mass housing, deploying health systems, and raising the efficiency of energy generation and usage. Dyer concludes, however, that despite TRW's optimistic effort most of its civil ventures "lost money or eked out meager returns during brief lifespans" (p. 360). He also delineates the troubles that TRW, especially its Systems Group, encountered in beating swords into ploughshares. These difficulties, however, did provide positive learning experiences.

Simon Ramo, a founder of Ramo-Wooldridge and a developer and articulator of the systems approach, as TRW executive vice president spurred on the firm's technology and management transfer efforts. He became, according to Dyer, "a kind of self-monitored radar to anticipate coming perils or opportunities" associated with "the imbalance between accelerating technology and [the nation's] lagging social maturity" (p. 363). Accelerating technological change brought the computer, space exploration, and the breaking of the genetic code; lagging social maturity resulted in crowded cities, a polluted environment, and depletion of energy sources. The systems approach, Ramo contended, offers a happy combination of solutions for society and profits for corporate enterprise.

California state and local governments awarded contracts to TRW to plan regional development, organize waste management, rationalize transportation systems, and process information about crime. The federal War on Poverty Program brought contracts for the company to train government personnel in systems management techniques. Small by defense contract standards, these civil contracts nevertheless stimulated a bold civil-systems initiative. With defense and NASA contracts on the wane in the 1970s, TRW formed joint ventures with other corporations to do community planning, housing, pollution control, data processing, and financial-investment analysis.

But TRW found civil governments far more difficult to deal with than the Office of the Secretary of Defense. With numerous initiatives showing losses, the company in the late 1970s began to liquidate its ventures in civil systems.

Dyer provides several overarching explanations for TRW's frustrations and disappointing performances in transferring the systems approach. The approach proved neither nimble nor flexible enough to deal with small projects, especially those involving disputatious and jurisdiction-guarding local governments. Entrenched bureaucracy resisted the deskilling impact of the systems approach. And some TRW managers and engineers experienced in aerospace projects just found the civil realm too messy. Recently the company has drawn upon its systems-approach experience to move once again into the civil sector with its "enormous, complex problems where there's a highly litigious and emotional set of characters who are influencing the particular problem" (p. 380). These characters inhabit a realm commonly called "politics."

OTHER NATIONS

Even though the conference concentrated on the spread of the systems approach in the United States, the approach transferred to other countries, too. The systems approach took root in Japan immediately after World War II when the U.S. occupational authority encouraged the transfer of American engineering and managerial techniques. In France, the Commissariat à l'Énergie Atomique and the Électricité de France encouraged the deployment of a systems approach to preside over the construction of nuclear power plants. The influence of the systems approach can also be traced in Germany and Sweden, especially in the design and construction of large-scale projects.

Arne Kaijser and Joar Tiberg chronicle one case of transfer in "From Operations Research to Futures Studies: The Establishment, Diffusion, and Transformation of the Systems Approach in Sweden, 1945–1980." Immediately after World War II the Swedish military cultivated operations research, depending on physicists and mathematicians, especially those in academia, to serve as consultants. In time, the military established small systems-research units, placing emphasis on systems analysis as a means of weighing the merits of prospective weapons systems. They also decided, with some guidance from visiting RAND researchers, that the quantitative emphasis of the physicists and mathematicians needed broadening to include a social science approach to messy problems—as had been done at RAND. A core of systems people, most of them civilians, located at the Sweden's National Defense Research Institute, acquired so much influence with policy makers that Kaijser and Tiberg call them the "spider" in Sweden's defense network.

As in the United States, operations research and systems analysis spread from the military to the civil realm. The systems approach took root because of a scientific management tradition in Swedish industry. Engineers and management experts used operations research techniques to rationalize existing production processes and the organizational structure of industrial corporations. The Royal Institute of Engineers (IVA) set up a section for operations research, which promoted the spread of the techniques throughout the industrial sector.

The parallels in development of the systems approach in the United States and Sweden are striking, in part because of the long-standing Swedish reliance on technology transfer. Much as Jardini and Dyer in their essays on RAND and TRW document a move of systems experts into public policy, so Kaijser and Tiberg recount the increasing emphasis by Swedish systems experts in the 1960s and 1970s on civil public-policy issues, such as those arising in the health sector. The systems approach became politicized when the governing Social Democratic party established in the prime minister's office a Secretariat for Future Studies staffed by systems experts who applied their tools to highly controversial issues, such as the future of nuclear energy. Kaijser and Tiberg suggest that seeming political bias reduced the expert status of the practitioners. They conclude that in Sweden the systems approach has now been incorporated into broader academic disciplines and that systems thinking is taken for granted. "The systems approach community," they believe, "has dispersed and lost its identity" (p. 407).

In his commentary on the Kaijser and Tiberg essay, John Staudenmaier argued that the systems-approach community lost influence because of overreaching. When the experts began to advise policy makers on broad social issues such as health care, their advice focused on technical and economic factors that did not really reflect the messy complexity of the issues. Staudenmaier drew a parallel between the loss of momentum of the systems approach, both in Sweden and the United States, and the comparable loss of influence of Frederick W. Taylor's scientific management approach early in this century. Both schools of management proved, according to Staudenmaier, "too rigid, too abstract, too clean."

INTERNATIONAL FRAMEWORK

The systems approach also embedded itself in an international framework. A conference panel on the history of the International Institute of

Applied Systems Analysis (IIASA) included members—Harvey Brooks, William C. Clark, Roger E. Levien, Alan McDonald, and Howard Raiffa—who had participated in the establishment and operations of IIASA. The essays by Brooks, McDonald, and Levien summarize the panel's presentations and enlarge upon their remarks.

Brooks and McDonald, in "The International Institute for Applied Systems Analysis, the TAP Project, and the RAINS Model," note that scientific organizations from twelve countries, including the United States and the Soviet Union, founded IIASA in 1972. The founders wanted IIASA to apply techniques of systems analysis to solve urban, industrial, and environmental problems that transcend international boundaries. Those who presided over the birth of IIASA also saw it as a "bridge building" initiative to reduce East-West tensions. Located near Vienna, IIASA invited researchers to stay in residence for varying periods, the average being about two years. They came primarily from the countries of the scientific organizations supporting IIASA.

IIASA's governing council, usually headed by a Russian, and its director, customarily an American, often chose problems associated with population, economic development, global warming, and other global environmental issues. This focus brought social scientists as well as physical scientists to Vienna. Brooks and McDonald believe that IIASA has succeeded in advancing both the theory and application of systems analysis as responses to these problems. This achievement reminds us of the comparable success of the RAND Corporation in cultivating the systems approach, a comparison explored by Levien in his essay.

Brooks and McDonald discuss five hallmarks of the IIASA approach to problem solving, which include an international and interdisciplinary emphasis and the maintenance of credibility with both scientists and decision makers. To illustrate this approach, Brooks and McDonald offer a case history of one of the organization's most successful ventures, a transboundary air-pollution project using the RAINS (Regional Acidification Information and Simulation) model of the impact of acidification in Europe. This informative case history focuses mostly on the changing nature, validity, and limitations of the RAINS computer model. Brooks and McDonald stress the ways in which the model has been adapted to the realities of policy making without losing its scientific credibility. The case history recalls Paul Edwards' analysis of global systems models.

Levien, in "RAND, IIASA, and the Conduct of Systems Analysis," provides a comparison of the systems approaches of these organizations. Having held leadership positions in research and management at RAND from 1956 to 1974 and having been director of the IIASA from 1975 to 1981, Levien is well qualified to discuss conditions at RAND conducive to "first class" systems analysis and to draw lessons from its history. He is also well positioned to analyze the reasons why he and others found the challenge of nurturing impressive systems analysis at IIASA so much more difficult than at RAND.

At RAND an interdisciplinary approach impressively met the standards of the academic community while also responding to the need of governmental agencies—especially the air force—for realistic operational and policy recommendations. A major reason for its success was an environment at RAND that supported a fruitful exchange among people with diverse disciplinary commitments, including some who felt most comfortable with traditional scholarly research and others who preferred to focus on problem-oriented projects and interactions with clients.

Levien's account of RAND provides an enlightening context for the essays by Hounshell and Jardini. He follows the early evolution of operations analysis, systems analysis during the Cold War, and policy analysis during the 1960s. Like Jardini, Levien suggests some of the reasons for the limited success of RAND's ambitious effort at knowledge transfer. Levien also carries his overview through the rise of an international systems approach and policy analysis. The lessons he draws from RAND's history help explain its successes. He stresses the advantages resulting from an interdisciplinary approach and from the granting of free choice to researchers in choosing problems. Levien contrasts this history with the problems encountered by IIASA leadership, including diverse points of view with regard to problem choice and expectations, as well as variations in competence among transient researchers.

Spread by Analogy

Two books by Norbert Wiener, an MIT mathematician, accelerated the spread of the systems approach to fields other than engineering and project management. *Cybernetics, or Control and Communication in the Animal and the Machine* (1948) influenced a professional audience, while *The Human Use of Human Beings: Cybernetics and Society* (1950) reached a general audience. Wiener's views stimulated scientists in various disciplines to see analogies between the systems they studied and the feedback systems he described. Within a decade, seeing the natural and human-built worlds as communication and control, or information, systems became commonplace among social and natural scientists, especially among those exploring molecular and developmental biology.

The information theory of Claude Shannon, an electrical engineer who worked at Bell Laboratories, complemented Wiener's ideas. Shannon's "Mathematical Theory of Communication," published in 1948, also encouraged scientists and social scientists to find information and control analogies. During World War II, both Shannon and Wiener had drawn on technological practice, the former in telephone communications and cryptography and the latter in gunfire control, in formulating theoretical concepts of information and control. After the war, increasing understanding of the workings of digital computers reinforced their closely related theories of information, communication, and control.

Lily E. Kay, in "How a Genetic Code Became an Information System," explains the ways in which concepts borrowed from Wiener and Shannon spread by analogy into molecular biology. The transfer took place mainly on the level of theoretical concepts, ones that were incorporated into laboratory practice by 1960. In search of theory with greater power to explain life processes, especially heredity, scientists used a host of metaphors from communications/information theory. Molecular biology began to represent itself as "a communication science, allied to cybernetics, information theory, and computers" (p. 463). Hereditary transmission came to be thought of as behaving in ways analogous to a guidance and control system. Such vocabulary as information, feedback, messages, codes, words, alphabets, and texts soon constituted the discourse of molecular biology.

Nobel laureate Jacques Monod argued that the human organism had a cybernetic, feedback system controlling its chemical processes; he even saw an analogy between gene-enzyme regulations and the automation circuitry of ballistic missiles. Another Nobel laureate, François Jacob, compared the genetic code to a computer program. Heredity functions, he believed, like "the memory of a computer." In 1952 another advocate of information-science-based biology offered a scriptural representation of the protein paradigm of heredity. Proteins became for some scientists messages, amino acid residues an alphabet.

James Watson and Francis Crick's 1953 description of DNA structure stimulated scientists cultivating an information-based biology to attempt to break the DNA code by using cryptoanalytic techniques and mainframe digital computers. Communication engineers, biologists, mathematicians, and physicists attacked the DNA problem over the next decade. Their persistent and ingenious code-breaking endeavors, however, failed to unravel the enormously complex behavior of DNA. But the decade of metaphoric transformation of molecular biology, Kay concludes, provided a shaping context in which subsequent research in the field and the human genome project developed.

In his commentary on the Kay paper, Timothy Lenoir, argued, like Kay, that information and cybernetic theory as articulated by Shannon and Wiener fell short of explaining the coded messages controlling hereditary because information theory is about syntax and not semiotics, or meaning. He also agreed with Kay that the early information models representing molecular biology amounted to little more than metaphors, analogies, and tropes. On the other hand, Lenoir faulted Kay's paper for not coping with the problem of explaining how biology has become, nonetheless, an information science.[16]

Conclusion

Taken together, the essays in this volume provide a history of the early decades of an innovative style of management and complex process analysis that took root after World War II. The systems style transferred from the military into the civil realm where it was often frustrated by political considerations and other factors that can be subsumed in the category "messy complexity."

Nevertheless, a more flexible systems approach to management, especially of technological systems, has, as Arne Kaijser and Joar Tiberg suggest, now been incorporated into a general repertoire of managerial practice. The influence of the systems approach is comparable to that of Frederick W. Taylor's scientific management earlier in this century.

Acknowledgment

We are indebted to Fred Quivik of the University of Pennsylvania for the information on citations.

Notes

1. The resulting compromise, called "black-boxing," allows local research and development contractor teams to fulfill system specifications for components in

various ways. When, for example, systems engineers specify the thrust required from a missile propulsion engine, the propulsion design team has leeway in choosing the mechanical, electrical, and chemical means for achieving these specifications. The systems engineers need not look into the "black box" within which the team is working and micromanage the means used to achieve the specified ends. Overarching systems management coordinates and schedules the numerous black-boxed subprojects.

2. Most of the commentators made informal oral presentations that have not been included in this volume.

3. Russell L. Ackoff, "The Future of Operations Research Is Past," *Journal of Operational Research Society* 30.2 (1979): 93–104; and "Resurrecting the Future of Operational Research," *Journal of the Operational Research Society* 30.3 (1979): 189–199. For the earlier views of Ackoff, see "The Development of Operations Research as a Science," *Operations Research: The Journal of the Operations Research Society of America* 4.3 (1956): 265–295.

4. Ackoff, "The Future of Operations Research Is Past," 93.

5. Ibid., passim.

6. Russell Ackoff, interview, 4 November 1993, in Philadelphia, Pennsylvania. See Russell L. Ackoff, " 'The Art and Science of Mess Management,' " *Interfaces* 17 (1981): 20–26.

7. Erik Rau, "Polaris," unpublished essay, University of Pennsylvania, 1993.

8. Harvey Sapolsky, *The Polaris System Development* (Cambridge: Harvard University Press, 1972), 112.

9. Author's interview with Robert Fuhrman, 25 June 1995, at the headquarters of the Lockheed Martin Corporation in Crystal City, Arlington, Virginia.

10. In this section we draw on a chapter entitled "Spread of the Systems Approach" in Thomas P. Hughes, *Rescuing Prometheus* (New York: Pantheon Books, 1998), 141–195. We are especially indebted to Fred Quivik, Atsushi Akera, and Erik Rau, research assistants on the Mellon Systems Project at the University of Pennsylvania, for research, essays, and insights helpful to us in following the spread of the systems approach. David Foster, formerly an undergraduate student at the University of Pennsylvania, and Elliott Fishman, a graduate student at the University of Pennsylvania, also provided valuable research assistance.

11. See the introduction by Merritt Roe Smith and the essays in *Military Enterprise and Technological Change: Perspectives on the American Experience* (Cambridge: MIT Press, 1985). See also the discussion of the American systems of production in David A. Hounshell, *From the American System to Mass Production, 1800–1932: The Development of Manufacturing Technology in the United States* (Baltimore: Johns Hopkins University Press, 1984); and David W. Noble, *Forces of Production: A Social History of Industrial Automation* (New York: Alfred A. Knopf, 1984). For a detailed military-history bibliography emphasizing technology in the military, see Barton

C. Hacker, "Military Institution, Weapons, and Social Change: Toward a New History of Military Technology," *Technology and Culture* 35 (October 1994): 768–834. On military influences on science and engineering, see Paul Forman, "Behind Quantum Electronics: National Security as Basis for Physical Research in the United States, 1940–1960," *Historical Studies in the Physical and Biological Sciences* 18.1 (1987): 149–229, and Stuart Leslie, *The Cold War and American Science* (New York: Columbia University Press, 1992). See also the perceptive remarks in Alex Roland, "Science and War," in Sally Kohlstedt and Margaret Rossiter, eds., *Historical Writing on American Science* (1985).

12. Author's interview with Alexander Kossiakoff, 3 April 1995, at Johns Hopkins University Applied Physics Laboratory.

13. John N. Warfield and J. Douglas Hill, *A Unified Systems Engineering Concept*, Battelle Monographs, ed. Benjamin Gordon (Columbus, Ohio: Battelle, 1972), 1.

14. These remarks on the spread of the systems approach are taken from Hughes, *Rescuing Prometheus*, 141–195.

15. Quoted in Melvin S. Day, "Space Technology for Non-Aerospace Applications," *1968 Wescon Technical Papers*, vol. 12, pt. 5, session 23, paper 4, 1968, 5–6.

16. Kay responds that molecular biology became an information science only in the colloquial sense of the word, as, for example, referring to library sciences as information sciences. Molecular biology, she continues, has never become an information science in the 1950s' sense of the word.

1

Automation's Finest Hour: Radar and System Integration in World War II

David A. Mindell

> It is nearly as hard for practitioners in the servo art to agree on the definition of a servo as it is for a group of theologians to agree on sin.
>
> —Ivan Getting, 1945

> At first thought it may seem curious that it was a Bell Telephone Laboratories group which came forward with new ideas and techniques to apply to the AA problems. But for two reasons this was natural. First, this group not only had long and highly expert experience with a wide variety of electrical techniques . . . Second, there are surprisingly close and valid analogies between the fire control prediction problem and certain basic problems in communications engineering.
>
> —Warren Weaver, 1945[1]

Examining, as this volume does, "the spread of the systems approach," suggests that some coherent approach to systems emerged within engineering before it diffused into other disciplines such as social policy and urban planning. While Thomas Hughes has chronicled a consciousness of systems in electrical power engineering early in the century, the historical literature overall has little to say about systems engineering, and what it meant technically and politically, in the period just before it began colonizing other fields.[2] This chapter examines a particular set of technical and institutional developments during World War II, to show how a new instrument of perception—radar—gave rise to a new approach to engineering systems. Combining servo-controlled gun directors with new radar sets raised problems of a system's response to noise, the dynamics of radar tracking, and jittery echoes. Engineers from Bell Laboratories, in conjunction with their rivals and collaborators at the MIT's Radiation Laboratory, learned to engineer the entire system's behavior from the beginning, rather than just connecting individual, separately designed components.

This new system logic reflected institutional relationships and evolved to suit their shifts. To the Radiation Lab it meant designing the system around its most critical and sensitive component—the radar—and not the director, computer, or gun. By the end of the war, the Radiation Laboratory, in competition with a number of other research labs, assumed control of system design. The Rad Lab ran the war's only successful effort to design a fully automatic radar-controlled fire control system, the Mark 56 Gun Fire Control System. Still, the existing tangle of arrangements between the Rad Lab, Section D-2 (later renamed Division 7) of the National Defense Research Committee (in charge of fire control), and the Navy Bureau of Ordnance did not give the Rad Lab the responsibility it sought. Ivan Getting, director of the Mark 56 project, redefined his organizational role and created the new job of system integrator, a technical, institutional, and epistemological position.

RADAR: AUTOMATING PERCEPTION

During the 1930s, the Army Signal Corps tried to incorporate new "radio ranging" devices into existing mechanical gun directors. In 1937, this work produced the SCR-268 radar (which Western Electric began producing in 1940), designed to supply fire control data to Sperry's M-4 mechanical gun director for directing antiaircraft fire toward attacking bombers.[3] (See figure 1.1.) The SCR-268, although deployed in large numbers, imperfectly matched the M-4, which was designed for use with optical tracking equipment (i.e., telescopes). These early radar sets performed similar to the old sound-ranging equipment they replaced: useful for detecting incoming aircraft and providing an idea where they were, but not as precision inputs to fire control systems. The SCR-268, however, worked much better than acoustic devices, and could direct searchlights to track a target.[4]

The SCR-268's poor accuracy derived in part from its relatively low frequency/long wavelength (1.5 meters). Existing vacuum tubes could not generate higher frequency (shorter-wavelength) signals at high enough powers for aircraft detection. So in 1940, shorter wavelengths, or "microwaves," were part of Vannevar Bush's solution to what he called the "antiaircraft problem," which was then proving critical in the Battle of Britain. When Bush's National Defense Research Committee (NDRC) began operations in 1940, it included microwave research, under Section D-1, the "Microwave Committee."

Figure 1.1
Army SCR-268 fire control radar.

It also included a division for "fire control," Section D-2, under the leadership of Warren Weaver, Director of the Natural Sciences Division of the Rockefeller Foundation. During the summer of 1940, Weaver and D-2 toured the field and learned about fire control, and the Microwave Committee did the same for radar. Both groups realized neither the army nor the navy were aware of each other's work. They found very little research on tubes capable of producing waves below one meter, and none for "microwaves," with wavelengths below ten centimeters.[5]

American radar radically changed in September of 1940, when a British technical mission, the famous "Tizard Mission," came to the United States and met with the NDRC. In a remarkable act of technology transfer, the Tizard Mission revealed the "cavity magnetron" to the Microwave Committee.[6] The device could produce microwave pulses with peak powers of ten kilowatts at a wavelength of ten centimeters. Not only did high frequencies produce more accurate echoes, but their small antennas could be carried aboard aircraft. Bush and the NDRC set up a central laboratory for microwave research at MIT, the "Radiation Laboratory," or Rad Lab. It become the NDRC's largest project.

The Rad Lab had three initial projects. Top priority was airborne radar for intercepting bombers, known as Project I. Project II sought automatic fire control. Harvard physicist Kenneth T. Bainbridge joined the Rad Lab and brought a young physicist named Ivan Getting. Getting, the son of Czeckoslovakian diplomats, had grown up in Europe and Washington, D.C. He attended MIT on scholarship and did an undergraduate thesis in physics under Karl Compton in 1934.[7] After completing graduate work in physics as a Rhodes Scholar at Oxford, he returned to the United States as a member of the Harvard Society of Fellows. In November 1940, Getting joined Project II, "to demonstrate automatic tracking of aircraft by microwave radar of accuracy sufficient to provide data input to gunnery computers for effective fire control of ninety-millimeter guns."[8] The intense, blue-eyed Getting was put in charge of the "synchronizer," the master timing device "which tied the system's operation together."[9] The group also included electrical engineers Henry Abajian and George Harris, and physicists Lee Davenport and Leo Sullivan.

At this time, tracking targets with radar remained a manual activity; it required "pip matching." The operator viewed radar return signals on an oscilloscope screen and used a handwheel-controlled blip to select which radar echo was indeed the target. Then the blip or "pip" and not the actual radar signal went on as the valid range. The operators worked as the "human servomechanisms" did in earlier Sperry directors: they distinguished signals from noise. Bowles and Loomis, aware of MIT's strength in servomechanisms automatic control, suggested Project II mechanize this task for "automatic tracking." If the radar signal itself could drive servos to move the antenna, the radar would follow the target as it moved. Project II set out to automate the work of the radar operator.

To solve this problem, the Rad Lab developed "conical scan," which rotated an off-center beam thirty times per second to make a precise "pencil-beam" for tracking. If the target was "off axis," that is, off the centerline of the beam, a feedback loop moved the antenna to return the target "on axis" to the center. If the target was moving, like an airplane, the antenna would thus track its motion. The Rad Lab obtained a machine-gun mount from General Electric to move the antenna, and G.E. engineer Sidney Godet to adapt the amplidyne servos for tracking. Godet became an informal Rad Lab member and taught the group about G.E.'s experience with servos. They first tested conical scanning at the end of May 1941 on the roof of one of MIT's engineering buildings.[10] By February 1942 the Rad Lab built a prototype, the XT-1; they bought a truck and modified the radar to fit inside.

The truck added more than mobility; it added enclosure. Earlier army radar sets (the SCR-268) mounted displays and operators directly on the rotating antenna platform, much as the Sperry directors had in the 1930s. This arrangement reflected the army's conception of the radar operators: they were soldiers on the battlefield operating a piece of equipment like a radio. To Getting it seemed foolish; the operators' eyes could not adjust to see the cathode-ray displays in bright sunlight; exposed to rain and snow, their hands got too cold to precisely tune the equipment.[11] Getting and his engineers saw the operators as technicians more than soldiers, reading and manipulating representations of the world. The XT-1 truck brought the operators inside a darkened, air-conditioned trailer: a control room, a laboratory.

Enclosure allowed their eyes to adjust to the delicate blips on the CRT; it freed their hands from cold; it isolated their ears from the sounds of battle. Glowing radar screens presented a captivating simulacrum of the world outside. Earlier oscilloscope displays, including the XT-1, showed a single horizontal trace of the radar echo over time. These were soon replaced, on the production version of the machine, with a "plan position indicator" or PPI: a round tube displaying a rotating beam tracing out a virtual map of the area being scanned. Now radar operators and their commanders could perceive and manipulate the field of battle as a map and not as electrical reflections. Radar created an analog of the world, collecting data from a broad area and representing them in compressed form. These systems were among the first in which an operator controlled a machine based on visual input from a cathode-ray tube—an act akin to today's interaction with computers.

After testing, the army reported, "The Radio Set XT-1 is superior to any radio direction finding equipment yet tested by the Coast Artillery or Anti-aircraft Artillery Boards for the purpose of furnishing present position data to an anti-aircraft director."[12] In April 1942 the XT-1 was standardized, or accepted by the army for production, as the SCR-584 radar system; the army ordered more than a thousand units from General Electric, Westinghouse, and Chrysler. (See figures 1.2 and 1.3.) As an "early warning system" it could scan the skies up to 90,000 yards and then track an aircraft to one-twentieth of a degree to a range of 32,000 yards. It provided output signals for azimuth, elevation, and range that could feed into the Sperry M-4 or M-7 directors,

Figures 1.2 and 1.3
SCR-584 Fire control radar with control van. Note tracking operator's console at left in van, and range operator's console at right. [From Louis Ridenour, *Radar System Engineering* (New York: McGraw Hill: 1947), 209.]

or the BTL M-9 director. The SCR-584 became the most successful ground radar of the war, with nearly 1,700 units eventually produced.[13]

The SCR-584 by itself was a remarkable device, "the answer to the antiaircraft artilleryman's prayer."[14] Rad Lab Project II, however, aimed at more than a tracking radar: it sought automatic fire control. Marching toward that goal, however, tread on D-2's terrain. Early on, Warren Weaver recognized the potential for overlap. He wrote to Loomis of his desire for "a reasonably definite understanding of the location of the fence between our two regions of activity ... a wire fence, through which both sides can look and a fence with convenient and frequent gates." Weaver proposed the relationship between the organizations mirror that of radar to a computer, of perception to integration: "The boundary between the activities between the two sections I would suppose to be fairly well defined by saying that your output (three parameters obtained from microwave equipment) was our input (input to a computer or predictor)."[15] Karl Compton, in charge of Division D, agreed and set up a special committee, known as D-1.5 to represent its liaison between D-1 (radar) and D-2 (fire control). It consisted of Bowles of D-1, Ridenour and Getting of the Rad Lab, and Caldwell and Fry of D-2. This group, in existence for only about a year, conducted a comprehensive survey of all radar development in the United States and Canada.

Where did the Sperry Gyroscope Company fit into this new domain? With a strong background in fire control and new work in radar, the company should have been the obvious choice to build new integrated control systems. The army, however, distrusted the company and requested Sperry only integrate its existing M4 director with the SCR-268 radar, both of which the army already possessed in large numbers.[16] But both the army and the NDRC drew on Sperry corporate knowledge in another way. Sperry's fire control director, Earl Chafee, joined the Ordnance Department and was assigned to survey existing technology and propose "the best all-around fire control system which could be put together out of equipment on which the basic research is now completed." Chafee was to work with D-2 and not only examine individual components, but "the emphasis is to be placed on the over-all aspects of the *system* ... on the role which radar should play in such a unified system."[17] The so-called Chafee Inquiry did not lead to a new development program but it clarified the systems nature of the problems involved in automating traditional instruments

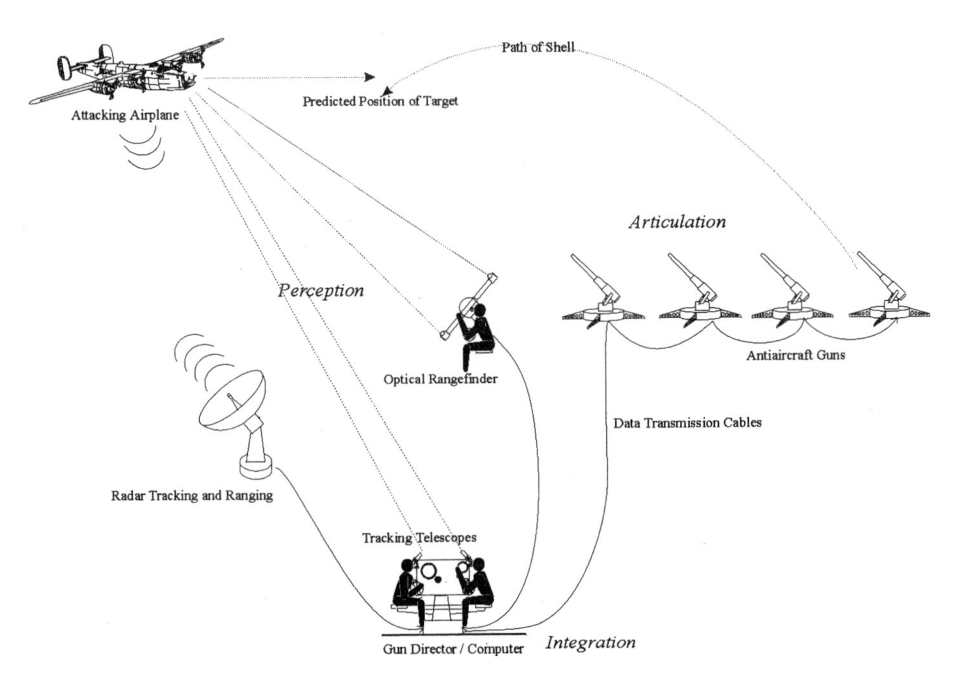

Figure 1.4
The antiaircraft problem. A tracking device (optical or radar) follows the target; the computer calculates its speed and direction, and then extrapolates that velocity into the future to choose an aiming point for the guns. A ballistics calculation turns this information into angles of elevation and azimuth for the guns.

of perception with microwave radar, problems Bell Labs and the Radiation Lab already faced.[18]

Bell Labs (BTL) was building an electrical computer, or gun director, under the NDRC's fire control section, D-2. The device would track and predict target positions for antiaircraft guns. (See figure 1.4.) It used the same algorithms as the mechanical Sperry directors, but implemented them with electromechanical servo-driven computing mechanisms and feedback amplifiers. Tracking input came from optical telescopes, but BTL built in provisions for radar inputs. The project got under way as soon as the NDRC began letting contracts in 1940, and the first prototype was delivered to the army a few days after Pearl Harbor. The army immediately ordered several hundred of the units, designated T-10 in development, for production, and "standardized" the machine (i.e., accepted it as operational) as the M-9 gun director. (See figures 1.5 and 1.6.)

Figure 1.5
M-9 gun director, tracking head with operators. One follows the target in elevation, the other in azimuth. The unit and the operators rotate with azimuth tracking. (AT&T Archives)

Figure 1.6
M-9 antiaircraft director with power supply, computer, tracking head, and servo-driven 90mm gun. With the SCR-584 radar, this machine fought the V-1. (AT&T Archives)

Ivan Getting learned of the Bell Labs director project during the D-1.5 survey; he began working with BTL to connect his XT-1 tracking radar to the M-9. Weaver's "wire fence" worked fairly well in this case. The T-10 and XT-1 designs proceeded together, and throughout BTL stayed in touch with the MIT group. Ridenour and Getting of the Rad Lab and Stibitz and Lovell of BTL visited back and forth, exchanging information and discussing interfaces between the machines. Getting was particularly interested in "time constants," measures of how quickly the T-10 could respond to inputs. When designing his antenna and tracking unit, he had to know how fast the T-10 could keep up with incoming data—its frequency response.[19] The T-10 final report touted the value of coordinated work: "Close liaison should be maintained between director designers and designers of radars and other tracking equipment. The specifications on each unit should be written with full consideration of the features and capabilities of the other."[20] During this project, the idea emerged that a system might be more than the sum of its parts; the added element was noise.

What difficulties did the Rad Lab and BTL face in trying to connect their instruments? Noise posed the biggest problem. Servos worked fine as calculators when input data was smooth and ideal. Errors in tracking, however, "would produce prediction errors of dominating proportions"; differentiating the prediction signal tended to emphasize high-frequency noise.[21] Radar signals had several sources of noise, making the problem especially bad. For example, as a radar beam reflected off an airplane, it would shift from one part of the plane to another (analogous to the airplane "twinkling" in the sun). A data smoother could eliminate short, high-frequency perturbations from the input data, but with trade-offs. Smoothers introduced time lag, so the smoothed data was no longer current when sent into the predictor.

How could one determine the optimal smoothing versus time lag for a network? Could one reduce the time lag for a given network? How did the smoother distinguish proper tracking data from erroneous inputs? What effect did the time lag of a smoother have on the dynamics of a feedback loop? Would smoothing avoid or induce instability? These questions resembled those telephone engineers had been asking for at least a decade. The work of BTL engineers Harry Nyquist and Hendrik Bode showed the answers depended on the frequency response of the system's components. Warren Weaver put it best when he observed that building radar-controlled systems raised "certain basic problems in communications engineering . . . if one applies the term *signal* to the variables which describe the actual true motion of the target; and the term *noise* to the inevitable tracking errors, then the purpose of a smoothing circuit (just as in communications engineering) is to minimize the noise and at the same time distort the signal as little as possible."[22] At BTL and the Rad Lab, just as at MIT's Servo Lab, building control systems meant rethinking the nature of electronic information. Using radar to close a feedback loop required paying attention to connections as well as to components. With radar, control engineering became a practice of transmission, of signals, of communications.

Neither the Bell Labs director nor the Rad Lab's radar had been designed from the first with such a practice of "systems engineering." Rather, the two groups tried to connect two separate machines, neither having formal responsibility for coordination. Still, the cooperation paid off. In the fall of 1942, the army held a competitive test of radar-controlled "blind firing."[23] The XT-1 was matched against two other radars, all connected to a T-10 director and Sperry power drives on a 90mm gun. The XT-1 performed best and competing programs were canceled. Although problems remained, particularly extraneous electrical noise in the cables, the system demonstrated that a radar-controlled director could track a target, figure a firing solution, and aim the guns (although it still required human input for target selection, pip matching, and a number of other tasks). By late 1943, the M-9/SCR-584 combination entered service in the European theater as an automatic antiaircraft fire control system.

The T-10/XT-1 program gave Getting new ideas for engineering systems. Technical success brought him new responsibility and the opportunity to articulate his vision: the Radiation Lab reorganized into a number of divisions for components, support, research, and "systems." Ivan Getting took charge of Division 8, responsible for all army ground radar and naval fire control. Nathaniel Nichols headed a special subsection for servos that included a theoretical section headed by Ralph Phillips and had on staff Walter Pitts and economist Paul Samuelson. While this group seemed to violate Weaver's cordial fence between division between D-1 and D-2, Getting believed system design orbited around radar; under his direction the Rad Lab would become the center of gravity for integrated systems.

THE DIFFICULT STEPCHILD: RADAR AND FIRE CONTROL IN THE NAVY

The source of that gravity, however, would not be the army but the world's fire control expert, the navy's Bureau of Ordnance (BuOrd). Between the world wars, BuOrd had been the world's leader in fire control systems, at first for heavy guns and then for antiaircraft, and had developed a closely knit, secret set of contractors, including the Ford Instrument Company, the Arma Engineering Company, and General Electric.[24] But by 1943, the M-9/SCR-584 combination gave the army the most automated fire control system in the war, leapfrogging the navy with help from the NDRC. BuOrd, for its part, had done little work with D-2, Division 7, or the Radiation Lab. Still, the navy was pushing radar because automated perception radically altered naval fire control. Naval control systems, especially for heavier guns, changed more slowly than equivalent army technology, because they depended on modifying ships instead of just sending systems into the field on trucks. This momentum, combined with the conservatism of BuOrd and its contractors and their failure to take immediate advantage of the NDRC, meant that the bureau came to Division 7 and the Rad Lab for help designing a new automated system. Before examining Getting's handling of this project, however, and hence his definition of system engineering, we must understand BuOrd's difficult cultivation of fire control radar.

Despite the navy's early work with the technology, in the words of an official BuOrd history, radar was "a stepchild slow to win affection." Typically it augmented existing fire control equipment not designed for electronic inputs. During the war, BuOrd's tough love spawned twenty-seven different fire control radar designs, only ten entered production, seven actually saw action, and only three (Marks 3, 4, and 8) became widely available.[25] They had problems with reliability, maintenance, short ranges, and target discrimination. Only intense human mediation—similar to the old "human servomechanisms"—could produce high-quality electronic inputs for rangekeepers (the mechanical computers that calculated solutions for naval guns). Operators needed to "pip match" to eliminate noise, and to manually follow the target with the antenna, much as with traditional optical rangefinders and telescopes. They routinely switched between optical and radar tracking, and the combination threatened to overload their attention. Optical tracking remained necessary, because tracking radars frequently jittered between closely spaced targets; they had particular trouble locking onto airplanes attacking low across the water—a weakness Japanese pilots exploited for tactical advantage. Radar underscored the navy's problems with antiaircraft fire control in general; it worked fairly well against high, straight-flying targets, but broke down when confronting fast, maneuverable, close-in attacks. Still the navy dreamed about fully automatic "blind firing," which could accurately shoot at night or through overcast (the anthropomorphic "blind firing" echoes the early use of radar for "blind landing" of airplanes).

Since 1941, BuOrd had attempted several projects to adapt existing control systems for blind firing, including several at the Rad Lab, all of which were terminated. Radar still played the frustrating stepchild. BuOrd and its established clique of secretive contractors simply could not produce a director and a radar at the same time. Blind firing remained an elusive goal.

Ivan Getting and Coordinated Design

Ivan Getting believed he could bring the stepchild into the family and make blind firing a reality. He redefined the system: no longer a set of separate components connected together, but a single, dynamic entity. Signals, dynamics, time constants, and feedback needed to be specified first—this *was* the system. The physical equipment and mechanical components merely solidified these relations. Beyond the technical relations, Getting's vision entailed a new role for his laboratory. BuOrd's earlier attempts at blind firing had failed, he argued, because they lacked a central, coordinating technical body that could oversee the integration of the system:

1. There was no attempt made to integrate the radar and the computer into a functioning whole.
2. The gross engineering was done by the Bureau of Ordnance, whereas the detailed engineering was done by the company [i.e., contractor], who was not informed of the problem as a whole.[26]

The fire control clique still saw the computer and the radar as comprising the "functioning whole." But to Getting they were subsidiary to a more abstract notion of the system. Similarly BuOrd, with its highly specified and compartmentalized contracting, still believed it could break the fire control problem into component parts, technically and contractually ("gross engineering" versus "detailed engineering").

Getting wanted to redefine the boundaries between components and between organizations in "a totally integrated effort starting from basic principles."[27]

Getting found willing allies in the NDRC and BuOrd. The NDRC reorganized in the end of 1942, and Section D-2 for fire control became Division 7, now headed by Harold Hazen, MIT's Department Head in Electrical Engineering who had made fundamental contributions to servomechanism theory in the early 1930s. Warren Weaver, though he remained an adviser to Division 7, left to head the newly created NDRC Applied Mathematics Panel. Section D-1, for radar, now became Division 14. Hazen, head of the new Division 7, recognized the value of coordinating radar and fire control design (he had grappled with similar systems problems ten years before with the Differential Analyzer). Among Division 7's priorities, Hazen announced, would be "the overall design of fire control systems and the optimum use of radar on navy directors."[28] To smooth relations with the Rad Lab, he invited Getting to join. Soon thereafter, Division 7 began discussing a blind firing director for the navy's 5-inch 38-caliber guns with Emerson Murphy, head of fire control research at BuOrd.[29] Getting proposed "a joint project under Division 14 and Division 7 . . . [for] compact blind firing director for heavy machine guns, 3-inch guns, and 5-inch guns for the U.S. Navy." Murphy, attending a Division 7 meeting, endorsed the idea. BuOrd chief William Blandy concurred, designating the project Gun Fire Control System Mark 56.[30]

Now Getting could start from scratch, defining the machine and defining his position. The NDRC would go one step beyond its usual role of designing equipment, building prototypes, and preparing drawings for production. It would now oversee the selection and preparation of manufacturers, and oversee a production run. This arrangement would allow the NDRC complete technical control of all phases of the project. But which part of the NDRC? A radar-driven fire control device fell within two domains: Division 14 (the Radiation Lab) and Division 7. Division 7 members argued the Radiation Laboratory didn't have sufficient experience with fire control, and that the project should use M-9 director technology developed for the army (BTL was then building for BuOrd the naval equivalent of its electrical director, an electronic rangekeeper).[31] Getting's idea for the new system, however, had radar at its core.

To connect radar and fire control, Hazen created a special section of Division 7, dubbed 7.6, "Navy Fire Control with Radar." Ivan Getting would head Section 7.6 as a member of both Division 7 and the Radiation Lab's systems division. He described the new section as "an attempt by Dr. H. L. Hazen to bring together the necessary elements which had been more or less artificially separated by organization, personality, and history."[32] Getting questioned the traditional lines between subunits: the NDRC's divisions dated from a time, just a few years before, when fire control and radar comprised separate technologies. For earlier projects, such as the M-9/SCR-584 combination, the arrangement worked well, given a high degree of communication between Bell Labs and the Radiation Lab. From that experience, however, Getting learned the value of coordination at the design stages and all the way through production—and the value of controlling that coordination. Section 7.6 absorbed a few other Division 7 projects relating to navy fire control and undertook a number of small contracts, but the Mark 56 formed its major work. Getting called the project, "the first fully integrated radar fire control system that was not restricted by history or by prejudices."[33]

Yet Getting took advantage of history. For the new section, and for the Mark 56, Getting tapped members of BuOrd's fire control clique. He included vice presidents from Ford Instrument and Arma, Al Ruiz of General Electric, MIT's Charles Stark Draper, and Robert M. Page, who had done the early radar work at the Naval Research Lab.[34] The committee did not actually meet until January of 1944, by which time the Mark 56 project was well under way. Section 7.6's primary function then became "supplying a forum where communications between the principals, including the Bureau of Ordnance, could be provided openly."[35] By this date, most 7.6 members were already overloaded with other work. Those from industry were further constrained: they had other contracts with BuOrd and could not discuss status or technical details. Nor did they wish to share such information in a forum in which their commercial competitors participated. The world of naval fire control, with its multilayered secrecy and its seeming archaism, frustrated Getting, used to the heady and open world of the early days of microwave radar.[36]

Despite Getting's vision, nothing inherent in "coordinated design" dictated it should be a radar group to capture and hold the terrain. He and Division 7 confronted not only BuOrd's fire control establishment, but also other centers of technical expertise. "Blind firing" became the high prestige project for BuOrd, and several groups vied for the technical spotlight. Others argued that Draper's "gyro culture" was best positioned for system engineering, or Bell Labs, where research shared a corporate umbrella with Western Electric's manufacturing (and "System Engineering" had been an established department since the 1920s). Getting bitterly opposed bringing Western Electric in even as a manufacturer; he disparaged his earlier work with the telephone company—"In fact the Radiation Laboratory and Bell Telephone Laboratories are not complementary but rather the same type of laboratories," he wrote to Karl Compton—and threatened to resign from the Mark 56 project if production contracts were given to Western Electric.[37] The contracts, instead, went to General Electric, the established navy fire control supplier with whom Getting had worked so successfully on the SCR-584.

Beginning in 1943 the Rad Lab undertook the Mark 56 program. (See figure 1.7.) Its conical-scan, X-band (3cm wavelength) radar could search broadly for targets, and then automatically track them, even at low angles. A "line of sight gyro" in the Mark 56 established a reference as the line between gun and target. Radar operations took place below decks; two sailors in the director itself could acquire and track targets optically. For the computers, the Rad Lab did not defer to prior experience, over Division 7 objections. Instead, Czech exile and fire control expert Tony Svoboda in the Rad Lab designed a wholly new type of mechanical computer, using innovative four-bar linkages. The MIT Servo Lab modified their Vickers servo to drive the director, but the devices were never used. In August 1943 Division 7 let a contract with General Electric's Aero and Marine Division in Schenectady for the gyro assembly. General Electric contracted to do production design on the radar based on a Rad Lab prototype. The Librascope Corporation of California (chosen over a competing proposal from Ford Instrument) produced the ballistic computer. The device was first tested on a specially constructed rolling platform at Fort Heath north of Boston in the spring of 1944. The first full-up test, including guns, took place the following December.[38]

The project's radical character adversely affected its timing. BuOrd, tuned for wartime production and deployment, allocated its priorities solely by anticipated delivery date. The long-term Mark 56 fell low on the list and its schedule suffered. Still, Getting saw his "ultimate" system as a crash program to get blind firing to the fleet as

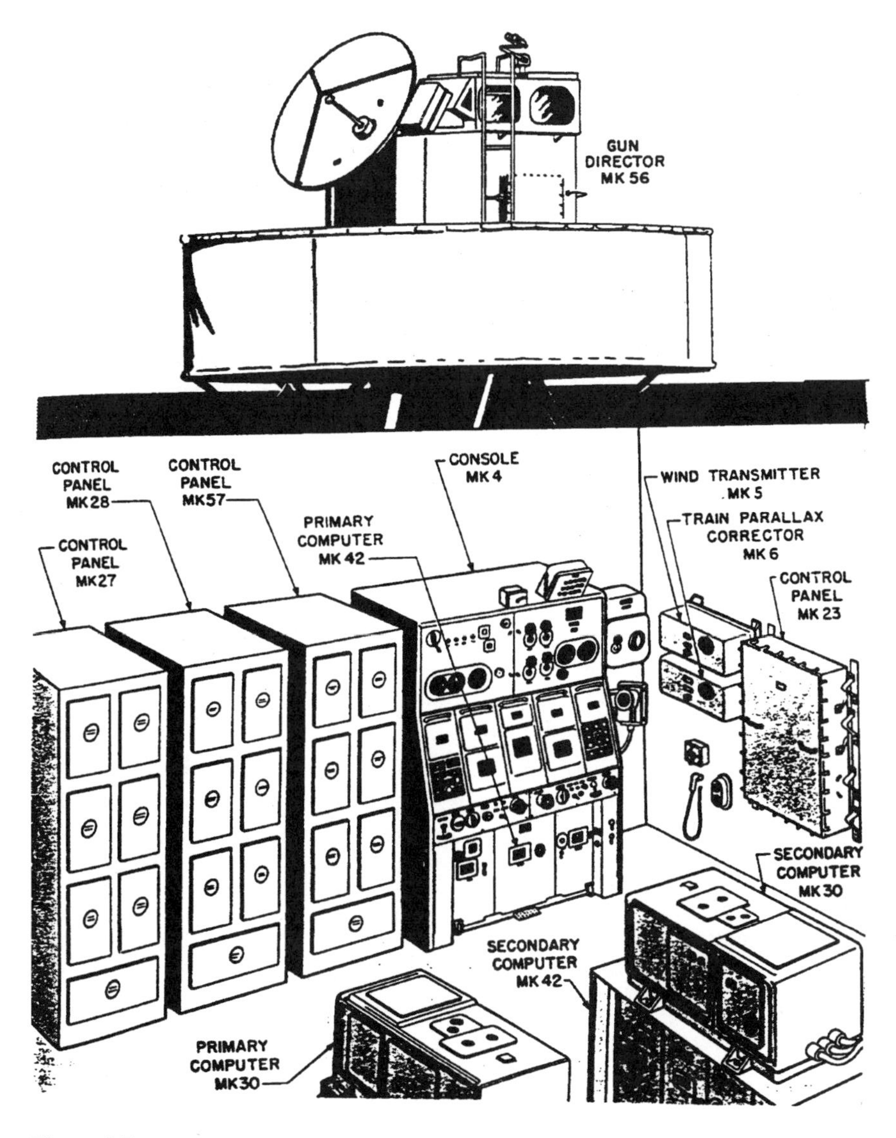

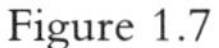

Figure 1.7
Layout of Mark 56 Gun Fire Control System. Two operators track optically from the deck positions, and two more work at the console in the control room below deck. [From *Naval Ordnance and Gunnery: Volume 2, Fire Control* (U.S. Navy, Bureau of Personnel, NavPers 10798, 1955, 319.]

soon as possible. He was greatly helped by Admiral King, Chief of Naval Operations, who was impressed by the Mark 56 and pushed BuOrd to let production contracts. But King voiced the fleet's frustration with previous automatic tracking radars and demanded the new system include optical as well as radar tracking—a further source of delay. When the war ended, Division 7 had five prototypes on order from General Electric, two of which neared completion. When the NDRC closed down in 1945, it transferred the contract to BuOrd in October 1945, which ordered 100 systems. Further problems, delays, and changes by the Bureau delayed Mark 56 production models from reaching the fleet until 1947. The device did, however, proliferate widely in the fleet and remained in service through the 1970s (never firing a shot in anger).

Throughout the Mark 56 project, Getting continued to redefine the work of building control systems. It entailed two parallel moves: transforming the Rad Lab from a radar group to a system integrator, and transforming the human operator into a dynamic component. For the first, Getting elaborated the Rad Lab's earlier position between the government and its contractors as a coordinating technical body. Earlier in the war, the urgency of the antiaircraft situation tended to smooth over political problems, and the NDRC's novelty provided a certain temporary authority. Furthermore, a new field like radar had no established expertise to resist the scientists' designs, so Getting had "complete technical control." Late in the war, however, as things became more established, routine, and industrial, they also became more complicated. Getting was used to dealing with the army, a low-tech service still awed by electronics; now he took on the Bureau of Ordnance, among the most technically sophisticated—and entrenched—groups in the services. Getting wanted to control not only engineering but production; otherwise the role of the Rad Lab would evaporate as the Mark 56 design neared completion. Toward this goal, Getting continued to cross established boundaries. He had joined Division 7, he had merged it with the Rad Lab (7.6), now he reached into the belly of the beast and sought to place a liaison within BuOrd. Warren Weaver, by now experienced at compromise with the services, thought the plans too ambitious, "discussed in over-pretentious terms," and suggested "the way to work with the BuOrd is, so to speak, to work with the BuOrd."[39] Still, in March 1945, Radiation Lab Director Loomis ordered that Getting be assigned to the Bureau of Ordnance, "to devote your time and efforts to technical problems on fire control and their application to radar."[40]

Getting now acquired the long-sought authority to delineate the role of the Radiation Lab. He formalized the Rad Lab's job of system integrator, which had previously been merely informal. Now the Rad Lab would

1. Make all technical information available to GE and the navy
2. Check and criticize designs at all stages of development
3. Send skilled representatives to participate in conferences
4. Report to the BuOrd on the progress of the project
5. Participate in testing of prototypes
6. Test preproduction models
7. Assist in establishing test and alignment procedures for manufacturing and acceptance tests
8. Assist in training programs

Engineering, production, testing, alignment, training: these activities comprised Getting's systems vision as much as time constants and signal spectra. To carry out these functions, the lab would have the following privileges:

1. To receive copies of correspondence between the navy and contractors
2. To receive copies of drawings and specifications prepared by contractors
3. To be notified when significant tests are carried out so representatives of the laboratory may participate
4. To be notified of technical conferences and conferences where technical decisions are to be made so that representatives of the laboratory may be present
5. To be given the opportunity to examine and criticize production designs or models before final design specifications are frozen
6. To have access to the establishments of the contractor and subcontractor by appointment, to confer with engineers, or to inspect equipment
7. To receive one of the first production models for test and study if directed by the navy[41]

Correspondence, drawings, specifications, tests, conferences, inspections: these embodied the relations between institutions. Getting needed to control them as much as the signal flows between components.

These remarkable lists reflect the experience Getting had acquired in a few years of doing research and managing contracts for the NDRC.

Each point seems to correspond to a particular episode where he lacked necessary authority: being excluded from meetings, not receiving correspondence, not having access to factory facilities. Getting redefined control engineering as an organizational as well as a technical task, and he vehemently argued BuOrd by itself was not up to it. Rather, Getting argued, the Radiation Laboratory had the best overall view of automatic control.

Where Getting appropriated authority from contractors, designers, and manufacturers, he also appropriated the work of human operators. Unlike system integrators who organized and collated different types of data, Getting's operators functioned purely mechanically, like "human servomechanisms." In 1945, while fighting for his project's priority, Getting wrote to Admiral Furer, the navy's Coordinator of Research and Development, connecting his ideas for designing new integrated systems with the principle of "automatic operation." Getting argued wartime experience had demonstrated the value of automation:

1. Human judgment introduced wrong guesses.
2. Human operators succumbed to battle fever.
3. The human mind reacts slowly compared to modern servo equipment.
4. The intellectual processes were incapable of utilizing most efficiently all the observable data.[42]

Radar burdened rather than relieved the operator by radically increasing the amount of information he had to sort through. Radar brought such complexity to military control that it strained human attention to hold the system together. Getting's automation would rein in that human involvement—a strategy that resonated with plans for demobilization, when men left the services but the machines remained.

To make his point, Getting invoked the success of the army's automated antiaircraft fire control. The M-9/SCR-584 system he helped design had entered the field, and Getting used the authority he gained by its success to sharply criticize the Navy's lack of automation: "In short the navy is an order of magnitude behind the army in heavy antiaircraft fire control and radar." The solution, of course, was to grant highest priority to Getting's Mark 56, "a wholly integrated operational system." But to what experience did he refer? How did automatic control perform in combat? What had been the experience of the human operators, whose behavior Getting now used to make his claim for automation? The M-9/SCR-584 combination did see service in the war. What were its successes? Where were its limitations?

Automatic Control's Finest Hour

As Getting promoted and composed his new project, the first automated antiaircraft system, the Radiation Lab's SCR-584 combined with Bell Labs' M-9 gun director, made its way off the production line and onto the battlefield. It was first successful at the beachhead in Anzio, Italy, in March 1944, when two of the radars and sixteen directors systems were deployed on the beach to cover the landing force. Together the SCR-584 and the M-9, combined with Sperry power drives to move the 90mm guns, shot down enemy aircraft that had been harassing the stalled landings.[43]

The M-9 still maintained the "constant altitude assumption" of the prewar Sperry directors. Rushed into production in 1942, it did not incorporate the latest results on predicting curved flight from work at BTL and MIT (being done by Norbert Wiener and others). The M-9 worked best, then, against attackers that flew straight and level—a tactic enemy bombers quickly learned to avoid. In June 1944, however, a new threat emerged from Nazi engineers, which perfectly matched the constant altitude assumption, exactly because it had no human operator. This threat itself relied on an automatic control system to fly, and hence was the perfect target for the automatic antiaircraft gun: the first operational robot bomb, the V-1.

Germany unleashed the "V-1 blitz" against London in mid-1944, and launched almost 7,500 "buzz bombs" against the English capital during the following eighty days. In the words of the British commander of the Antiaircraft Command, "It seemed to us that the obvious answer to the robot target of the flying bomb . . . was a robot defense."[44] Here the M-9/SCR-584 combination, to paraphrase Churchill, saw its finest hour. In anticipation of the V-1 blitz, and in response to a special request by Churchill, Radiation Lab engineers rushed systems out of production, onto ships and accompanied them to England. Members from the original SCR-584 design group (Davenport, Abajian, and Harris) and other Rad Lab staff members traveled along the English coast from battery to battery, aligning equipment, training crews, and tuning the radars—conveying tacit laboratory knowledge to crews in the field.[45]

One other technology completed the system: the proximity fuze, developed by Merle Tuve's Division T before their own foray into fire control. The proximity fuze (known as VT or variable-time fuze) placed a miniature radar in each shell which sensed when it neared the target airplane and set off the explosion.[46] Until then, antiaircraft, with all its feedbacks and controls, remained an open-loop system once the shell left the gun. The proximity fuze closed the loop—making each shell a one-dimensional guided missile, capable of reacting to its environment.

Buzz bombs posed no easy targets. Smaller than a typical airplane, they flew faster than bombers of the day (380 mph), and at low altitudes, averaging about 2,000 feet (indeed fast and low would become the classic radar-evading strategy). And they proved remarkably robust to shellfire, sometimes taking several hits before falling. Still, between 18 June and 17 July 1944, the automated guns shot down 343 V-1's, or 10 percent of the total attack, and 22 percent of those shot down (the others were hit by aircraft, barrage balloons, and ships). During this period the AA batteries were deployed in a ring south of London; and their ability to fire was limited to avoid hitting fighters that also pursued the buzz bombs. The guns could fire only on positive identification of the target and if no fighter were in pursuit, giving aircraft the first chance to shoot down the missiles. In mid-July, the AA batteries moved to the coast, where they could fire without limit over the channel. From 17 July to 31 August, the automated guns accounted for 1,286 V-1 kills, or 34 percent of the attack, 55 percent of those shot down (the improved success rate probably also reflects the effects of the Rad Lab members' assistance).[47] That October, the M-9/SCR-584/VT-Fuze combination defended Antwerp from the V-1 with similar success. In this tense confrontation of robot weapons, the automated battlefield, which even today remains a dream of military technologists, began to take shape.

Despite its success, the system had seams in its automation. Radar's new way of seeing did not immediately replace ocular vision. Throughout the war, automatic and manual perception had an uneasy coexistence—translating between the two proved difficult, error-prone, and fatiguing. A detailed assessment of these issues came not from Ivan Getting but from his rivals and former collaborators at Bell Labs. In July and August of 1944, a group of four army officers and two BTL employees, including Clarence A. Lovell (who headed the T-10/M-9 design team), traveled to Europe to tour antiaircraft batteries and observe their operation against the V-1's. This group's report set out requirements for future antiaircraft systems. Unsurprisingly, the BTL report criticized the Rad Lab radar because the SCR-584 could not search and track simultaneously (BTL's rival SCR-545 could).[48] BTL also reported the system demanded unreasonable concentration from its operators—"there are too many sources of present position data for the computer"—because it allowed radar, optical trackers, and a rangefinder, or a combination. Operators had to judge and juggle these alternate instruments. Manual tracking, for example, was still necessary because of interfering ground echoes (for targets low on the horizon), closely spaced targets that a radar might not be able to distinguish, and the possibility of jamming.[49]

The M-9/SCR-584 was more a combination of two separate units (the BTL director and the Rad Lab radar) than an integrated system. Radar trackers sat inside a trailer while optical trackers and rangefinders (on the director) sat outside. BTL's report proposed adding a means for switching between radar and optical tracking. Ultimately, it argued, any new system should mount optical instruments right at the radar station so operators could "track either optically or by radar without changing their positions or the controls which they employ." BTL's report recommended combining tracking and computing in a single unit, similar to the integrated, blind-firing system Ivan Getting proposed to the navy in 1944.

Getting built that case on the success of the SCR-584/M-9 combination, and on the seeming inability of human operators to keep up with the data flow. Much of the trouble, of course, arose not from the limits of human performance, but from relationships between design organizations divided among perception of inputs, integration of several sources of data, and articulation or output of results. Getting's Mark 56, the "wholly integrated, operational system," proposed to overcome these difficulties by defining a new institutional role, the system integrator, supervising tighter coupling of radar and computer, design and production, operator and machine.

More Than the Sum of Its Component Parts: Dynamic Systems and Military Contracting

Radar's new subtlety accompanied new expertise; the Radiation Lab staked out a role as a system integrator. Organizational relationships solidified as technical systems, at first the partially integrated but combat-tested SCR-584 radar, and then the integrated Mark 56 Gun Fire Control System. The Rad Lab also embodied its claims as knowledge, among its most lasting contributions. After the war, the laboratory, with OSRD funding, published a twenty-seven-volume series on radar to distribute the results of its wartime work. Getting proposed a volume on fire control but series editor Ridenour turned him down because of security restrictions. Still, three of the twenty-seven volumes emerged from the work of Getting and his associates: Louis Ridenour's *Radar System Engineering*, Tony Svoboda's *Computing Mechanisms and Linkages*, and *Theory of Servomechanisms* by physicist Hubert M. James, Rad Lab Division 8 servo engineer Nathaniel B. Nichols, and Division 8 mathematician Ralph S. Phillips.[50] Along with similar volumes from Bell Labs and the Servo Lab, "James, Nichols, and Phillips" became a canonical postwar text of control engineering, introducing a generation of engineers to newly constituted discipline.[51]

For the Rad Lab scientists and engineers, the boundaries of this knowledge derived from the boundaries of radar-driven fire control. The book opens, "The work on servomechanisms in the Radiation Laboratory grew out of its need for automatic-tracking radar systems." Ivan Getting introduces the volume and reviews the basic definitions of servomechanisms and the history of design techniques. Noting the field's lack of stable epistemology, Getting observes, "It is nearly as hard for practitioners in the servo art to agree on the definition of a servo as it is for a group of theologians to agree on sin." Getting and his co-authors certainly acknowledged their predecessors; the twenty-page introduction cites earlier pioneers in servo design and theory: Hazen, Bush, Minorsky, Nyquist, Harris, Brown, Hall, Wiener, and Bode. Still, the book reflects Radiation Lab culture: design examples include the SCR-584 radar, numerous automatic and manual tracking schemes, filters for radar signals, and methods for dealing with noisy echoes. The Rad Lab volume, while stabilizing control systems as a coherent body of knowledge, defined that stability by the systems vision of radar scientists.

Their notion of the system as a dynamic entity, however, conflicted with the prewar vision, which saw a system as a "sum of component parts." In the 1930s, for example, Harold Hazen defined the modular blocks of the differential analyzer so he could manipulate and recombine them ad infinitum. In this world, the whole was exactly the sum of its parts. But radar, noise, and feedback complicated that simplicity. Hazen articulated the newer approach in his 1945 preface to Division 7's "Summary Technical Report":

> One must always remember that a fire-control system is more than the sum of component parts. It is an integrated whole with inter-

> related functioning of all its parts and one is safe in considering parts separately only if one always keeps in mind their relation to the whole.[52]

In a dynamic control system, each component affected the others. Computer design, for example, depended on the bandwidth of the radar, its noise spectrum, and the capabilities of the human operator. But in the early 1940s, the political economy of military technology was built on the older model where systems were decomposable. BuOrd divided up problems, assigned pieces to separate contractors, and assembled the pieces into systems. That approach only worked, however, if a system really was the sum of component parts; noise proved it was more, and pushed systems to a higher level of complexity. The NDRC's fire control division, and then the Radiation Lab's Ivan Getting, reconfigured the structure of contracting to suit a dynamic, noisy, error-prone model of a system. To embody their model in working systems, however, they needed a set of engineering techniques to complement institutional relationships. In parallel developments, those techniques began to emerge during the war as well, driven by similar problems of radar noise and feedback loops, gradually defining a general quantity to flow through the new integrated systems: information.

NOTES

1. Ivan Getting, "Introduction," in Hubert M. James, Nathaniel B. Nichols, and Ralph S. Phillips, *Theory of Servomechanisms* (New York: McGraw Hill, 1947), Radiation Laboratory Series #25. Warren Weaver, foreword to "Final Report: D-2 Project #2, Study of Errors in T-10 Gun Director," National Archives RG-227, Office of Scientific Research and Development, Division 7 (hereafter referred to as OSRD7) Office Files of Warren Weaver, 3.

2. Thomas P. Hughes, *Networks of Power* (Baltimore: Johns Hopkins University Press, 1983).

3. Roger B. Colton, "Radar in the United States Army: History and Early Development at the Signal Corps Laboratories, Fort Monmouth, N.J.," *Proc. I.R.E.* (November 1945): 740–753.

4. Henry Guerlac, *Radar in World War II* (New York: Tomash Publishers/American Institute of Physics), 103–110. A similar device developed by the army, the SCR-270, with a wavelength of 2.5 meters, was designed as an early warning and search system. It was deployed in Hawaii in August 1940, and detected the attack on Pearl Harbor at a distance of over 100 miles.

5. Guerlac, *Radar in World War II*, 243–250.

6. David Zimmerman, *Top Secret Exchange: The Tizard Commission and the Scientific War* (London: A. Sutton Publishing/McGill Queen's, 1996).

7. Ivan Getting, *All in a Lifetime: Science in the Defense of Democracy* (New York: Vantage Press, 1989), 37.

8. Getting, *All in a Lifetime*, 107.

9. Ivan Getting, "SCR-584 Radar and the Mark 56 Naval Gun Fire Control System," *IEEE Trans. Aerospace and Electronic Systems*, AES-11 no. 5 (September 1975): 924.

10. Getting, "SCR-584 Radar." For a technical discussion of the 584 servos, see Hubert M. James, Nathaniel B. Nichols, and Ralph S. Phillips, *Theory of Servomechanisms* (New York: McGraw Hill, 1947), Radiation Laboratory Series #25, 212–224, Stuart Bennett, *A History of Control Engineering: 1930–1955* (London: Peter Peregrinus, 1993), 143–146, which includes Godet's servo.

11. Ivan Getting, interview with author, 3 March 1996, Coronado, California. Notes and recording in author's possession. Getting oral history interview by Frederik Nebeker, 11 June 1991. IEEE Center for the History of Electrical Engineering, Radiation Lab Oral Histories, available on the World Wide Web at http://www.ieee.org:80/history_center/oral_histories/oh_rad_lab_menu.html

12. "Report of A.A.B. Test on XT-1 at Fort Monroe, Virginia, February, 1942," Radiation Laboratory Report no. 359. For firsthand accounts of the XT-1/SCR-584 development and its field deployment, see Henry Abajian oral history interview by Frederik Nebeker, 11 June 1991; Lee Davenport oral history interview by John Bryant, 12 June 1991; and Leo Sullivan oral history interview by Frekerik Nebeker, 14 June 1991. IEEE Center for the History of Electrical Engineering, Radiation Lab Oral Histories.

13. For a summary of SCR-584 projects, including a number of modifications, see National Defense Research Committee, *NDRC Division 14 Final Project Report*, MIT Archives, 2-41 to 2-68.

14. The SCR-584 proved no simple device to manufacture. It required 140 tubes and a host of specialized electronics parts, weighed ten tons total, and cost about $100,000. It did not go into full production until mid-1943. For the difficulties of producing the SCR-584 see George Raynor Thompson, Dixie R. Harris, Pauline M. Oakes, and Dulany Terrett, *The United States Army in World War II: The Technical Services, The Signal Corps: The Test (December 1941 to July 1943)* (Washington, D.C.: Office of the Chief of Military History, United States Army, 1957), 265–274; Getting, *All in a Lifetime*, 121–127; Guerlac, *Radar in World War II*, 481–483; Getting, Harris, Abajian, Davenport oral histories. For the operational history of the SCR-584, see George Raynor Thompson and Dixie R. Harris, *The United States Army in World War II: The Technical Services, The Signal Corps: The Outcome (Mid-1943 through 1945)* (Washington, D.C.: Office of the Chief of Military

History, United States Army, 1966), 474–477. Guerlac, *Radar in World War II*, 480–496, 853–862, 882–897, 1018–1025. Field commanders employed the rugged and versatile SCR-584 for numerous uses beyond the one originally envisioned. It could track mortar shells back to their source, so army units could attack mortar positions. It tracked V-2 trajectories, so American bombers could go after their launch facilities. In combination with automatic plotting boards, it enabled air controllers to "talk" fighter planes to their targets—prefiguring the automated air defense systems of the Cold War and the air traffic control systems of today. During testing at the Aberdeen Proving Ground, it tracked shell fired from the army's 90mm guns and revealed a significant error in their firing tables. The firing table had been calculated on a Bush differential analyzer, but its operator had set up its gearing incorrectly. These errors had then been built into all the Sperry M-7 directors, but since the T-10 was still in development, it could be properly corrected. The army used it during the Battle of the Bulge for tracking enemy vehicles as well. It was also used to track remote-controlled planes for automated bombing attacks (like the one in which Joe Kennedy, Jr., was killed). A number were given to the Soviet Union, and for many years Soviet radars incorporated many of the SCR-584's design features. Getting, *All in a Lifetime*, 130–135. Also see Abajian, Davenport, Harris, Getting oral histories.

15. WW (Warren Weaver) diary, 5 December 1940, meeting with Loomis. WW to Loomis, 10 December 1940. WW diary, 13 December 1940. OSRD7 General Project Files, box 70, collected diaries, vol. 1.

16. See TCF (Thornton C. Fry) diary of meeting with Col. Bowen, 3 July 1941, OSRD7 General Project Files, box 70, collected diaries, vol. 2, and Earl W. Chafee, "Memorandum of Conference in Fire Control Department," 24 September 1942. OSRD7, E-83 Office Files of Warren Weaver, box 4, Sperry Gyroscope folder.

17. Emphasis original, WW diary, 12 November 1942. ORSD7 General Project Files, box 72, collected diaries, vol. 5. WW to Lovell, 23 November 1942. WW to Chafee, 1 December 1942. OSRD7 E-82 Office Files of Harold Hazen, box 9, Rad Lab folder; see other correspondence to Weaver from Fry, Hazen, and Caldwell as input for Chafee's report, many of which are more informative on issues of "coordination" between system elements than the report itself.

18. Earl W. Chafee, "Study of the Requirements for a Satisfactory Antiaircraft Fire Control System," 15 February 1943. Sperry Gyroscope Company Papers, box 33, Hagley Museum and Library. The report can also be found in OSRD7 E-82 Office Files of Harold Hazen, box 9, Rad Lab folder. The Chafee Report includes the most comprehensive history of Sperry's prewar antiaircraft development program in the historical record. A meeting held at Sperry Gyroscope in February 1943 covers similar issues, with input from Rad Lab officials (Ridenour, Griggs), the Ford Instrument Company (Tear, Jahn), and Sperry (Draper, Bassett, Holschuh, Willis, White). John B. Russell diary, OSRD7 General Project Files, box 70, collected diaries, vol. 3.

19. WW to Fletcher, 28 February 1941. Project file 23140, ATT. Ridenour to Lovell, 24 September 1941. Project file 23140, ATT. GRS diary, 21 May 1941.

Ridenour to Lovell, 6 August 1941. Lovell to Ridenour, 23 September 1941. OSRD7 General Project Files, Project #2. Getting, interview with author.

20. "Final Report: D-2 Project #2, Study of Errors in T-10 Gun Director."

21. "Study of Errors in T-10 Gun Director," 72. For a Rad Lab study of jitter in a tracking servo from radar data, see "Data Smoothing," Radiation Laboratory Report, no. 673.

22. Warren Weaver, foreword to "Final Report: D-2 Project #2, Study of Errors in T-10 Gun Director," OSRD7 Office Files of Warren Weaver, 3.

23. The competitors were a similar Bell Labs radar, the SCR-545 (which was produced in limited numbers), and the Canadian GL-III-C, which had been designed in response to Tizard's initial assignment for gunlaying. The SCR-545 was the closest rival to the Radiation Lab set, and included a long-wave search radar along with its microwave tracker. SHC (Samuel H. Caldwell) diary, 26 March 1942, and J. B. Ridenour Diary, 4 April 1942. OSRD7 General Project Files, Project #2.

24. For a detailed account of BuOrd and its "fire control clique," see David A. Mindell, "Datum for Its Own Annihilation: Feedback, Control, and Computing, 1916–1945" (Ph.D. diss., MIT, 1996), chap. 2.

25. Rowland and Boyd, *The U.S. Navy Bureau of Ordnance*, 421, 429.

26. IAG (Ivan A. Getting) to KTC (Karl Taylor Compton), "U.S.N. AA Director Mk. 56," 29 December 1943. OSRD E-39, Office Files of Karl Taylor Compton, box 51, Division 7 folder.

27. Getting, "SCR-584 Radar," 932; interview with author.

28. Division 7 meeting minutes, 3 February 1943. OSRD7 General Project Files box 72, Division 7 meetings folder.

29. HLH (Harold Locke Hazen) diary, 20 and 21 April 1943. OSRD7 General Project Files, office files of Harold Hazen, box 70.

30. Division 7 meeting minutes, 28 April 1943. Guerlac mistakenly recounts these events as the summer of 1942, in *Radar in World War II*, 490, based on a misunderstanding of Getting's letter to Compton of 29 December 1943.

31. Getting, "History of Division 7.6," 7. See Fagan, ed., *History of Engineering and Science in the Bell System*, 158–162. BTL built a prototype of this computer, designated Mark 8, which directly replaced the Ford Instrument Mark I, but it was never put into production.

32. Getting, "History of Division 7.6," 7.

33. Getting, oral history interview.

34. The Complete 7.6 membership was George Agins, vice president, Arma Corporation; R. F. Cooke, vice president, Ford Instrument Company; C. S.

Draper, MIT; A. W. Horton, Bell Telephone Laboratories; R. M. Page, Naval Research Laboratory; E. J. Poitras, Division 7 (Ford Instrument Company); R. B. Roberts, Section T, OSRD; A. L. Ruiz, Division 7 (General Electric).

35. Getting, *All in a Lifetime*, 201.

36. Getting, "History of Division 7.6," 10.

37. IAG to KTC, "U.S.N. AA Director Mk. 56," 29 December 1943. OSRD7, E-39, Office Files of Karl Taylor Compton, box 51, Division 7 folder.

38. For the design history of the Mark 56, see IAG diary, "Conference on Mark 56 Director," 10 June 1943, "Mk 56 Radar Discussions at Bureau of Ordnance," 15 July 1943, "Mk 56," 2 July 1943, "Mk 56," 26 July 1943, OSRD7 General Project Files, box 72, IAG diary folder. Division 7 "Minutes of Rochester Meeting," 5 January 1944, OSRD7 General Project Files, box 72, Division 7 Meetings folder. Getting, *All in a Lifetime*, 177–181. For an operating description of the system, see *Naval Ordnance and Gunnery, Volume 2: Fire Control* (U.S. Navy Bureau of Personnel, NavPers 10798), 318–340. For project history, see *Division 14 Final Report*, 4-55 to 4-63. For Svoboda's relay computers, see "Eloge: Antonin Svoboda, 1907–1980," *Annals of the History of Computing* 2, no. 4 (October 1980): 284–292.

39. WW to IAG, 16 January 1945. OSRD7, office files of Ivan Getting, box 62.

40. Loomis to IAG, 9 March 1945. OSRD7, office files of Ivan Getting, box 62.

41. "Statement of Relationships between the Bureau of Ordnance, U.S. Navy and the National Defense Research Committee, OSRD, on the Development and Production of the Gunfire Control System Mark 56," reprinted in Getting, *All in a Lifetime*, 186.

42. Getting to Furer, 26 April 1945, reprinted in *All in a Lifetime*, 182–185.

43. Leo Sullivan from the Rad Lab accompanied the SCR-584 to Anzio. See Sullivan oral history.

44. General Sir Fredrick Pile, *Ack-Ack, Britain's Defence against Air Attack during the Second World War* (London: Harrap, 1949), 314–315. Also see Pile to George C. Marshall, quoted in Bush to Hazen, 31 August 1944. OSRD7 E-82 Office Files of Harold Hazen, box 9, Rad Lab folder.

45. Getting, Davenport, Abajian oral histories.

46. "Antiaircraft Artillery Fire Control," prepared by the Bell Telephone Laboratories for the Ordnance Department, U.S. Army in fulfillment of Contract W-30-069-Ord-1448, 1 May 1945. ATT, 14. Those manning the batteries were often slow to recognize the value of the fuze. If the VT fuzed shells didn't find a target, they exploded after some fixed time-out period due to a self-destruction mechanism. Because these explosions were likely to be far from the targets, the proximity fuze did not produce large numbers of explosions near the target like time fuzes did. Instead, gunners would see very few explosions near the target and

many explosions far beyond it. "To those used to seeing large numbers of bursts around the target from time fuzed ammunition, this distribution of bursts makes the performance of the battery look very poor," despite much improved accuracy.

47. Guerlac, *Radar in World War II*, 859. For a personal account of the automatic system versus the V-1, see Abajian interview.

48. "Antiaircraft Artillery Fire Control," 9.

49. Ibid., 10.

50. Hubert M. James, Nathaniel B. Nichols, and Ralph S. Phillips, *Theory of Servomechanisms* (New York: McGraw Hill, 1947), Radiation Laboratory Series #25. Antonin Svoboda, *Computing Mechanisms and Linkages* (New York: McGraw Hill, 1948), Radiation Laboratory Series #27. Louis B. Ridenour, *Radar System Engineering* (New York: McGraw Hill, 1948), Radiation Laboratory Series #1.

51. Gordon S. Brown and Donald P. Campbell, *Principles of Servomechanisms* (New York: Wiley, 1948). Leroy MacColl, *Fundamental Theory of Servomechanisms* (New York: Van Nostrand, 1945). See Chris Bissel, "Textbooks and Subtexts: A Sideways Look at the Postwar Control Engineering Textbooks, Which Appeared Half a Century Ago," *IEEE Control Systems* 16, no. 2 (April 1996): 71–78, for an account of the postwar publishing effort, and a comparative discussion of control textbooks. Comparing degrees of importance for these books is, of course, splitting hairs, although Bissel calls the Rad Lab volume "perhaps the most influential of all the American publications of the 1940s."

52. Harold Hazen, "Fire Control Activities of Division 7, NDRC," in *Summary Technical Report of Division 7, NDRC*, volume I: *Gunfire Control*, 4. Stuart Bennett has noted the "systems approach" in his comparison of British and American fire control work during the war in *A History of Control Engineering: 1930–1955*, 125.

2

The Adoption of Operations Research in the United States during World War II

Erik P. Rau

Operations research (OR) today has infiltrated such a diversity of activities, from military planning and logistical support to hospital management and airline scheduling, that even its practitioners struggle to define their field.[1] Its emphasis on mathematical modeling and computer simulation often obscures OR's relatively straightforward object: to understand and improve the use of complex socio-technological systems.[2]

Systems, as we know from the work of Thomas and Agatha Hughes, are often heterogeneous, consisting not only of technological components, but also individuals and institutions.[3] As a result, systems approaches like OR challenge existing boundaries and draw new ones. Thus, cybernetics dissolves the biological-technological divide, creating cyborgs; systems analysis blurs the distinction between present and future by attempting to show the future outcomes of actions taken now; and OR, at least for the period covered in this chapter, integrates the makers and users of technology into a conceptual whole.[4] The history of OR, therefore, is not limited to a discussion of evolving methods. It must also include the inevitable negotiations that arise whenever a conceptual framework is imposed on aspects of social life.[5] Systems approaches are, after all, discourses about control, social as well as technological.

This chapter examines the adoption of operations research in the United States during the Second World War. Despite the relative ease with which it developed in Britain, OR generated considerable controversy when scientists, engineers, and military officials attempted to introduce it into the United States. At the root of the conflict lay incompatible strategies for organizing research and development for the war effort. Vannevar Bush, director of the Office of Scientific Research and Development (OSRD), designed his organization not only to mobilize the nation's civilian research and development capacity, but also to shield that capacity from government influence.[6] Keeping the two apart, however, had the practical effect of separating the makers of new weapons systems from their users—the operational military commands. Advocates of OR anticipated using it to integrate the two, undermining the organizational boundaries Bush had painstakingly constructed around the OSRD.

OR's eventual success, despite Bush's recalcitrance, is emblematic of systems approaches' ascendancy in collaborations between the federal government and civilian technical expertise during and after World War II. Bush's approach represents the older conservative preference for a weak state harnessed to the needs and agendas of business and science, variously labeled by historians as corporatism or the associational state.[7] Often the individuals involved in implementing OR came to professional maturity either during or after the New Deal. They began the war with fewer reservations about government-directed research, and even these worries were often reduced by the rigors of the war effort. Systems approaches like OR seemed to promise a more robust alliance between the state and civilian expertise. OR offers a prehistory for the collaboration President Dwight D. Eisenhower later described as the "military-industrial complex," a relationship explored elsewhere in this volume. The adoption of operations research in the United States also illustrates the limitations of Bush's corporatist strategy and the emergence of what historian Brian Balogh has called the "proministrative state."[8]

It is perhaps no coincidence that operations research originated in Britain rather than in the United States. In the years immediately prior to the Second World War, the British scientific and engineering elite had fewer reservations about government-directed research than their American counterparts. In fact, many of those involved in OR were inclined toward the political left. After the war and the election of a Labour Government, wartime operations researchers helped manage nationalized industries, like coal and steel.[9] The immediate impetus behind OR's creation, however, was less political than pragmatic: it became essential for the effectiveness of Britain's radar air defense system.

In 1936, Henry Tizard, a leading figure in the British development of radar, devoted time to integrating the technology into an effective air defense system. Radar promised to maximize the protection of a relatively small number of fighter aircraft. To do so, however, required trained military users; smooth communication among radar stations, Royal Air Force (RAF) Fighter Command headquarters, and air bases; and new tactics. Machines, human operators, and institutions had to work in concert. Together with RAF officials, he initiated a series of experiments at the RAF's Biggin Hill Station that lasted nearly two years. In 1938, A. P. Rowe, an RAF official appointed to supervise the radar development at Bawdsey Research Station, continued and expanded this work. He dispatched scientists and engineers from Bawdsey to undertake these tasks. To differentiate them from the "developmental research" at Bawdsey, Rowe coined the term "operational research." RAF Fighter Command officers participated in the early OR experiments and soon instituted the practice of maintaining a group of scientists and engineers on hand to advise on the use and improvement of the air defense system. By the time the war started in 1939, these groups had became the earliest OR sections.[10]

As radar began being used in antiaircraft, antisubmarine, and bombing operations, OR extended beyond RAF Fighter Command to other parts of the RAF and finally to the navy and army. Tizard and other scientists, especially physicist Patrick Maynard Stuart Blackett, aggressively promoted OR as an essential part of radar systems.[11] After the Battle of Britain, in which radar played a critical role, OR groups began to tackle problems other than radar. At the Ministry of Home Security's Research and Experiment Station at Princes Risborough, for example, researchers collected data on bomb damage inflicted by the Luftwaffe. They later used this information to recommend the kinds of bombs and fuses to be used against German targets.[12]

By the end of 1941, OR had become an important part of every major military community. While some staff officers seem to have resisted the civilian outsiders, most did not. OR not only addressed the technical problems a command might experience with new military hardware, it also educated civilian researchers in the military considerations—technical requirements, tactical and strategic consideration—that officers faced. It helped the makers and users of advanced military technology organize themselves into systems. By early 1942, immediately after the United States entered the war, British scientists encouraged their American counterparts to begin their own OR program.[13]

British researchers may not have appreciated the political culture in which American scientists and engineers lived, one significantly different from their own. Through the 1930s, the attitude of the American scientific and engineering elite toward government involvement in research was, at best, ambivalent. As recently as 1930, the National Research Council (NRC), the operating arm of the National Academy of Sciences, harbored virulent opposition to government participation in civilian research. Only with the Great Depression did the NRC change its tune, when a younger generation of scientific statesmen, led by Massachusetts Institute of Technology president Karl T. Compton and Johns Hopkins University president Isaiah Bowman, adopted a position akin to that of American business leaders.[14]

Sometimes called corporate liberalism, or simply corporatism, this political stance updated the earlier antistatist, laissez-faire orientation of American capitalism. Basically, the business elite sought to turn government regulatory power to serve their interests.[15] Mining engineer Herbert Hoover had developed some of these ideas into a philosophy he pursued as president. Many among America's civilian researchers believed Hoover's approach to be descriptive of their own ideas.[16] Compton and Bowman followed similar convictions as they tried to forge a new relationship between the federal government and the research community.

However, by the time Compton, Bowman, and other research elites took the initiative, Hoover had lost the White House to Franklin D. Roosevelt. The New Deal was anathema to many of the research elite, although Compton and Bowman cautiously proceeded with their courtship of several agencies. New Dealers in the administration regarded these attempts as self-serving. Mutual mistrust prevented any meaningful change in the relationship between the two sides for the remainder of the 1930s.[17]

The stagnation ended when rumors of war crossed the Atlantic. During 1939, the Massachusetts Institute of Technology's former engineering dean, Vannevar Bush, moved to Washington to take up the presidency of the Carnegie Institute of Washington and chair the National Advisory Committee for Aeronautics. Bush grew concerned about the state of readiness among the armed forces and considered ways of mobilizing American civilian research for the impending war. Bush persuaded Roosevelt of the necessity of mobilizing the nation's research and development capacity and in June 1940 he was made the chairman of the National Defense Research Committee (NDRC). A year later, Bush won the approval to establish another organization, the Office of Scientific Research and Development (OSRD). The OSRD provided support to the NDRC as well as well as several other research-related functions.[18]

Bush intended the NDRC not only to mobilize America's civilian research capacity, but also to provide corporate and university research facilities with a bulwark against government influence. Although the NDRC was itself a government agency, Bush staffed it with members of the scientific elite, including Compton, National Academy of Science president Frank Jewett, and Harvard University president James B. Conant. Moreover, Bush intended that the NDRC be dissolved at war's end, thereby burning the bridge between the government and civilian research.[19]

The central feature of the NDRC was its development of the research contract. Contracts limit the obligation of the co-signers. Bush believed that properly devised research contracts could put private research institutions in harness for the war and release them from further obligation after specified tasks were completed, protecting the institutions from future government interference. Historian Larry Owens has remarked upon the strengths and weaknesses of this novel strategy. Contracts were limited to research and development *only*. Responsibility for production of devices once NDRC contractors had designed them fell to the military procurement agencies. And although NDRC's various divisions might provide some support, the military departments held responsibility for training, maintenance, and other support.[20]

An unavoidable consequence of the NDRC's charter and its contract strategy was the separation of the makers and users of weapons systems developed by the NDRC. Moreover, Bush refused to bridge the gap if doing so committed the personnel of the NDRC or its contractors to nondevelopmental activities. Thus Bush's corporatist strategy imposed precisely the boundary that OR had been developed to dissolve.

Bush may have been aware of operations research as early as the late summer of 1940. Even after January 1941, when it is certain that he did know of it, Bush did nothing to encourage its adoption in the United States. Toward the end of August 1940, Tizard led a British diplomatic and scientific delegation to the United States whose goal was to secure greater scientific exchange between the United Kingdom and the United States.[21] Most of the excitement generated by the visit, for those with the proper security clearances, was British radar technology, specifically the cavity magnetron, a device capable of generating strong microwaves, critical for radar.[22] The NDRC had already created a division to research microwave generation, later known as Division 14 (Radar),[23] which soon contracted with MIT to organize the Radiation Laboratory. Please refer to David A. Mindell's contribution to this volume.) Because of the centrality of OR to the British radar program, it seems unlikely that Tizard would not have described it to Bush. Bush took no action on the matter.

Nor did he do so after January 1941, when Conant reciprocated Tizard's visit with an American scientific delegation to Britain and discovered OR. For several weeks, Conant and a small band of NDRC scientists visited British research installations and reviewed British research organization. An NDRC liaison office was established in the American embassy, known as the London Mission, which began using U.S. State Department communication channels to relay technical information between the two allies. Selected NDRC staff members spent tours of duty at the London Mission and acted as liaisons on related Anglo-American research efforts.[24] During their tours, the Americans also made note of their British colleagues' involvement in military operations. Princeton physicist H. P. Robertson and MIT architect John Burchard, for instance, were especially impressed with the OR work on blast damage being compiled by J. D. Bernal, Solly Zuckerman, and others at Princes Risborough.[25] Robertson and others urged the adoption of British OR practices, but Bush maintained his silence on the issue.

Bush had other means of keeping the NDRC—and, later, the OSRD—coordinated with the military, but none of these brought NDRC researchers in direct contact with military commands in the field in the way OR did. Bush established formal liaisons with both the navy and the army, but these were a far cry from the exchange capable through OR. For instance, the navy's coordinator of research and development, Rear Admiral Julius A. Furer, had no authority over naval matters beyond his office. To make matters worse, Bush's relationship with the chief naval officer, Admiral Ernest J. King (also the commander-in-chief, U.S. Fleet) was stormy throughout the war.[26]

At the War Department, Bush enjoyed cordial relations with Secretary Henry L. Stimson and his special assistant Harvey H. Bundy. What was decided in Washington, however, sometimes bore no relationship with what occurred at the front. In April 1942, Bush convinced Stimson to take on a technical advisor, Edward L. Bowles of the MIT electrical engineering department, with whom Bush had strained relations. Bowles, familiar with the Radiation Laboratory's radar research, was particularly interested in the effective use of new technology in warfare, especially radar, for he had been working at Radiation Laboratory. Soon after his appointment, Bowles grew aware of OR and became quite excited by it. In Bowles's opinion, Bush had a talent for identifying the strategic implications of new weapons, but paid too little attention to the details of how new devices would be deployed in the field.[27]

Finally, Bush gained access to the Joint Chiefs of Staff when he persuaded Stimson to use his leverage and create a staff panel where the services and the OSRD could keep one another abreast of military requirements and technological capabilities. Bowles later criticized Bush's use of the Joint Committee on New Weapons and Equipment (JNW), as the panel was called. Bowles complained that Bush, who chaired the JNW, dwelled too much in details.[28]

Even into the early months of the war, Bush continued to avoid taking a position on operations research, trying alternative strategies instead. None of these, however, addressed the central issue of overcoming the boundary between the makers and users of new weapons as OR did.

Despite its two-year warning, the United States was ill prepared to enter the war in December 1941. In three areas, the unpreparedness led to the introduction of operations research to the military services, specifically the Army Signal Corps, the Army Air Forces, and the navy. These areas were radar air defense, strategic bombing, and antisubmarine warfare. In all three, American forces either borrowed heavily from, or were forced to collaborate closely with, their British counterparts. Doing so inevitably exposed U.S. forces to OR. Before America's first year in the war had elapsed, its military departments had founded OR groups and pressured the OSRD for assistance.

The event that catapulted the United States into the Second World War, the Japanese attack on Pearl Harbor, also triggered the first official interest in OR. The surprise raid on Hawaii had been successful, in part, because of the local air defense system's failure. The contrast with Britain's radar experience could not have been more striking. During the Battle of Britain and the subsequent Blitz during the summer and fall of 1940, Britain's radar network had played a critical, even decisive, role in turning back the German Luftwaffe. Stimson immediately demanded that American air defenses be reviewed. Importantly, he requested that Great Britain contribute one of its radar pioneers to lead the study. Before the end of December 1941, Robert Watson-Watt had accepted this assignment and settled in at the British embassy in Washington.[29]

Watson-Watt filed his report in January. Far from sanguine, he noted the "insufficient organization applied to technically inadequate equipment used in exceptionally difficult conditions."[30] Watson-Watt delivered a similarly critical review of the Panama Canal Zone's radar defenses, an area of considerable concern among American leaders. The OSRD's radar program was still months away from putting units into production; in any event, the inadequate use of radar equipment was at least of as much concern as the hardware itself. Watson-Watt suggested that air defense training centers use British methods and equipment. Among the British methods, of course, was operations research.

Watson-Watt's report significantly embarrassed several services in the War Department. One was the Army Signal Corps, under the leadership of Chief Signal Officer Major General Dawson Olmstead. Within weeks of the report's release, Olmstead ordered his Director of Planning, Colonel Frank C. Meade, to establish an OR group directly under Olmstead's command.[31]

The report's impact on the Army Air Forces (AAF)[32] had even greater consequences for the development of OR in the United States. Watson-Watt's review was no less embarrassing for the AAF's commanding officer, Lieutenant General Henry H. Arnold. Arnold immediately appointed Colonel Gordon P. Saville to his staff as Director of Air Defense. Saville was no newcomer to radar air defense, nor to OR. Under Arnold's orders, he had spent much of 1941 constructing radar networks on both coasts modeled on the British example. His knowledge of British radar and OR came from his tenure as an AAF military observer in Britain, which lasted well over a year. With Watson-Watt's help, Saville established an interceptor training center in Orlando, Florida. The center, eventually known as the AAF School of Applied Tactics, closely followed Watson-Watt's recommendations. Radar units of British design were acquired via Canada. British trainers arrived soon thereafter.[33] Given Saville's and Watson-Watt's background, it is not surprising that the training and experiments carried out at the center closely resembled those of the RAF Biggin Hill Station nearly six years before. From the center's early planning stages, Saville began assembling his own OR section.

In itself, however, Saville's commitment to OR had little effect. Air defense had little priority in the AAF. The AAF was far more heavily invested in strategic bombing. The reasons for this lay in AAF officers' ambitions to break their service loose from army control and to achieve equal status with the army and navy. The RAF had accomplished this immediately after the First World War. Throughout the interwar years, the U.S. Army's aviators argued their case in congressional hearings, interviews, even books—all without success. Arnold was a veteran of these public relations battles, as was General Ira C. Eaker, with whom Arnold published several books on the strategic importance of an air force.[34]

Army aviators argued that an air force had strategic potential, that it could destroy enemy war-making potential. Army officials and their public allies countered that the Army Air Corps (as it was known before World War II) should provide tactical support to the army. As a tactical service, the Air Corps did not merit equal status with the army and navy. Throughout the interwar period, army aviators tried to prove otherwise. Many believed that the public's negative reaction to the horrors of trench warfare augured well for the air force bid. As a result Air Corps leaders spent much of the war developing the doctrine of precision bombing, which held that an air force had the capability to penetrate enemy air space and eliminate targets of military and industrial importance. Enemy war-making potential would be eliminated without the horrors of trench warfare or indiscriminate bombing of civilian populations.[35]

Although air force proponents spent over a decade developing precision bombing doctrine, they had done virtually nothing in translating this doctrine into concrete operational plans and objectives before the summer of 1941, when Roosevelt called upon the services to forward their military plans to him. Arnold found the first attempts unsatisfactory.[36] When the AAF began sending units to Britain to participate in the European bombing campaign, they had little in the way of guidance.

The early American bombing effort in Europe was organized as the Eighth Army Air Force. General Carl Spaatz commanded the Eighth, which had been formed in early 1942. The core of the Eighth was its Bomber Command, under Eaker's leadership. Eaker traveled to Britain in that spring to establish his headquarters and work with RAF officials in coordinating the bombing campaign. Eaker's RAF opposite was Air Marshal Sir Arthur Harris, of RAF Bomber Command. Eaker established his headquarters at Wycombe Abbey, close to Harris's.[37]

Not long thereafter, Eaker noticed Harris's OR Section, directed by B. G. Dickins. The OR Section carried out a diverse array of studies for Harris, ranging from bomb selection to bombing accuracy. Eaker persuaded Spaatz, headquartered in London, to request an OR group for the Eighth. Arnold received Spaatz's request before spring had melted into summer.[38] Although he took no immediate action, OR's cause was taken up by several of his staff. Saville, of course, was one; another was General Muir Fairchild; another was Brigadier General Harold M. McClelland, the Director of Technical Services, who was also concerned with radar. Brigadier General Muir Fairchild, Arnold's Director of Bombardment and an important architect of the AAF's precision bombing doctrine, also argued for an OR program for the AAF. Along with Bowles, who spent much of his time in the War Department advising Arnold, these men began to persuade the AAF's commander of OR's merits.

If the AAF had been caught unprepared for fighting Germany, the navy was no less so.[39] German U-boats found easy prey within sight of American shores as early as December 1941. The navy's anti-submarine doctrine was out of date, its complement of destroyer escorts for convoys meager. Responsibility for antisubmarine training and doctrine was far removed from active escort crews most familiar with German submarine attacks. Captain Wilder D. Baker of Destroyer Command persuaded Admiral Royal E. Ingersoll, Commander-in-Chief of the Atlantic Fleet, to consolidate the fleet's various antisubmarine activities. Baker was made the chief of a new Antisubmarine Warfare Unit (AWU), charged with monitoring all antisubmarine operations, evaluating their effectiveness, and coordinating training and doctrine.[40]

As with the bombing campaign, Britain already had over two years' experience with fighting the enemy. Baker crossed the Atlantic to take notes on British methods, inevitably encountering Blackett's OR sections at RAF Coastal Command and the Admiralty. Baker returned to the United States determined to incorporate OR in the American war against the U-boat.[41]

The branches of the military mentioned above all lagged significantly behind their British counterparts. Like the British, American commanders endeavored to take advantage of new weapons technology wherever possible. The most serious gap between the allies involved not the weapons themselves, but using them effectively. As the Americans, most often field commanders rather than Washington chiefs, turned to the British for advice and examples, they inevitably encountered OR. Throughout 1942, the military services organized OR groups. To staff these groups, they turned to the most obvious source of technical expertise: the OSRD.

The requests began arriving in January, shortly after Watson-Watt's report had been distributed. By the end of the month, Meade, with Saville not far behind, formally requested assistance from the OSRD in setting up their operations research groups. Bush could no longer hope to ignore the issue. His efforts to deflect OR away from the OSRD, however, were often undermined by his own staff.

Bush addressed the issue of OR for the first time at the 30 January 1942 meeting of the OSRD's Advisory Council. The Council included, among others, Bush, Conant from the NDRC, Furer, and Bundy. Bush persuaded those in attendance that OR was not within the OSRD's domain; the National Academy of Sciences could better provide assistance, he argued. Bush then took a precautionary rear-guard action; immediately after the meeting, he dashed off a memorandum to all NDRC chiefs, summarizing his argument and asking the chiefs to "please consider that this matter [OR] for the present is in Dr. Jewett's hands, and if you hear of any moves looking toward the establishment of such sections let him have the information in regard to the matter promptly."[42]

Jewett uncomplainingly located two civilians to lead and develop the OR groups for Meade and Saville. He chose seasoned engineers for the job, not scientists. Both William L. Everitt and C. M. Jansky had been professors in radio engineering and both currently headed successful consulting firms. Considering that both groups would be heavily involved with improving the use of radar, Jewett's choices were completely appropriate. Both men eagerly prepared themselves for their assignments.[43]

Ironically, Meade and Saville ran afoul of Civil Service regulations in establishing military units staffed by civilian personnel. Low salaries and hiring restrictions took their toll on Meade's program especially. Everitt suffered a 50 percent cut in income for his duty and found many highly qualified engineers unwilling to do likewise. Meeting hiring protocols extended the process of recruitment. It seems that Everitt was unable to recruit the required quantity of personnel throughout the war. Saville and Jansky tried to bypass Civil Service codes by categorizing Jansky and his recruits as temporary consultants, paid on a per diem basis. But this scheme also had drawbacks. Jansky still suffered a large pay cut, if not quite as bad as Everitt's. And bypassing the onerous hiring procedures only delayed trouble, for civilian consultants could be kept on only for a total of 180 days.[44]

Bush's memorandum, however, seemed of little avail. A number of NDRC officials continued their interest in OR. Besides Burchard and Robertson, Warren Weaver began praising the British OR groups. As a member of NDRC Division 7 (Fire Control), Weaver first encountered OR during his 1941 stay in Britain. In November 1942, he became the chief of the NDRC's Applied Mathematics Panel, charged with finding solutions to military problems. The quantitative field data OR groups produced, such as statistics on bombing or gun accuracy, were absolutely critical to the AMP's work.[45] Even NDRC divisions that nominally shared Bush's position seemed to have ambivalent personnel. Karl Compton, influential in Division 14 (Radar), noted that although OR was "not directly in the field of responsibility assigned to me," as per Bush's wishes, "its successful handling is probably of more concern to [Division 14] than any other civilian group."[46]

Compton's missive underscores the reason why these men lacked Bush's conviction. Unlike Bush, they recognized the benefits accruing to the OSRD through an OR program. OR not only served military commanders, by improving the effectiveness of military units using new technologies, it also gave researchers a peek at the conditions under which the weapons they designed would have to operate.

The first NDRC chief to openly defy Bush's wishes was John T. Tate, a physicist from the University of Minnesota and the chairman of Division 6 (Sub-Surface Warfare). Captain Baker approached Tate in mid-March 1942 on the advice of Admiral Furer.[47] After hearing Baker describe his intentions, Tate heartily agreed to fund an OR group for naval warfare. An OR section in the navy's antisubmarine command might benefit Division 6 as much as the navy. Tate drew up a contract with Columbia University to provide the administrative support. He also advised Baker whom to choose to lead the group: MIT physicist Philip M. Morse.

Morse, an expert in acoustics, had already worked in the Radiation Laboratory and on several army and navy sound studies. Yet he was bored and believed that his contributions to date were rather insignificant to the war effort. He had already vented his frustrations to Tate, hence Tate's recommendation. After a brief meeting in which Baker and Morse sized up one another, a deal was struck. Morse began work on 1 April, recruiting members for the Antisubmarine Warfare Operations Research Group (ASWORG), officially formed a month later.[48]

Bush's position compromised, it came under additional pressure through Bundy and Bowles in the War Department, who had begun urging field commanders to attach OR groups to their outfits. Bowles had immersed himself in the problem of providing air support to navy ships hunting U-boats off the American shoreline and began collaborating with Morse. Bundy sought the application of OR to army problems more generally. After returning from an inspection of the Panama Canal Zone defenses that April, Bundy asked Bush for assistance in setting up an Army OR program.[49]

Bush chose to address the issue at the 16 June meeting of the JNW. He prepared carefully for the meeting, seeking a way to turn it to his advantage. In late May, he sent letters to his army and navy counterparts on the JNW, Brigadier General R. G. Moses and Rear Admiral W. A. Lee, Jr. In it, Bush argued that "different conditions" existed in the United States than in Britain. The British had no equivalent to the OSRD and OR helped to organize scientists and engineers for the war effort. British methods of organization, therefore, were inappropriate for the American context. Granting that military commands might find useful the kind of support OR groups were capable of, Bush nevertheless opined that "if something is done here it ought to be under a different name" to prevent "identify[ing] matters that should be kept distinct."[50] In particular, the "NDRC is concerned with the development of equipment for military use, whereas these groups are concerned with the analysis of its performance, and the two points of view do not, I believe, often mix to advantage." The "type of man to be used in such work is very different from the type needed for experimental development. Analysts, statisticians, and the like are not usually good developmental research men."[51]

Bush's observations plainly flew in the face of the British experience, where "developmental men" had been among the first and most successful operations researchers. Whether or not Bush actually believed his own argument is uncertain. It seems more clear that Bush's main object was protecting the NDRC and OSRD from attrition to OR groups. OR did not fit within his understanding of the OSRD's research and development mission. In fact, OR had the potential to wreak havoc on it. Bush hoped to deflect responsibility by suggesting that the services administer and staff their own program. Advising on the "utilization of new weapons," outside of OSRD's charter, should be conducted by "groups that are a part of the armed services themselves, on account of the intimate interconnection with [military] operations."[52]

Bush's JNW counterparts were staff officers and may not have been aware of the concerns occupying field officers like Baker, Saville, and Eaker. All of the services had yet to develop a policy regarding OR. Therefore, when the JNW met on 16 June, Moses and Lee were easily persuaded by Bush. It was noted at the meeting that British scientists had already contacted Bush and suggested that OSRD scientists should be sent for apprenticeships in British OR units, continuing to serve in those units until American forces had need of them. Bush encountered little difficulty in securing a rejection of the proposal. As for the path the American armed forces should take, Bush suggested that a thorough study of British and American conditions be undertaken. The committee agreed and formed a study panel.[53]

As its representative on the panel, the army, apparently at Bundy's behest, selected Major Walter Barton Leach of the General Staff. Leach had been a reserve officer before the war; in peacetime, he had been a professor at Harvard Law School. Leach worked closely with Bundy, also a Boston lawyer, and it is possible that they were acquainted before the war. Bush chose Conant's deputy executive officer, engineer Ward F. Davidson, to represent the OSRD. For unclear reasons, the navy never volunteered a representative for the study.[54]

Bush lost little time in attempting to influence Leach. In his official directive to the lawyer, Bush encouraged him to investigate other potential sources for OR personnel than the OSRD. Bush even orchestrated a meeting between Leach and his longtime legal associate Elihu Root.[55] The aging Root, a partner of a prominent New York law firm and former Secretary of War, had collaborated with Bush at several points in the past and shared many of Bush's corporatist views on organizing research. Bush's efforts paid off; by the end of the summer, Leach was persuaded that lawyers, not scientists, were best able to lead OR groups because of the "exacting diplomatic" demands of dealing with the military brass. "This is work which lawyers can do better than anyone else," Leach wrote Bundy. "You and I know it and Dr. Bush and Dr. Davidson are convinced."[56]

Although Davidson helped draft the report, Leach seems to have done much of the legwork, traveling to Britain to interview officers and scientists and uncovering areas of latent OR activity in the United States. Besides ASWORG, Jansky's group, and Everitt's section, physicist Ellis A. Johnson, a colleague of Bush's from the Carnegie Institute of Washington, had started a small and informal OR group at the Naval Ordnance Laboratory shortly after the Pearl Harbor attack. By summer, its leadership had fallen to MIT physicist Francis Bitter.

The duo completed the so-called Leach-Davidson Report in mid-August and submitted it to JNW. Leach and Davidson acknowledged the soundness of many British practices, especially that of maintaining OR sections as staff organizations, with no authority in the chain of command. Second, the report agreed that the groups should be staffed with civilian personnel to ensure objectivity. Leach and Davidson also recommended some deviations from British practice, and it is here that Bush's influence is most apparent. The report advised that the War and Navy Departments jointly arrange a contract with a civilian organization, such as the National Research Council, to recruit and administer OR personnel.[57] Such an arrangement would allow the military departments to circumvent cumbersome Civil Service regulations, while simultaneously excusing the OSRD from any further involvement in operations research.

The JNW drafted its recommendations to the Joint Chiefs of Staff based on the Leach-Davidson Report. Leach, in fact, seems to have been its author.[58] Just as the committee prepared to vote on their recommendations, activity came to a grinding halt. Moses and Lee seemed to lack the mandate of their services to pass along the recommendations. As the momentum behind the Leach-Davidson Report threatened to stall, Bush frantically sought some means to reach closure on the OR issue. Bush needed somehow to circulate some kind of policy statement, even if it did not have official sanction. General McClelland had already taken up Spaatz's request for an OR section and it seemed likely that whatever means were used to organize it would become the pattern for staffing future groups for the AAF. Bush recommended that the report be given to military "personnel who may be concerned with operations analysis" provided it was made clear that the report carried no sanction from the Joint Chiefs.[59] The JNW approved this request. Apparently the Leach-Davidson never reached the Joint Chiefs for final approval, but now that it was circulating in AAF and navy headquarters, the issue had become moot.

Or maybe not. Bush may have thought he could rely on Admiral King's antipathy toward outsiders to prevent the proliferation of

ASWORG. In fact, King had only become aware of ASWORG in June, when he ordered Baker's AWU to join his staff in Washington. Baker requested that ASWORG come along. King at first refused to bring a civilian group into the inner sanctum of navy command. At great professional risk, Baker insisted on ASWORG's transfer. King finally relented, but only after demanding a written statement from Tate that NDRC relinquished all control over ASWORG and had no claim over the information it produced.[60] The NDRC now paid for a rapidly growing ASWORG, but derived no real benefit for its expenditures.

Bundy, while perfectly willing to implement the recommendations of the Leach-Davidson Report, nevertheless had difficulties doing so. In September, he deputized Leach to negotiate an OR contract with the National Academy of Sciences and its National Research Council. When Leach proposed the contract to Jewett and NRC executive secretary George Barrows, however, the two balked. Jewett had never intended his favor to Saville and Meade to escalate into a long-term commitment. On the other hand, Jewett was also an NDRC member and had that institution's interests to consider as well. After deliberating for five days, Jewett declined. The spurned Leach fired a note to Bundy, commenting acidly that "this does not break my heart ... I think the National Academy ... would have been an inept instrumentality."[61] Other potential sponsors existed, of course. Both the Institute of Advanced Studies at Princeton and Columbia University were willing to take on the work. Columbia would have been particularly suitable because of its ASWORG support. Leach and Bundy, however, felt New York and Princeton too distant from Washington for effective oversight. Leach went as far as to persuade one of his lawyer friends to form a new corporation, Special Services, Incorporated, to support the OR program. Bundy seems to have intervened just before Leach's friend filed the legal papers.[62]

Equally frustrating was the army's lack of enthusiasm for OR. Officers in the Army's Operations Division announced interest in December, but it suddenly evaporated for unclear reasons.[63] The principal support of OR in the War Department remained the AAF. In mid-October, Fairchild, Saville, and McClelland finally coaxed Arnold's approval for an OR program and Leach transferred to the AAF to set it up.[64] The Operations Analysis Division (OAD), as Leach's program office came to be known, opened in December and fell in the jurisdiction of the Office of Management Control, the portion of Arnold's staff dedicated to logistical support.[65] The problem of recruitment remained, however. Leach was an outsider to the scientific community. He knew only the OSRD. Failing to find an OR recruiting contractor left Leach little choice but to return to Bush for help.

Bush's bid to deflect operations research away from the OSRD had failed. Although he denied that "developmental men" were appropriate for OR, it was precisely such researchers that military officers often wanted. Moreover, NDRC personnel began to see the usefulness of OR to their research and development responsibilities. Bush's attempts to coax the army and navy into taking responsibility for a joint OR program also bore no fruit. The navy's ASWORG remained subsidized by the OSRD and the War Department endeavor to find an appropriate organization to recruit and administer its program faltered. Leach would be back.

For Bush, OR posed a threat to the OSRD, in part, because it could divert technical expertise away from research and development. OR also did not fit well into the OSRD's mission as Bush envisioned it. But the boundary between makers and users of new weapons, which the OSRD represented, would cause much discussion in the months ahead as Americans gained experience with OR.

By the end of 1942, senior OSRD staff began to question Bush's judgment on the operations research issue. They did so through both actions and words. Division 2, Division 14, and AMP began training programs in OR for Leach's program, although they may have had mixed motives for doing so. John Burchard, for instance, commented later that his Division 2 began an OR training program "frankly, because I foresaw from things which were happening that if we were not prepared to furnish men to the Air Force whom we had specifically selected for that purpose we might be forced by the needs of the war to give up men who would be a great loss to our other projects."[66]

Yet Burchard was equally aware that OR could be of use to the OSRD. For that reason he had made Robertson available for a six-week tour with the OR section Leach had created for the Eighth Army Air Force in Britain. Moreover, Division 2 needed air strike information to prepare data sheets indicating the appropriate bomb and fuse combinations for specific targets. At the Applied Mathematics Panel, Warren Weaver complained to Bush in February 1943 that NDRC divisions "[a]re sorely in need of better contact with the Operational Research Groups. At the present time the former simply are not getting the data from the latter."[67] In December, physicist Alan T. Waterman, head of Division 14's Manpower Division, reached similar conclusions and began assembling an OR training program for Division 14. "I confess I was mystified to find that nothing was apparently known about these [operational radar problems], at least on this side [of the Atlantic]," Waterman admitted to Leach in March.[68] In December, Division 14 requested authorization to spend $100,000 toward the development of the OR training program.[69]

By the beginning of 1943, then, two main perspectives dominated OSRD thinking about operations research. Bush considered it essential to avoid OR in order to preserve the integrity of the OSRD's research mission. Bush was by no means alone. Waterman notwithstanding, a number of leading radar personnel in Division 14 and the Radiation Laboratory also regarded OR, especially the AAF's program, as a net loss to OSRD. Louis Ridenour, the Radiation Laboratory's assistant director, told Bush at one point that "Colonel Leach's operational research groups are rather inflexibly set up on a basis of permanent personnel." If radar specialists were to be sent into the field as OR personnel, Ridenour felt that NDRC divisions should at least be able to recall them to research and development tasks when the need arose. Ridenour also believed that the OR groups should report directly to theater commanders, like Army General Dwight D. Eisenhower; their talent was squandered on mere field commanders. That Ridenour felt strongly on the issue of recall is no surprise. The Radiation Laboratory had lost about 100 scientists and engineers to the military in the first year of the war.[70]

The second perspective regarded OR as a potential gain, not a net loss. Burchard, Weaver, Robertson, and Morse became persuaded early in the war that understanding what happened at the front was critical to the success of the research and development effort behind the lines. If some, like Burchard and Waterman, appreciated Bush's organizational concerns, they also believed that Bush's position inadequately reflected wartime realities by the end of 1942.

Robertson was especially adamant. Operations research was a scientific activity and the OSRD had a stake it, he wrote Bush in January 1943. "In view of the technological nature of modern war, I am flatly of the conviction that the Chief [of an OR group] should be a man of scientific or engineering training," he continued. "In this I dissent sharply from the position taken in Leach and Davidson's report ... [I]n view of the role now played by OSRD in the war machine, it has a special responsibility toward and interest in Operational Analysis." OSRD's responsibility toward OR arose because it represented the "biggest scientific man-power bloc in the country," and it had an interest in OR's fate "because [OR groups] form nuclei through which its developments can be brought to the attention [of the military], and through which relevant operational factors can . . . be brought to bear" on its research and development programs.[71]

Already, the level of OSRD involvement in OR, both officially (funding ASWORG, training future OR personnel) and unofficially (helping Leach recruit) had helped ruin Bush's earlier strategy of deflection. As Davidson pointed out, the OSRD had little to show for its involvement at that point. ASWORG's "formal monthly reports that so-and-so many men were employed scarcely are adequate basis . . . for justifying the expenditures."[72] Even without the lack of return on investment, the OSRD funding of ASWORG was problematic from a legal standpoint. OSRD was essentially aiding and abetting the navy in an end-run around Civil Service regulations. Bush had already put the OSRD in the bad graces of the Bureau of the Budget. Its director, the New Dealer Harold Smith, had caught Bush in some free-wheeling fund transfers with the military in order to overcome cash-flow problems.[73] Bush did not need additional aggravation with the Civil Service. "I think I'm out on a bit of a limb in continuing to carry Morse," Bush confided to his assistant, Carroll L. Wilson.[74]

Obviously, a new policy was necessary. It seemed, however, that any viable position involved OSRD participation in OR, a prospect Bush still resisted. Bush stalled until the end of February. He complained to Compton that OR "is certainly of great importance and I hope that it will go well. On the other hand, I hesitate to become too much involved with it."[75] Davidson countered, "It is an obvious mistake to assume that the proper operations of OSRD are restricted to laboratory investigations and the development of devices." He pointed to examples in which ASWORG had made sure that OSRD technologies were being used effectively and troubleshooting whenever they were not. OR activity, Davidson opined, "should be greatly expanded [within the OSRD] rather than curtailed." Why not devise a means for letting OR contracts and supervising the work?[76]

Bush finally relented. On 2 March 1943, he let Furer and Leach know that despite his own inclinations, "there are certain auxiliary or related matters carried on within OSRD, and appropriate interrelations are desirable."[77] He instructed Wilson to investigate the legal aspects of absorbing ASWORG and administering the other OR programs by establishing an OR division within the OSRD.

Deciding that the OSRD should involve itself in operations research was one thing; deciding precisely what that role should be was quite another. For one thing, it seemed that everyone in the OSRD with an interest in OR had something different in mind. While most researchers seemed to favor Morse's arrangement over Leach's, differences remained over the kind of people to employ in OR, how to train them, and what authority the OSRD should have over its OR practitioners once they were in the field.

Leach wanted to secure highly trained specialists for his OR groups. To his thinking, radar experts from the Radiation Laboratory, for instance, and ballistics experts from Division 2's Princeton University Station would be perfect for the OR sections in the European air forces. Needless to say, these were precisely the men Ridenour and Burchard were trying to keep for themselves. On more than one occasion Leach and the NDRC leadership butted heads over this issue.[78] In contrast, Morse insisted that "generalists" were better suited to OR work than specialists. NDRC division chiefs eagerly latched onto this argument, which allowed them to substitute personnel they would not otherwise have employed for researchers they wanted to retain. Of course this was not what Morse intended. He suspected that division chiefs were keeping the most promising people for themselves. "If this is the case we [ASWORG] might as well resign ourselves to taking what dregs the other NDRC laboratories will see fit to give us," he complained.[79]

Training was the second contentious issue. By March 1943, Divisions 2 and 14 and the AMP had developed their own training programs, although Burchard and Weaver also exchanged trainees and set up orientation trips to other NDRC divisions. Confusion arose over who had responsibility for recruitment. NDRC technical aide Lincoln R. Thiesmeyer attempted to smooth ruffled feathers by organizing a comprehensive training program, which provided uniform training from each of the major divisions. But Thiesmeyer's plans would have required an excessively long training period. This dilemma was overshadowed by another. By the end of 1943, OR groups in the AAF had proliferated to all field commands. Their rapacity for OSRD-trained OR practitioners resulted in an increasingly anemic curriculum, as trainees were rushed through the program so that they could be shipped overseas to their new units. Waterman also observed that the Selective Service often ignored personnel exemptions for trainees. Most trainees became nervous as their numbers came up and preferred to bail out of the training programs and volunteer in the service while they could still bargain for good assignments.[80]

Finally, the issue of OSRD's authority over OR practitioners in the field soon transformed the original logic behind the OSRD's OR division. Since Bush's acquiescence on the issue, the question of control over the OR personnel plagued OSRD and NDRC leaders. Robertson tacitly endorsed Leach's approach, which Division 14 and its contractors adamantly opposed. Waterman, like Ridenour, countered that the OSRD should be able to recall its personnel from OR groups at will. As a compromise, he suggested that two classes of operations researchers be defined. The trainees not essential to development could be handed over to the army and navy. "High-power" groups, by which Waterman meant specialists, would operate under OSRD's jurisdiction.[81] Weaver also proposed differentiating OR groups, but in less hierarchical terms. "Operations research," Weaver suggested, would be reserved for field groups, while teams stateside would conduct less urgent and more complex "warfare analysis."[82] Obviously, NDRC division chiefs' suggestions reflected their own organization's goals and work methods. Some of their expectations bordered on unrealistic. Davidson undercut Waterman's arguments by pointing out that the "high-power" groups the latter suggested, intended to be free of military control, would not be tolerated by commanders.

Through the spring and early summer, Wilson attempted to synthesize these disparate points of view into a plan for the planned Operations Research Division. Wilson defined operations research as being "concerned with ... developing tactics for the most effective use of equipment and forces," as well as conducting "equipment analysis which is concerned with ... obtaining maximum performance from equipment through improvement of installation, maintenance, calibration, and testing procedures." Wilson had combined Weaver's and Waterman's positions. The OSRD would still have the power to assess military needs and send OR personnel on a temporary loan basis. Wilson, looking ahead to the winding down of research toward the end of the war, saw OR as a means of keeping the staffs of the OSRD and its contractors busy and under OSRD control until the very end of the war. "[B]y the end of the war, a substantial proportion of those now engaged in laboratory research ... who are capable of doing Operational Research will have shifted to this activity." Wilson's proposal found widespread approval in Division 14.[83] If Weaver recorded his thoughts on the matter, they do not seem to have survived.

With the basic plan in place, Wilson, with Waterman's and James Conant's help, proceeded to set up a charter and budget for the new OR Division. It would be parallel and equal to the NDRC within the OSRD framework. Some fencing with Harold Smith at the Budget Bureau occurred as expected. More interestingly, the ability to let contracts to individuals needed congressional approval; securing this took the better part of the spring and summer.[84]

Finally came the matter of choosing a leader. Nearly unanimously, everyone with a stake in the program nominated Morse. Robertson was considered briefly, but the hot-foot incident disqualified him. Samuel Goudsmit, a Dutch émigré scientist working in Division 14, was another candidate, but as Compton remarked to Bush, "his foreign accent might be prejudicial in circles where he is not personally known." Clearly, the man for the job had to be a skilled diplomat and yet hold the respect of the scientific personnel. Compton, along with many others, supported Morse.[85] Ironically, Morse himself did not. Morse had long-standing reservations over the OSRD's administration, believing it to be cumbersome and disruptive, top-heavy with "secretaries and stooge workers." He preferred his current situation and felt no enthusiasm for ASWORG's impending transfer to the OSRD, let alone for a job as an OSRD administrator.[86]

Bush believed university presidents to be ideally suited for the post, capable of impressing both officers and scientists. Compton, upon being asked for suggestions, could name only a few.[87] Compton was, of course, the university president Bush had in mind and eventually he accepted the position. Bush selected Waterman to serve as Compton's deputy. The OR Division would therefore be dominated by the Division 14 perspective: a willingness to sacrifice information and cooperation for control over personnel.

Over the course of the summer, Bush found a way to resurrect his original agenda, but with a twist. The OSRD would proceed with Wilson's plans to create a new division, but the new division would not bridge the gap between makers and users, as OR promised to do. Rather, it would extend the emphasis on research and development—privileging the making over the using of technology—into the field. OR would actually become a minor emphasis in the OR Division. On his return from an antisubmarine warfare conference in Britain during July, Bush praised Wilson's plans and spadework, and promptly modified it.

> The scheme is, as I see it, much broader than merely the furnishing of men to operational research groups for particular groups. It can include the furnishing of men outside of these groups for particular purposes ... More to the point, I think that the sending of men to operating theaters in connection with particular weapons, as temporary visitors concerned with a particular subject, may well grow to be of much importance.[88]

To reflect the change, Bush discarded "Operations Research Division" in favor of the Office of Field Service (OFS).

Bush marginalized operations research because it was not amenable to OSRD's control. Successful OR required the negotiation, even participation, of both makers and users. Bush preferred focusing the OSRD's attention on matters requiring less bargaining, in which its personnel could regard the military services as clients and not partners. By appointing researchers with similar views to head the Office of Field Service, Bush hoped to minimize the OSRD's commitment to an activity that defied his ambitions for his organization.

By the early fall of 1943, planning for the Office of Field Service took on some urgency. The reason for this lay in the Pacific theater. Around the time of his return from Britain, Bush realized that the Pacific conflict would gradually absorb more of the military's attention. In the European theater, the OSRD had enjoyed tremendous institutional support. It had relatively good relations with the officers in the theater. Through its mission in the London embassy it maintained excellent communications with its agents. In the Pacific, the OSRD had none of these amenities. It had few contacts with officers and virtually no working history with any of them. The military controlled almost all of the communications channels. Bush and Compton intended the OFS to tame this landscape and make it amenable to the needs of research and development. Just as the London Mission had extended the OSRD's reach to Europe, the OFS would do the same for it in the Pacific.

Bush announced the establishment of the OFS on 15 October 1943. The OFS had the blessing of the Navy and War Departments, but this was not the audience that needed to be impressed. Within a month, Compton was planning a whirlwind tour of the Southwest and Central Pacific to meet with the top theater commanders, General Douglas MacArthur, commander of the Southwest Pacific Area, and Navy Admiral Chester Nimitz, commander of the Central Pacific Area.[89] Essentially, the purpose of the trip was to sell MacArthur and Nimitz on the OSRD.

Compton left shortly before Christmas 1943, and returned after a hard month of diplomatic work.[90] The results were mixed. Nimitz, headquartered in Hawaii, had been cordial, but noncommittal. He welcomed the offer of help, even suggested potential projects, but had no interest in hosting an OSRD field office. The local army commander (the army was subordinate to the navy in the Central Pacific Area), Lieutenant General Robert C. Richardson, seemed genuinely cooperative. Compton found the situation in Southwest Pacific more hopeful. MacArthur approved the establishment of an OSRD branch office at his Australian headquarters in Brisbane.[91] As soon as Compton returned stateside, he started to organize the branch offices. George Harrison, MIT's Dean of Science and an important manager of the radar program, agreed to direct it. On 10 March 1944, Compton outlined Harrison's duties to him and on 28 March MacArthur approved Harrison's selection. Two days later Harrison arrived in Brisbane. The establishment of a branch office for Richardson trailed by several months, but such an office was finally established in May, with physicist Lauriston C. Marshall as director.[92]

Nevertheless, Compton's carefully laid plans proved largely futile. In its primary mission, extending the OSRD's research agenda to the front, the OFS proved an utter failure. The reasons for this can be attributed to what Davidson had foreseen: field commanders might welcome technical help, but would rarely tolerate civilian groups in the midst that were not under their control. Harrison's group, named the Research Section, illustrates some of the intractable problems.

Among the most serious was that of communication. Upon his arrival, Harrison found that MacArthur's staff forbade all mention of the OSRD in correspondence. Communications, which in Europe had easily flowed over State Department channels, were subjected to the time-consuming censorship. Several OFS personnel resorted to sending updates to the OSRD via personal mail although this practice was discouraged. Harrison began making frequent use of the diplomatic pouch until MacArthur's chief of staff, Lieutenant General Richard K. Sutherland, upbraided him.[93]

Harrison also found that, instead of reporting directly to MacArthur, as he had expected, the Research Section had been assigned to a mere major far down the chain of command. The very title of the group, "Research Section," had been imposed rather than chosen. But most humiliating of all was MacArthur's abandonment of the Research Section at Brisbane as American forces moved north to retake the Philippines. Only through Waterman's efforts in Washington was it finally relocated. Harrison resigned in a huff after only four months on the job. At his exit interview he sputtered that he had taken the job only as a favor to Compton. With some vindictiveness, he groused that he considered his most important achievement in the Southwest Pacific to have been getting his commanding officer, whom he reportedly described as a "neuropsychiatric and homosexual," dismissed from the Army.[94] Harrison realized in the end that he had been a casualty in the war over organizational control over technical expertise in the Pacific theater of war. MacArthur reserved absolute organizational control, which Bush, Compton, Waterman and others at the OSRD found difficult to accept.

Compton and Waterman worked furiously over the next few months to reach a compromise solution, with Waterman himself traveling to MacArthur's command, but found MacArthur and his subordinates opposed to any scheme that allowed the OSRD freedom of movement in the theater. Exploiting the OSRD's influence with Stimson and the AAF, Bush and Compton finally circumvented MacArthur's authority and secured authorization in Washington for an independent field station in the Philippines, the Pacific Branch of the OSRD (PBOSRD).[95] It was a Pyrrhic victory. It had taken until the spring of 1945 to attain the approval and several months were spent acquiring staffing, funding, and equipment for the organization.[96] Compton did not arrived in Manila to take control of the PBOSRD until 5 August 1945, the day before the atomic bombing of Hiroshima.

Marshall's group on Hawaii fared somewhat better, but hardly ideally from the OSRD's perspective. General Richardson's command constituted the rump of the army's Pacific presence and was effectively under navy control. As the war in the Pacific progressed, this control was transferred back to MacArthur. By this time, the army's Hawaiian command post became increasingly irrelevant to the war effort.[97] Nimitz, meanwhile, continued to rebuff OSRD advances for the rest of the war. Only through Morse's group, technically part of the OSRD but divorced from its authority, did the OSRD enjoy success with the navy.[98]

None of this should suggest that the OFS failed to provide important technical assistance to the military during the last two years of the war. Most famous of the successes is ALSOS, the scientific intelligence mission Samuel Goudsmit led in Europe which uncovered—among other things—the German atomic bomb project. Numerous, and more anonymous, short-term projects were undertaken by small teams temporarily assigned to military commands for specific purposes, such as controlling insect infestations, advising on medical procedures and technical requirements, and troubleshooting equipment.[99]

Nevertheless, the intent behind the OFS was to extend control over the OSRD's research and development mission up to the front. When Waterman offered Bush his opinions of the OFS's wartime performance, his nine-page brief focuses almost exclusively on maintaining control over research and development in the field. The small projects receive scant attention and when Waterman mentions Leach's OR program it is primarily to complain about it. Rather than reevaluate the OSRD's attempts to have a free hand in technical matters, Waterman instead fantasized on what might have been, if only the war had lasted longer: "On looking back over the activities of OFS, it seems to the members of its staff that the final arrangement planned in the Pacific Branch, OSRD, would have worked out in a most satisfactory manner."[100] Some found it difficult to look beyond Bush's ideal.

Ironically, if not surprisingly, the military services most valued the OFS as a source of men for its OR programs. ASWORG remained, by far, the largest and longest lasting OFS project. Although transferred to the OFS, ASWORG (later simply named the Operations Research Group when it began to expand its program to cover pro-submarine, amphibious, and antiaircraft warfare) remained as much under navy control as it ever had, perhaps even more so. Morse found all his worst fears regarding OSRD management realized. His personnel experienced great difficulty with the OSRD bureaucracy, especially with military exemptions, travel money, and equipment. Morse finally found it more expedient to work through navy supply channels and have the OSRD reimburse the navy, even though he felt this taxed his relationship with top navy brass. His frustration continued to mount in February 1944, when Morse fired off a vitriolic letter to Bush.

> I would feel badly enough if operations research had failed because of the Navy's unwillingness to take it on. I believe this hurdle has

> been passed already and the Navy has accepted us. It certainly would be a very much more terrible thing to contemplate to have to confess that the failure of operations research in the Navy was due entirely to the inability of OSRD to service an organization of this sort.[101]

Thiesmeyer, now the OFS's technical aide, intercepted the letter before it reached Bush, dismissing it as "static."

The Army Air Forces refused to alter its OR program, and the OSRD had to be content in its role as talent scout and trainer. For this, Bush had no one but himself to blame, for he had been instrumental in the design of Leach's program. The navy and AAF OR programs, the largest consumers of OFS talent, effectively removed this expertise from the OSRD's control. By itself, this fact illustrates the failure of the OFS and the limits of Bush's strategy for the OSRD.

Operations research's wartime adoption in the United States provides a great irony. The OR community today is rightly proud of OR's contributions during the Second World War, but often seeks to associate their field's past with an individual and an organization hardly worthy of the acknowledgment. In professional histories, OR practitioners are fond of invoking Bush and the OSRD as central characters in the development of their field.[102] Bush and the OSRD did, in fact, play a central role, but not the one most practitioners (or historians, for that matter) suspect. OR was adopted by the American civilian research community in spite of, not because of, the intervention of Bush, who consistently attempted to distance the OSRD from OR throughout the war. Bush acquiesced and agreed to have the OSRD administer an OR program only when no other options remained available to him.

The story of OR's arrival in the United States helps to illustrate the limits of Bush's strategy for mobilizing America's civilian research and development capacity. His was a corporatist approach, which had been favored by his colleagues during the Progressive Era and the New Deal, when the captains of industry and research argued for a weak government in the service of business and science. Bush attempted to use OSRD's painstakingly constructed institutional boundaries to protect civilian research from government influence. Ironically, the OSRD's success in research and development midwifed the military-industrial-academy complex with which corporatist ideals were incompatible.

Instead, the military-industrial-academy complex fostered the systems approach. The technological systems so strongly identified with the Second World War—radar, the atomic bomb, the proximity fuse, antisubmarine warfare—were heterogeneous in character. They included not only the artifacts designed by OSRD scientists and engineers, but also the military personnel who were responsible for using them. Officers and civilian researchers directly involved with the technologies in question—Morse and Baker, Robertson and Eaker—were fully aware of their interdependency. Their systems could only be encumbered by institutional boundaries like those of the OSRD. Operations research provided a way to circumvent those boundaries and to integrate makers and users, men and machines. After the war, OR, systems engineering, and other systems approaches would sustain this sensibility, exhibiting Schumpeterian disregard for preexisting institutional boundaries even as they created their own. Bush's OSRD strategy illustrated the stagnation of one approach; OR's adoption heralded the ascendancy of another.

Glossary

AAF	U.S. Army Air Forces
ASWORG	Antisubmarine Warfare Operations Research Group
AWU	Antisubmarine Warfare Unit (U.S. Navy)
JCS	Joint Chiefs of Staff
JNW	Joint Committee on New Weapons and Research (Joint Chiefs of Staff)
MIT	Massachusetts Institute of Technology
NDRC	National Defense Research Committee
NRC	National Research Council
OAD	Operations Analysis Division (U.S. Army Air Forces)
OFS	Office of Field Service
OR	Operations Research (in the United Kingdom, Operational Research)
OSRD	Office of Scientific Research and Development
PBOSRD	Pacific Branch of the Office of Scientific Research and Development
RAF	Royal Air Force (United Kingdom)

Notes

1. Part of the difficulties practitioners have had is drawing professional boundaries. For a useful perspective on this discussion, see Hugh J. Miser, "The Need for Inclusion," *OR/MS Today* (December 1993): 38–42.

2. Philip Morse, introduced later in this account, was an early proponent of this view. See Philip M. Morse, *In at the Beginnings: A Physicist's Life* (Cambridge: MIT Press, 1977), 263–264.

3. The Hughes' influence is present in each of the chapters in this volume. For a concise statement of this approach, see Thomas P. Hughes, "The Evolution of Large Technological Systems," in *The Social Construction of Technological Systems: New Directions in the Sociology and History of Technology*, ed. Wiebe E. Bijker, Thomas P. Hughes, and Trevor Pinch (Cambridge: MIT Press, 1987).

4. For a discussion of cyborgs see Donna J. Haraway, "A Manifesto for Cyborgs," *Socialist Review* 15, no. 2 (1985): 65–107; and Paul N. Edwards, *The Closed World: Computers and the Politics of Discourse in Cold War America* (Cambridge: MIT Press, 1996). Steve J. Heims explores the early impact of cybernetics on the social sciences in his *The Cybernetics Group* (Cambridge: MIT Press, 1991). For critiques of systems analysis (and systems approaches generally) see Ida R. Hoos, *Systems Analysis in Public Policy: A Critique* (Berkeley: University of California Press, 1972); and Robert Lilienfeld, *The Rise of Systems Theory: An Ideological Analysis* (New York: Wiley, 1978). The idea of using quantitative methods to understand risk and "colonize the future" here recently attracted the attention of historians and social theorists; see, for example, Anthony Giddens, *Modernity and Self-Identity: Self and Society in the Late Modern Age* (Cambridge: Polity Press, 1991); and Theodore M. Porter, *Trust in Numbers: The Pursuit of Objectivity in Science and Public Life* (Princeton: Princeton University Press, 1995).

5. The history of operations research is only beginning to be addressed by historians. See M. Fortun and S. S. Schweber, "Scientists and the Legacy of World War II: The Case of Operations Research (OR)," *Social Studies of Science* 23 (November 1993): 595–642; Robin A. Rider, "Operations Research and Game Theory: Early Connections," *History of Political Economy* 2 (1992): S225–239; Stephan P. Waring, "Cold War Calculus: The Cold War and Operations Research," *Radical History Review* 63 (Fall 1995): 28–52; and M. Kirby and R. Capey, "The Air Defence of Great Britain, 1920–1940: An Operational Research Perspective," *Journal of the Operational Research Society* 48 (1997): 555–568. Practitioners' accounts for the Second World War abound. Most helpful are Charles W. McArthur, *Operations Analysis in the U.S. Army Eighth Air Force in World War II* (Providence: American Mathematical Society, 1990); Joseph F. McCloskey, "The Beginning of Operations Research: 1934–1941," *Operations Research* 35 (1987): 143–152; "British Operational Research in World War II," *Operations Research* 35 (1987): 453–470; and "U.S. Operations Research in World War II," *Operations Research* 35 (1987): 910–925. Also useful is Keith R. Tidman, *The Operations Evaluation Group: A History of Naval Operations Analysis* (Annapolis: Naval Institute Press, 1989); and

Air Ministry, *The Origins and Development of Operations Research in the Royal Air Force* (London: H. M. Stationery Office, 1963).

6. For excellent overviews of the OSRD's creation, see Daniel J. Kevles, *The Physicists: The History of a Scientific Community in Modern America* (New York: Vintage, 1979), 287–301; and Larry Owens, "The Counterproductive Management of Science in the Second World War: Vannevar Bush and the Office of Scientific Research and Development," *Business History Review* 68 (Winter 1994): 515–576.

7. Robert Kargon and Elizabeth Hodes, "Karl Compton, Isaiah Bowman, and the Politics of Science in the Great Depression," *Isis* 76 (1985): 301–318. For explanations of corporatism and corporate liberalism, see Louis Galambos, "Technology, Political Economy, and Professionalization: Central Themes of the Organizational Synthesis," *Business History Review* 57 (1983): 471–493; Ellis Hawley, "The Discovery and Study of a 'Corporate Liberalism,'" *Business History Review* 52 (1978): 309–320; and James Weinstein, *The Corporate Ideal in the Liberal State, 1900–1918* (Boston: Beacon Press, 1968). For a masterful summary of this literature, with implications for historical research for the postwar era, see Brian Balogh, "Reorganizing the Organizational Synthesis: Federal-Professional Relations in Modern America," *Studies in American Political Development* 5 (Spring 1991): 119–172.

8. See Balogh, "Reorganizing," and his *Chain Reaction: Expert Debate and Public Participation in American Commercial Nuclear Power, 1945–1975* (New York: Cambridge University Press, 1991).

9. J. Rosenhead and C. Thunhurst, "A Materialist Analysis of Operations Research," *Journal of the Operational Research Society* 33 (1982): 111–122; and Jonathan Rosenhead, "Operational Research at the Crossroads: Cecil Gordon and the Development of Post-war OR," *Journal of the Operational Research Society* 40 (1989): 3–28. For a celebratory account of OR in the British coal industry, see Rolfe C. Tomlinson, ed., *OR Comes of Age: A Review of the Work of the Operational Research Branch of the National Coal Board 1948–1969* (London: Tavistock Publications, 1971).

10. Kirby and Capey, "The Air Defence of Great Britain," 560–562. See also Air Ministry, *The Origins and Development of Operational Research in the Royal Air Force* (London: Her Majesty's Stationary Office, 1963).

11. McCloskey, "The Beginnings of Operations Research," 148–149. For Blackett's wartime formulation of OR, see his *Studies of War: Nuclear and Conventional* (New York: Hill and Wang, 1962), 171–198.

12. Solly Zuckerman recalls this work in *From Apes to Warlords* (New York: Harper & Row, 1978). See also John E. Burchard, *Rockets Guns and Targets: Rockets, Target Information, Erosion Information, and Hypervelocity Guns Developed during World War II by the Office of Scientific Research and Development* (Boston: Little, Brown, 1948), 247 and 296–300.

13. U.S. Military Attaché to Bush (draft), 28 February 1942, Office of Scientific Research and Development Papers (RG 227), OSRD Papers, Records of the OFS—Manuscripts, Histories, and Summaries (E-177), OFS Records, Indoctrination Programs folder, National Archives and Records Administration (NARA), Washington, D.C.; and Darwin to Bush, 12 May 1942, Joint Chiefs of Staff Papers (RG 218), JCS Papers, Joint Committee for New Weapons and Equipment—Subject File 1942–45 (E-343A), JNW Records, Operations Research 5/1942 through 12/1942 folder, NARA.

14. Kargon and Hodes, "Politics of Science," 301–302.

15. See note 7.

16. Even during the administration of Franklin D. Roosevelt, Bush and National Academy of Science president Frank Jewett meant Hoover when they talked about "the Chief." See Nathan Reingold, "Vannevar Bush's New Deal for Research: or The Triumph of the Old Order," *Historical Studies of the Physical Sciences* 17 (1987): 302.

17. Kargon and Hodes, "Politics of Science," 309–318; and Larry Owens, "MIT and the Federal 'Angel': Academic R & D and Federal-Private Cooperation before World War II," *Isis* 81 (1990): 188–213, especially 206–211.

18. Owens, "The Counterproductive Strategy," 521–530; and Kevles, *Physicists*, 296–300.

19. Owens, "The Counterproductive Strategy," 559–560.

20. Ibid., 523–537.

21. For more information on the Tizard Mission, see David Zimmerman, *Top Secret Exchange: The Tizard Mission and the Scientific War* (Buffalo: McGill-Queen's University Press, 1996).

22. Zimmerman, *Top Secret Exchange*, 90–91. For a detailed discussion of radar technology in the Second World War, see Henry Guerlac, *Radar in World War II* (Los Angeles: Tomash Publishers, 1987).

23. To avoid confusion, the NDRC division titles used in 1945 have been used consistently throughout the article, regardless of earlier titles. The NDRC's divisional structure was revised at several times during the war. The structure and personnel of the divisions examined in this chapter were stable enough that this should not cause confusion. For an explanation of the OSRD's and NDRC's structure throughout the war, see Irvin Stewart, *Organizing Scientific Research for War: The Administrative History of the Office of Scientific Research and Development* (Boston: Little, Brown, 1948).

24. The papers of the London Mission survive in the National Archives and Records Administration, among the OSRD Papers (RG-227), as Records of the London Mission (E-176).

25. Burchard, *Rockets Guns and Targets*, 247.

26. Kevles, *Physicists*, 312–315; and Harvey M. Sapolsky, *Science and the Navy: The History of the Office of Naval Research* (Princeton: Princeton University Press, 1990), 19–23.

27. Kevles, *Physicists*, 309–311.

28. Ibid.

29. Wesley Frank Craven and James Lea Cate, *The Army Air Forces in World War II* (Chicago: University of Chicago Press, 1948), I:291.

30. Quoted in Crave and Cate, *Army Air Forces*, I:291–292.

31. George Raynor Thompson and Dixie R. Harris, *The Signal Corps: The Outcome* (Washington, D.C.: Office of the Chief of Military History, United States Army, 1966), 5–6.

32. The AAF was officially considered the aerial tactical support arm of the army until the passage of the 1947 National Security Act, which created the United States Air Force. As we shall see, the AAF had nevertheless begun the fight for its independence long before that time.

33. Compton to Jewett, 7 February, OSRD Papers, OFS Records, OSRD in Operation Analysis Division folder.

34. Michael S. Sherry, *The Rise of American Air Power: The Creation of Armageddon* (New Haven: Yale University Press, 1987), 47–61.

35. Ibid.

36. A helpful supplement to Sherry for the development of American air war plans is Alfred C. Mierzejewski, *The Collapse of the German War Economy, 1944–1945: Allied Air Power and the German National Railway* (Chapel Hill: University of North Carolina Press, 1988).

37. For a detailed account of life in the U.S. Eighth Army Air Force, see Roger A. Freeman, *The Mighty Eighth: A History of the Units, Men, and Machines of the U.S. 8th Air Force* (New York: Orion Books, 1989).

38. McArthur, *Operations Analysis*, 6.

39. For an account of the Battle of the Atlantic, as seen from Admiral King's perspective, see Thomas B. Buell, *Master of Sea Power: A Biography of Fleet Admiral Ernest J. King* (Boston: Little, Brown, 1980), 282–299.

40. United States Navy, Commander-in-Chief, Atlantic Fleet, *United States Naval Administration in World War II*, no. 143, vol. 8, *Commander Fleet Operational Training Command* (1946), 247–260, Rare Book Room, Navy Department Library, Naval Historical Center, Washington, D.C. For an account of German submarine operations in the western Atlantic, see Michael Gannon, *Operation Drumbeat: The Dramatic True Story of Germany's First U-Boat Attacks along the American Coast in World War II* (New York: Harper and Row, 1990). United States Navy, Commander-in-Chief, Atlantic Fleet, *Commander Fleet Operational Training Com-*

mand, no. 143, vol. 8 of *United States Naval Administration in World War II* [1946], 247–258, Rare Book Room, Navy Department Library, Naval Historical Center, Washington, D.C.; Keith R. Tidman, *The Operations Evaluation Group: A History of Naval Operations Analysis* (Annapolis, Maryland: Naval Institute Press, 1984), 31–32.

41. Tidman, *Operations Evaluation Group*, 31–32; and Morse, *In at the Beginnings*, 173.

42. Bush to Chairman of NDRC Division, 30 January 1942, OSRD Papers, OFS Records, OSRD in OAD folder.

43. "Memorandum No. 2: Operations Analysis in the United States Army and Navy," JCS Papers, JNW Records, Operation Analysis in England, USN, USA Memos Nos. 1 and 2 folder. Collectively, memoranda 1 and 2 are known as the Leach-Davidson Report and will henceforth be cited as such.

44. Ibid.

45. "Report of Other Activities of the OSRD London Mission, Part Fifteen," 23, OSRD Papers, London Mission Records, Part Fifteen folder. Evidence of AMP's extensive involvement with OR can be found in the OSRD Papers, Records of the Applied Mathematics Panel—General Records (E-153), AMP Records, box 4. For background on the AMP, see Larry Owens, "Mathematicians at War: Warren Weaver and the Applied Mathematics Panel, 1942–1945," in *The History of Modern Mathematics*, ed. David E. Rowe and John McCleary (Boston: Academic Press, 1989).

46. Compton to Jewett, 7 February 1942.

47. Tidman, *Operations Evaluation Group*, 32.

48. Morse, *In at the Beginnings*, 172–173.

49. "Operations Analysis, Headquarters, Army Air Forces, December 1942–June 1950," Air Force History Microfilm Collection 143.504, Air Force History Support Office, Bolling Air Force Base, Washington, D.C.

50. In subsequent months, Bush used the term "operations analysis" and encouraged its use in the OSRD. The AAF adopted this title as well, but it is not clear if Bush is the source.

51. Bush to Moses and Lee, 29 May 1942, JCS Papers, JNW Records, box 9, Operations Research 5/1942 through 12/1942 folder.

52. Bush to Moses and Lee, 29 May 1942.

53. Minutes of the Sixth Meeting of the JNW, JCS Papers, JNW Records, Minutes 1–20 folder.

54. Minutes of the Sixth, Seventh, and Eighth Meetings of the JNW, 16 June 1942, 23 June 1942, and 30 June 1942.

55. Bush to Root, 24 July 1942, JCS Papers, JNW Records, Operations Research 5/1942 through 12/1942 folder.

56. Bush to Root, 24 July, and Leach to Bundy, 2 September 1942, Office of the Secretary of War Papers (RG 107), OSW Papers, Records of Harvey H. Bundy (E-82), Bundy Records, Operations Analysis 1942 folder.

57. Leach-Davidson Report.

58. JNW to JCS (draft), August 1942, JCS Papers, JNW Records, Operations Research 5/1942 through 12/1942 folder. Leach alludes to authorship in Leach to Bundy, 2 September 1942.

59. Minutes of the Fifteenth Meeting of the JNW, 25 August 1942.

60. Morse, *In at the Beginnings*, 182–183.

61. Leach to Bundy, 30 September 1942, OSW Papers, Bundy Records, Operations Analysis 1942 folder.

62. Leach to Bundy, 30 September 1942.

63. Hall to Bundy, 1 December 1942, OSW Papers, Bundy Records, Operations Analysis 1942 folder.

64. Advisory Council to General Arnold, 17 October 1942, Henry H. Arnold Papers, Decimal File, 1940–45, box 115, 400.112 Research and Development, Library of Congress Manuscripts Division, Washington, D.C.

65. "Operations Analysis, Headquarters, Army Air Forces, December 1942–June 1950."

66. Burchard to Waterman, 1 March 1944, OSRD Papers, OFS Administration Records, box 287, Indoctrination—Plans and Policies—Special Operations Analysis Group folder.

67. Weaver to Bush, 25 February 1943, OSRD Papers, OFS Records, box 283, Definitions and Methodology of Operations Analysis folder.

68. Leach to Waterman, 23 March 1943, OSRD Papers, OFS Records, OSRD in Operations Analysis Division (OAD) folder.

69. Waterman to Bush, 17 December 1942, JCS Papers, JNW Records, Operations Research 5/1942 through 12/1942 folder; and Minutes of the 21st Meeting of the NDRC, 18 December 1942, OSRD Papers, NDRC Minutes of Meetings (E-3), 18 December 1942 folder.

70. Ridenour to Bush, 12 May 1943, OSRD Papers, OFS Records, OSRD in OAD folder.

71. H. P. Robertson, "Report of Activities on Operations Analysis Assignment," JCS Papers, JNW Records, Operations Analysis–Robertson Report folder.

72. Davidson to Bush, 24 February 1942, JCS Papers, JNW Records, Operational Analysis folder.

73. Owens, "The Counterproductive Management," 533–535.

74. Bush to Wilson, 1 March 1943 (attachment to Weaver to Bush, 25 February 1943).

75. Bush to Compton, 15 February 1943, OSRD Papers, OFS Records, OSRD in OAD folder.

76. Davidson to Bush, 24 February 1942.

77. Bush to Furer, 2 March 1943, JCS Papers, JNW Records, Operational Analysis folder.

78. For example, see Waterman to Wilson, 15 March 1943, JCS Papers, JNW Records, Operational Analysis folder.

79. Morse to Bush (unsent), 10 July 1943, OSRD Papers, OFS Records, box 284, ORG—Operations Research Group—PM Morse's Final Summary folder.

80. Waterman to Bush, 9 May 1946, OSRD Papers, OFS Records, "Critical Summary for Dr. Bush" folder.

81. Waterman to Bush, 26 May 1943, OSRD Papers, OFS Records, box 283, General Summary—Establishment of OFS folder.

82. Weaver to Bush, 25 February 1943.

83. Wilson to Bush, 30 July 1943, OSRD Papers, OFS Records, box 283, Steps Leading to the Establishment of OFS folder.

84. Wilson, Memorandum for Operational Research Division Files, 11 September 1943, OSRD Papers, OFS Records, box 283, Steps Leading to the Establishment of OFS folder; and Wilson to Bush, 30 July 1943.

85. Waterman to Wilson, 23 July 1943, OSRD Papers, OFS Records, box 283, Definitions and Methodology of Operation Analysis folder; and Compton to Bush, 23 August 1943, OSRD Papers, OFS Records, box 283, Steps Leading to the Establishment of OFS folder.

86. Morse to Wilson, 10 July 1943, OSRD Papers, OFS Records, box 283, Steps Leading to the Establishment of OFS folder.

87. Compton to Bush, 23 August 1943.

88. Bush to Wilson, 6 August 1943, JCS Papers, JNW Records, Operational Research 1/1943 through 7/1943 folder.

89. Bush to Bundy, 12 November 1943, OSRD Papers, OFS Records, box 284, R/S—SWPA folder.

90. For a celebratory account of Compton's mission see Lincoln R. Thiesmeyer and John E. Burchard, *Combat Scientists* (Boston: Little, Brown, 1947), 3–32.

91. Compton to MacArthur, 21 January 1944, OSRD Papers, OFS Records, box 284, R/S—SWPA folder; and Compton to Bush, 21 January 1944, OSRD Papers, OFS Records, box 284, R/S—SWPA folder.

92. Thiesmeyer and Burchard, *Combat Scientists*, 50–51.

93. "An Interview of Dean Harrison," OSRD Papers, OFS Records, box 284, R/S—SWPA—Misc. Notes folder.

94. Ibid.

95. The PBOSRD idea was modeled on the Radiation Laboratory's facilities in Britain, the British Branch of the Radiation Laboratory (BBRL).

96. The PBOSRD set up was also extremely complex. Apparently, MIT, through the Radiation Laboratory, managed it, the National Academy of Sciences administered it, and the OSRD paid for it.

97. Nevertheless, the OSRD branch office on Hawaii enjoyed much better relations with General Richardson, and its members appear to have been satisfied with the conditions under which they worked. Researchers and soldiers seem to have finessed the issue of authority in this case. See "Final Report of Activities, Operational Research Section, Headquarters, AFMIDPAC, 31 May 1944 to 2 September 1945," OSRD Papers, OFS Records, box 284, ORS-POA—LC Marshall and JK Howard's Final Reports folder. Ironically, the Army insisted on calling the branch office the "Operational Research Section."

98. Thiesmeyer and Burchard, *Combat Scientists*, 291–292.

99. For short summaries of the projects supported by OFS, see OSRD Papers, OFS Records, box 284, Reports Prepared by Mrs. Schneider folder.

100. Waterman to Bush, 9 May 1946.

101. Morse to Bush (unsent), 8 February 1944, OSRD Papers, OFS Records, box 284, ORG—Operations Research Group—PM Morse's Final Summary folder.

102. All, that is, except for Morse. In his autobiography, he avoids stating his opinion of OSRD and NDRC; Bush is mentioned as much for his work on the differential analyzer as for his role as OSRD director. In one revealing passage, Morse writes, "Of course, Bush and Compton and Conant, who were running NDRC, were in touch with the highest military authorities and were undoubtedly contributing to the overall plans. But there must be need for more detailed studies, for mathematicians and theoretical physicists to work at lower command levels, to forge multiple links between new technology and military requirements." Morse, *In at the Beginnings*, 170–171.

3

From Concurrency to Phased Planning: An Episode in the History of Systems Management

Stephen B. Johnson

> There can be no bold approaches when the program must be defined in explicit detail before initiation. If an approach must survive the review of a series of boards, panels, committees, advisors, etc., before it can be started, it is safe to say that the technology it employs will be the least common denominator among these many reviewers, and we risk the danger inherent in the story that "a camel is a horse designed by a committee."
>
> —Colonel Otto Glasser, USAF

As the United States Armed Forces struggled to develop large-scale, complex weapons systems in the 1950s and early 1960s, they developed new organizational structures and tools that collectively came to be known as "systems management." Project management and matrix management were its two fundamental "exemplars" used for organizational structures,[1] while the Program Evaluation and Research Technique (PERT) and the Critical Path Method (CPM) became the primary exemplars of tools from which many variations evolved. Within the air force, Air Force Systems Command and its predecessors led the development of new methods for the management of its weapons systems.

This chapter will discuss one significant change in the conceptions and processes of systems management: the change from the philosophy of "concurrency" to that of "phased planning." As we will see, this change was a managerial and organizational response to a deeper restructuring of priorities within the United States government. Priorities shifted from an overriding concern with rapid development and deployment of large weapons systems to a primary concern with their cost. It is a good example of how an organization translates upper-level decisions and priorities into changes in processes and procedures that in turn modify the products they create.

"Concurrent" and "sequential" methods can be modeled as two ends of a spectrum of military procurement methods. Concurrent procurement is a method by which a government agency builds a system at the same time that it develops its plans and designs. In sequential procurement, the "phases" of definition, design, and production are clearly separated, and are not allowed to overlap. In his book, *Flying Blind*,[2] Michael Brown uses this model to argue that sequential procurement leads to lower-cost, better-performing products when significant technological uncertainties are present. He argues that when a project needs little new development, concurrent approaches are more successful. Because engineers and technicians are familiar with existing technologies, they understand the design and production implications of using them at the start of a program. Conversely, if the project requires substantial new development, then sequential approaches work better, because they methodically develop the necessary design and production knowledge. In his view, the air force has often been "flying blind," because it has pushed for concurrent development, even in the face of large technological uncertainties.

"Phased planning" was the term used by the air force to describe one "sequential" method that they adopted in the early 1960s. This chapter describes how, under pressure from McNamara's regime in the Department of Defense, phased planning replaced concurrency as the standard procedure for Air Force Systems Command.[3] Phased planning represented a partial retreat from concurrent methods toward sequential techniques. This retreat became more significant than it would otherwise have been because McNamara embedded the new procedures in the Department of Defense's research and development regulations, giving them a long bureaucratic life that extends to the present. They have become a standard element of how the Department of Defense "does business" in military research and development. In this way, McNamara's cost-conscious values persist to the present in R&D, not only in the Department of Defense, but also in other organizations around the world.

Organizational Learning and Innovation

Large technological systems are built by individuals embedded in a much larger network of organizations and resources upon which they draw. Large weapons systems, such as missiles and submarines, are built by groups of engineers, scientists, and technicians, who are in turn guided by managers or by groups of engineers and scientists acting in management capacities. Scholars increasingly recognize that even the technological features of the final artifacts depend upon the specific characteristics and values of these groups.

Historians have shifted their analysis from the characteristics of artifacts to the characteristics of the individuals, groups, and social processes that produce them. In this shift, they draw upon existing traditions of historical and sociological research that focus upon how specific artifacts come to embody these values. Marxist, feminist, and constructivist approaches are only three of many possible avenues of investigating the development of technologies through social interactions.[4] Other scholars investigate the capacity of communities of practitioners, and how these communities interact to produce technical innovations. They emphasize the ability of communities to compete, cooperate, and learn as groups.[5] In addition, they interpret organizations as systems with characteristic internal and external dynamics that extend their scale and scope, and influence their capacities for innovation and change.[6]

All of these approaches assume that an organization can gather information internally and externally, can use this information to set goals, and can translate these goals into material artifacts. This in turn implies that internal methods exist to communicate ideas, decisions, and values within the organization. No organization can function without appropriate means of communication between the individuals within the organization. James Beniger in his 1986 work, *The Control Revolution*, and JoAnne Yates in her 1989 book, *Control through Communication*, investigated the means of control and communication within organizations.[7] Both view organizations as networks of communication, where management actively creates or promotes increasing speed and efficiency of communication to control their organizations.

Drawing from Beniger and Yates, this chapter assumes that communication techniques are the means by which managers and others control social structures and processes. These include both hardware technologies and the processes and procedures where humans play the role of information processors and transmission devices. Processes apply to organizations that create technologies, as well as manufacturing them or using them. I examine the genesis of one specific process that the Department of Defense used to control the creation of large-scale technological systems in the American military. "Phased planning" has become a standard model for procedures to control the development of technology as it transitions from research to development and operations. An investigation into the values and conflicts expressed at its creation yields insights into characteristics of the American R&D system.

Concurrency

Concurrent methods, where engineers develop components in parallel with each other, and then integrate them into aircraft, developed in the crisis atmosphere of World War II. However, these methods fell to the wayside upon the war's completion. The term "concurrency," as it came to be used in the air force, had its origins in the development of American ballistic missile systems in the 1950s. Ballistic missiles, where air force officers revived these methods, also first appeared in World War II. Usually based upon the German V-2 design, several American organizations began experimental development of their own rockets during and after the war. Soviet nuclear weapons development prompted the next burst of missile activity. Fear of Soviet fusion bomb development spurred Americans to create their own fusion weapons. The hydrogen bomb could potentially be made light enough to be carried on ballistic missiles. Even if missiles were not accurate, their payload would nonetheless destroy the target. Colonel Bernard Schriever, then charged with planning for the future of the air force, immediately recognized the implications. Within days he was talking with the scientists and pressing for intercontinental ballistic missile (ICBM) development. His campaign was successful, leading to a crash program to develop ICBMs.[8]

The air force was tasked with developing ICBMs as "rapidly as technology would allow." To ensure that the program moved quickly, the air force created the Western Development Division (WDD), and placed Brigadier General Bernard Schriever in charge. The WDD and its systems engineering contractor Ramo-Wooldridge concentrated on accelerating the Atlas missile program that had been progressing slowly under contract to General Dynamics Convair Division.[9] In 1954, the air force made ballistic missiles its top priority, allowing WDD to operate with its own special management and control procedures. In September of 1955, the Eisenhower administration placed ICBM development at the top of the nation's defense priority list, while a special civilian-military committee reviewed the management of the program to further reduce delays in the program. The Gillette Committee established separate managerial and financial procedures and organizations for ballistic missiles at the Office of Secretary of Defense level. Schriever won independence from the rest of the air force, reporting directly to the secretary of defense. In practice, he reported to Congress, which directly allocated the funds to the program.[10]

Despite the program's top priority and streamlined management, neither the air force nor anyone else knew for certain if ICBMs with adequate performance and reliability could be built. Substantial uncertainties remained regarding the feasibility of liquid rocket engines and missile guidance. Because of the uncertainties and perceived criticality of ICBMs, the WDD and Ramo-Wooldridge believed it unwise to rely on only one approach. If it did not work, then the entire Atlas program would be in vain. Consequently, the WDD developed separate technical solutions for the Atlas in parallel, so as to have an alternative if any one of them failed. It also developed an entirely separate missile system, the Titan. Although costly, this "concurrent" development drove system design and testing forward at a rapid pace while reducing technical risks. It continued this policy for the construction of launch sites and training for operations. Searching for a way to explain this parallel development, Schriever coined the term "concurrency" to describe his management strategy and process.

Although concurrency drove Atlas and Titan development rapidly forward, it came at considerable cost. Because designs were immature, design changes plagued the program. Changes in one subsystem frequently required changes in connected or related subsystems. Since many of the affected subsystems were already being built, they had to be redesigned and rebuilt, or retrofitted with the latest modifications. The ballooning costs and obvious inefficiency led in turn to congressional investigations. Citing the "missile gap," and the new "principle of concurrency," Schriever fended off criticism and maintained his own autonomy through the 1950s. While concurrent methods themselves were not all that new, the term "concurrency" helped him portray an image of managerial innovation.[11]

The perceived success of Schriever's methods led to the creation of new regulations to encapsulate and promote them. To govern large-scale projects such as ballistic missiles, the air force published the "375-Series" regulations in 1961. These regulations divided "system programs" into conceptual, acquisition, and operational phases, and provided for centralized planning, management, and funding of the acquisition phase, just as Schriever's group had done for ballistic missiles. With the establishment of Air Force Systems Command in 1961, and the 375-Series regulations that governed its large programs, systems management became a standard air force process. Concurrency therefore became the established policy of the air force.[12]

McNamara's Trombones and the Rubel Philosophy

The relative autonomy of Schriever's organization was short lived. The air force did not care for the independence of Schriever's group, and the Department of Defense soon clamped down on all of the services, including the air force. In the 1950s, President Eisenhower fought to control the individual services, whose constant bickering and competition he believed to be detrimental to national security. Stating that "separate ground, air, and sea warfare is gone forever," he lobbied Congress to give the secretary of defense fuller, more centralized authority over the services.[13] The Defense Reorganization Act of 1958, passed in the wake of the Sputnik crisis, gave the secretary of defense the authority to "transfer, reassign, abolish or consolidate" service functions, and control over the defense budget.[14] No longer could the separate services present their budget requests directly to Congress. The act also recognized the criticality of new technology by authorizing a new director for research and engineering to "supervise all research and engineering activities in the Department," and to control the research and development budget. This critical reorganization cleared the way for a strong manager to take control of the DOD.

Just such a man appeared in the next administration. Robert McNamara, John Kennedy's appointee as secretary of defense in 1961, was famous for centralized control, implemented through quantitative measurement. With the power to withhold funding from the services, he could exert his power effectively. He spent the spring of 1961 gathering information about the department, initiating over 100 studies known as "McNamara's 100 trombones," or in other circles the "92 Labors of Secretary McNamara."[15]

One of the major shortcomings in the department was the separation of planning from budgeting. This caused huge cost overruns. Each service could and did plan for far more than could ever be paid for, attempting to secure expanding budgets for current and future years. In the early phases of development, weapons systems cost far less per year than during their future procurement and long-term operation. Thus, getting a small appropriation today to develop a much larger system for tomorrow virtually guaranteed a large budget for the future. This was known as the "thin edge of the wedge" or the "foot in

the door" strategy. Lack of coordination among the services compounded the budget problems, since each service utilized "wedge" strategy to its own benefit, without having to negotiate with the other services for potentially overlapping systems. Only in 1956 was joint planning required of the military services, with an annual Joint Strategic Objectives Plan.[16] However, without any tie from plans to budgets, coordination had little meaning.

McNamara brought in Charles Hitch, the chief economist of RAND Corporation, to be the comptroller for the Department of Defense. In the early spring of 1961, after listening to Hitch's proposals for financial control system reform of the strategic nuclear forces, McNamara exclaimed "that's exactly what I want," but with ". . . one change. Do it for the entire defense program. And in less than a year." This decision and its implementation hit the Department of Defense so quickly that when the services finally realized what was happening, it was a fait accompli.[17]

Given McNamara's background as a financial manager at Ford, and Hitch's qualifications as an economist, it was not surprising that they considered economic criteria to be foremost in making decisions for future weapon systems. According to Hitch:

> The choice of an appropriate economic criterion is frequently the central problem in designing a systems analysis. In principle, the criterion we want is clear enough: the optimal system is the one which yields the greatest excess of positive values (objectives) over negative values (resources used up, or costs) . . . Military decisions, whether they specifically involve budgetary allocations or not, are in *one of their important aspects economic decisions*; and . . . unless the right questions are asked, the appropriate alternatives selected for comparison, and an *economic criterion* used for choosing, the most efficient military power and national security will suffer.[18] [Hitch's italics]

Hitch's Program Planning and Budgeting System required that life-cycle cost estimates be performed prior to the decision to develop and acquire a new weapon system. This agreed with the result of Project 102, entitled "Shortening Development Time and Reducing Development and Systems Cost." The new policy resulted in equating "performance" with "reducing lead time and cost." It deemphasized state-of-the-art performance: the policy stipulated that "real requirements, not technical feasibility, dictate developments." "Feasibility and effectiveness studies" were to calculate technical risks and cost-to-

effectiveness ratios. With this, cost became as important or more important than technical performance, which had heretofore dominated the thinking of armed forces leaders.[19]

In September 1961, McNamara assigned the task of improving the management of research and engineering to John Rubel, the deputy director of defense research and engineering.[20] Rubel began with the air force, to set up "model programs" for management that could then be used throughout all of the services. He started with Agena, the TFX fighter, the Titan III, and the medium range ballistic missile (MRBM) programs.[21]

Rubel likely selected the air force because it had been a traditional supporter of the Office of the Secretary of Defense since 1947 when Congress created both. Ever since that time, air force leaders had allied themselves with the secretary of defense, believing that arguments for economy and efficiency favored its relatively inexpensive strategic nuclear forces over the large, human-intensive conventional forces of the army and navy. In its relative youth, air force leaders were less bound by historical traditions than the army or the navy. They had relatively few ties to the civil service, preferring instead to work with industry and with special, nonprofit research and development organizations like the RAND and MITRE corporations. Such organizations were in the vanguard of systems analysis, systems engineering, and project management. In comparison to its fellow services, the air force favored efficiency and modern systems management methods. When McNamara and Rubel tried to set up model programs for the management of research and engineering, they found an ally in the air force.

In the Agena upper-stage program, Rubel proposed a "Phase I" effort to develop a preliminary design. This would ensure "that the cost estimates for the subsequent development effort are based on a solid foundation." Phase I was to generate "a set of drawings and specifications and descriptive documents" that could be subjected to an independent review. Arrangements for the long-term management of the program were also to be described, including schedules, milestones, tasks, objectives, and policies.[22]

Rubel also promoted the same "phased" approach for the Titan III program. He saw this Phase I effort to "correspond in principle" with that envisaged for Agena, and made approval of the Titan III program contingent upon substantial improvements and implementation of management methods. These included "a strong project

organization," and the use of PERT and other "accounting centers, accounting and auditing practices." Rubel required that the project's contractors also adopt the new standards. He stated "strong centralized project-type organization must be insisted upon for all major elements."[23] Rubel had no reservations about forcing industrial contractors to organize and manage their projects in the way he wanted them to. If they wanted the job, they had to conform. The Martin Company, one of the leaders in systems management since the early 1950s and manufacturer of the Titan III, readily complied.[24]

By early December 1961, Rubel and the air force had worked out the details of how to manage Phase I for Titan III. Operations Research Incorporated would help set up PERT networks, and also "serve as a useful technical link" between Rubel and "the Air Force groups endeavoring to synthesize and improve management procedures and operations." The Request for Proposal (RFP) made clear that a go-ahead for Phase I *did not constitute program approval.* Previously, award of a contract for preliminary design work would have constituted de facto project approval for system development and production. This was no longer true. Only the Secretary of Defense could approve a project, and he would not do so until Phase I was completed and the program reviewed.[25]

Rubel also authorized a Phase I Program Definition Phase for the mobile mid-range ballistic missile (MMRBM) program.[26] He stated that

> the fact that improved definition is required before larger-scale commitments are undertaken is neither surprising nor unique, although it is true that on most programs this definition phase has been less clearly identifiable since it has been stretched out in time and interwoven with other program activities such as development, model fabrication, testing and, in some cases, even production.

Rubel did not believe that a Program Definition Phase would slow down the high-priority program. "In fact," he wrote, "our real progress should be accelerated as the result of obtaining a better focusing of our efforts."[27]

The phased approach brought several benefits to upper management. Phase I ensured better cost, schedule, and technical definition. If program personnel did not provide the appropriate information, management could cancel the program, force it through another Phase I, make changes, or go ahead. This ensured that companies would make a strenuous effort to finalize a design and estimate the cost of its design, production, and operations. The preliminary design phase provided management with a decision point *before* commitment of large sums of money, making it easy to terminate. From this point on, the use of phased-planning approaches began to spread throughout the Department of Defense, and then to other organizations.[28]

On 9 October 1961, Rubel distributed a more general "discussion paper" outlining problems and approaches to management in development programs. Known in the air force as "The Rubel Philosophy," it became a lightning rod for debate on the Department of Defense's initiatives, and the air force's acquisition methods, including phased planning. While Rubel's objectives were hardly controversial—reduction of lead times and costs, better utilization of engineering and scientific talent, better quality, and so on—his proposed solutions were. He identified a number of problems in DOD R&D programs, including nonuniform, unrealistic requirements, erroneous and misleading plans, costly and inefficient contractor selection methods, excessive changes to new systems, excessive reporting, and using the term "research" to cover for ineptness in "development." To alleviate these problems, he proposed solutions that increased the power of DOD centralized management, at the expense of the services and their project officers. These included standardization of reporting in the DOD, improving the information in contract bids and evaluations, and collection of manpower and cost statistics.[29]

AFSC Reactions to Phased Planning and the Rubel Philosophy

Department of Defense criticism directly affected Schriever's Air Force Systems Command (AFSC), the newly formed successor to the Ballistic Missile Division and the Western Development Division. McNamara's Project 102 policy statement reached the air force on 25 September 1961 as "policy guidance." In response, Colonel James Stewart in the AFSC Systems Office described actions that had been taken in AFSC and the USAF that were independent of, but "in full accord with" McNamara's memorandum.[30] McNamara wanted to shorten development lead time. Stewart stated that this had been a primary objective for years, with the principle of "concurrency."

Cost was a more difficult issue. Stewart noted that AFSC had "established an in-house cost analysis capability" and was "working aggressively to train additional personnel in the techniques of cost estimating and analysis." AFSC had established financial procedures and formats to obtain annual cost data for all major systems. To ensure improved monitoring, they were also placing engineers in contractor facilities to "work with the contractor on time, performance, and cost tradeoffs." All proposed programs were to be "subjected to technical feasibility, study, analysis of effectiveness, and a cost estimation to determine a cost/effectiveness factor for the time period concerned." Schriever's officers placed responsibility in the System Program Office, as a single point of contact for all phases of the program cycle. McNamara, ever concerned to pinpoint accountability, now had one person to praise or blame for the performance of the program.

The 375-Series regulations required the submittal of a "preliminary development plan" providing information for management to decide whether or not to proceed.[31] Although Stewart felt that the 375-Series regulations dealt with this issue, this is precisely what Rubel deemed inadequate. Rubel added an additional program step to produce better information than was previously forthcoming.

Not all responses to McNamara's memorandum were so sanguine. In January 1962, Brockway McMillan, the assistant secretary of research and development in the Department of the Air Force, reported that notwithstanding McNamara's memo, "there is little evidence of good understanding or application of these principles." RFPs, including those for Titan III and MMRBM, had little or no mention of trade-offs of cost versus performance. Unless the secretary of defense's instructions were followed "at the lowest levels," they would have little effect. He stated that "it is difficult to obtain widespread recognition and application of the rather nebulous provisions of the Secretary of Defense's memorandum." Nonetheless, he felt that it was important to consider cost and performance trade-offs. He believed that "in the end, an active educational program will be required, carrying out the word to the working levels in AFSC, perhaps by means of seminars or briefings."[32]

Air force divisions also responded to the "Rubel Philosophy" paper.[33] The Minuteman program had inherited some of the management ideas used on Atlas and Titan, but was run on a somewhat more orderly basis than these two early "crash" programs. Their team, headed by Colonel Samuel Phillips, soon to be the manager of NASA's Apollo program, thought the Rubel discussion paper treated symptoms rather than causes of the problems in weapons system acquisition. The real question to be addressed was: "Is there an orderly, logical way early in a program to identify or to define the weapons system configuration?" Answering in the affirmative, they proceeded to proclaim that Minuteman used such a logical procedure. It required "precise definition by defense agencies of such matters as maintenance plans, supply plans, logistic plans, and operational plans," and could be extended to "cover all elements of weapon system hardware." To them it was "obvious that if the above is done, optimization can be achieved among dollars, quality or performance, and schedule or lead times. By the same token, it should be recognized that without the information proposed, no, or limited optimization is possible."[34]

Other air force officers saw the McNamara-Rubel initiatives as a threat, particularly to the principle of concurrency. Colonel Jewell Maxwell, Air Force Assistant for Management Policy, stated that the new process of "definition and development implies a tendency toward the 'fly-before-you-buy' concept which will obviate to some extent application of the concurrency principle during the earlier phases of urgent system acquisition programs."[35] Maxwell noted that "it has been conservatively estimated that an additional six to nine months will be required to complete the MMRBM development program under the principles established in the DDR&E [Defense Department Research and Engineering, Rubel] memorandum." Additional delays might occur "if too much emphasis is placed on achieving finality of definition during that phase of the program when data are most meager and decisions are most subjective." Maxwell saw that the earlier emphasis in the air force upon achieving rapid deployment of advanced systems, even if at high cost, was now being changed by the new emphasis on reducing cost to the detriment of the schedule.

A more serious problem was a tendency toward "creeping centralization." Maxwell saw in the new initiatives a "trend toward imposition of super-management organization at the top of current review and approval channels." The problem was subtle, "the more intangible consequence resulting from more detailed time-consuming control by the higher echelons outside of the Air Force." External controls removed "program control flexibilities" from "responsible operating levels in the field."[36]

General Schriever became increasingly concerned with this "creeping centralization." Schriever's management reforms, encapsulated in the 375-Series regulations, sought to decentralize and streamline the decision-making process to reduce weapon system lead times.[37] The Secretary of Defense's initiatives worked at cross-purposes to the intent of AFSC's regulations. In April 1962, Schriever wrote to his commanding officer, General Curtis LeMay:

> During the past several months a disturbing trend has developed toward increased review of program details at Department of Defense level, prior to the release of decisions required to proceed with approved programs. The extent and detail of information required in these reviews and the nature of the decisions being withheld is unprecedented. This trend is generating demands for large volumes of information and program data that is magnified at each succeeding organizational level. Decisions on matters that have never been previously reviewed are being withheld for inordinate lengths of time.[38]

Schriever noted that Office of Secretary of Defense reviews repeated those already conducted by AFSC. This caused delays in the program, and if continued, would require Schriever to increase AFSC upper-level staff to support the increasing demands for information. He refuted "the apparent belief that our procedures have not or will not produce the desired results," stating that the AFSC "concept of management is sound, provided we give it a chance to work."[39]

Colonel Otto Glasser, the former Atlas program manager, summarized the new management trends at a conference in 1962, stating, "Dominant among all of the external influences in systems management is the trend toward centralized civilian control. . . ."[40] Due in large degree to the complexity and cost of new weapon systems, the "validation of requirements [for a weapon system] can no longer be left to the possible parochial opinions of a single service . . . The requirements of today must be national and they must be validated in detail before a program can be seriously undertaken."[41] The DOD now evaluated proposed systems primarily on the basis of cost-effectiveness. Once they were approved, an upper-level manager subjected systems to continual review.[42] As Glasser put it, the reforms "seem to be in the direction that you would go in handling your own household budget and I believe that it is safe to say that is just how the average taxpayer expects us to employ the tax dollars." On the other hand, there were dangers. Glasser compared current dangers to the problems found in the 1930s with the old-style "efficiency experts":

> The efficiency expert is not a creator, but rather an improver on other people's creations. For a limited period, the benefits of his labors can be enormous, but care must be taken to avoid stifling all new ideas. There can be no bold approaches when the program must be defined in explicit detail before initiation. If an approach must survive the review of a series of boards, panels, committees, advisors, etc., before it can be started, it is safe to say that the technology it employs will be the least common denominator among these many reviewers, and we risk the danger inherent in the story that "a camel is a horse designed by a committee."[43]

To this refrain, Schriever agreed. "If we are to be held to this overly conservative approach, I fear the timid will replace the bold and we will not be able to provide the advanced weapons the future of the nation demands."[44]

Conclusion

Phased planning became a cornerstone of the DOD research and development "system," enshrined in regulations. Borrowing and modifying DOD concepts, NASA made "Phased Project Planning" the basis of its research and development management regulations. The European Space Agency ultimately adopted it from NASA as its standard for systems management.[45]

The shift from concurrency to phased planning signified a larger shift from concerns with performance and rapid deployment of state-of-the-art systems at virtually any cost, to a focus on reducing cost. Huge cost overruns inhered in the earlier "principle of concurrency."[46] While justifiable to Congress and the Eisenhower administration under the perceived national emergency in the 1950s, this emergency status could not be sustained forever. Phased planning was one element of correction to this unbalanced approach. Inherent in this shift was a slowdown in the pace of technological innovation driven by the government. Managers believed that to control costs, they needed to have an additional "checkpoint" where they could assess whether a program should move from research and preliminary design to full-scale development. To decide whether to continue or cancel a program, civilian managers wanted more information than the military or its contractors had formerly supplied. Additional information reduced program risks and made for better cost estimates, but also slowed down the development of the program. As perceived by Schriever and Glasser, the involvement of cost-conscious managers in the go-ahead decisions for programs could also stifle technological development if it appeared too risky or costly.

This shift in the meaning and goals of system management had consequences for those who adopted system management for their organizations, including NASA, ESA, their industrial contractors, and numerous other organizations both inside and outside the government. Systems management became a fad in the late 1960s through the 1970s, proclaimed as the means to efficient, cost-effective management. However, it came to acquire other characteristics as well, including tight centralization, a fetish for quantification, and a significant slowdown from research to development to deployment of technologies.

In his analysis of concurrency, Michael Brown stated that "concurrency flourished in the Pentagon under McNamara's stewardship in the 1960s," because McNamara believed that "paper studies could substitute for prototype testing."[47] On the other hand, Brown also found that civilians concerned with cost pressed for sequential strategies."[48] In this chapter, we find that McNamara promoted phased planning to consolidate his power over the services, and to better control costs. There is also good evidence that McNamara did not really understand the nuances of technological development, and believed in the value of systems analysis "paper studies."[49] With McNamara's background in statistics and logistics, and lack of experience with R&D, this should not be too surprising. We can paint a more complex picture of McNamara's regime, trying to achieve cost control through sequential methods by use of systems analysis in phased planning. This shows that in at least one aspect, contrary to Brown's assessment, this was not the "golden age of concurrency."

Procedural and structural changes are the means by which political and strategic forces modify organizational behavior and capabilities. Phased planning is one example of this kind of organizational change. Although upper-level management desired better cost control through phased planning, lower-level managers had many reasons to keep cost estimates ambiguous. As we have seen, Schriever appealed to the wider emergency to expand missile programs and to delegate power to his organization. This appeal had lost its force by 1961, and McNamara was therefore able to impose managerial control through phased planning, and then to ensure its continuation through bureaucratization of these procedures. They subsequently took on a life of their own, well beyond the careers of the original actors. The long-term ramifications have yet to be investigated, but are sure to hold surprises in the complex interplay of interests inherent in the development of large technological systems.

ACKNOWLEDGMENTS

I would like to thank Robert Seidel and Jackie Geier for their reading and editing of this chapter, Tom Hughes for the opportunity to present it, John Lonnquest for his insights on Schriever's use of concurrency, and Harvey Sapolsky for his commentary, which brought Brown's work to my attention.

NOTES

1. See Thomas Kuhn, *The Structure of Scientific Revolutions*, 2d ed. (Chicago: University of Chicago Press, 1970), 187–191; Kuhn's idea of exemplars is extended here to models for technologies and social structures.

2. Michael E. Brown, *Flying Blind: The Politics of the U.S. Strategic Bomber Program* (Ithaca, N.Y.: Cornell University Press, 1992). Brown also develops a set of eight characteristics that define concurrency. A highly concurrent program has a large number of these characteristics, whereas a highly concurrent program has few.

3. It is important to note that AFSC's usage of the term "phased planning" is not identical to Brown's use of the term "sequential." Phased planning introduced one sequential element in the overall procurement of a weapon system. It became a regular part of the "weapon system concept" that Brown characterizes as "highly concurrent." This point is argued by John Lonnquest in his recent dissertation on Bernard Schriever and the Atlas program. For Lonnquest, the air force weapon system concept was a necessary precursor to concurrency but is not identical with it. See John Lonnquest, "The Face of Atlas: General Bernard Schriever and the Development of the Atlas Intercontinental Ballistic Missile 1953–1960" (Ph.D. diss., Duke University, 1996).

4. Examples of this kind of analysis are Donald MacKenzie, *Inventing Accuracy: A Historical Sociology of Nuclear Missile Guidance* (Cambridge: MIT Press, 1990); Wiebe Bijker, Thomas P. Hughes, and Trevor Pinch, eds., *The Social Construction of Technological Systems* (Cambridge: MIT Press, 1987); David F. Noble, *America by Design: Science, Technology, and the Rise of Corporate Capitalism* (New York, Knopf, 1977); Judith Wajcman, *Feminism Confronts Technology* (University Park, Pa.: Pennsylvania State University Press, 1991).

5. See Edward Constant, *The Origins of the Turbojet Revolution* (Baltimore: Johns Hopkins University Press, 1980); Nathan Rosenberg, *Exploring the Black Box: Technology, Economics, and History* (Cambridge: Cambridge University Press, 1994); and Ross Thomson, "The Firm and Technological Change: From Managerial

Capitalism to Nineteenth Century Innovation and Back Again," *Business and Economic History* 22:2 (Winter 1993): 99–134.

6. See Thomas P. Hughes, *Networks of Power* (Baltimore: Johns Hopkins University Press, 1983); Bernward Joerges, "Large Technical Systems: Concepts and Issues," in Renate Mayntz and Thomas P. Hughes, eds., *The Development of Large Technical Systems* (Boulder, Co.: Westview Press, 1988), 9–36. See also Alfred Chandler, *The Visible Hand* (Cambridge: Harvard University Press, 1977), and *Scale and Scope* (Cambridge: Harvard University Press, 1990).

7. James R. Beniger, *The Control Revolution: Technological and Economic Origins of the Information Society* (Cambridge: Harvard University Press, 1986); and JoAnne Yates, *Control Through Communication: The Rise of System in American Management* (Baltimore: Johns Hopkins University Press, 1989).

8. MacKenzie, *Inventing Accuracy*, 105–113; Edmund Beard, *Developing the ICBM: A Study in Bureaucratic Politics* (New York: Columbia University Press, 1976), 134–157. Author's interview with Bernard A. Schriever, 25 March 1999.

9. Jacob Neufeld, *Ballistic Missiles in the United States Air Force, 1945–1960* (Washington, D.C.: Office of Air Force History, United States Air Force, 1990), 44–50, 68–79.

10. Neufeld, *Ballistic Missiles*, 133–137. See Lonnquest, "The Face of Atlas," 224–248. See also General Bernard A. Schriever, Oral History Interview with Dr. Edgar F. Puryear, Jr., 15 and 29 June 1977, Air Force Histoical Research Agency (AFHRA) K239.0512–1492, 3. As Schriever later noted in his interview with Puryear, he "loaded" the committee "pretty much with people who knew and who would come up with the right answers." That is to say, they would favor Schriever's independence and authority to assure rapid speed of development.

11. See Lonnquest's overall argument in "The Face of Atlas." See also Hearings before the Subcommittee of the Committee on Appropriations, Senate Department of Defense Appropriations for 1960, 86th Congress, 1st Session (Washington, D.C., 1960), part 5, "Procurement," 662–671. Stephen B. Johnson, "Insuring the Future: The Development and Diffusion of Systems Management in the American and European Space Programs" (Ph.D. diss., University of Minnesota, 1997), chap. 4.

12. Michael H. Gorn, *Vulcan's Forge: The Making of an Air Force Command for Weapons Acquisition (1950–1985)* (Washington, D.C.: Office of History, Air Force Systems Command, 1989), chap. 2.

13. James M. Roherty, *Decisions of Robert S. McNamara* (Coral Gables, Fl.: University of Miami Press, 1970), 56.

14. C. W. Borklund, *The Department of Defense* (New York: Frederick A. Praeger, 1968), 70–71.

15. Charles J. Hitch, *Decision-Making for Defense* (Berkeley: University of California Press, 1967), 31–32. McNamara started with 92 studies, which then expanded

to over 100. See F. T. Moore, memorandum to Research Council, subject: "The 92 Labors of Secretary McNamara," Date: 3–17–61, WM-663; University of Minnesota, Charles Babbage Institute Collection 90, Burroughs Collection, System Development Corporation Series, box 1, "RAND Corporation: Correspondence and Clippings" folder.

16. Hitch, *Decision-Making for Defense*, 25.

17. Deborah Shapley, *Promise and Power, The Life and Times of Robert McNamara* (Boston: Little, Brown, 1993), 99–101.

18. Hitch, *Decision-Making for Defense*, quoted from Roherty, 72–73.

19. "DOD System Acquisition Policies," an internal summary used for a meeting in the Air Force, ca. 1962, box 19. All subsequent memoranda and internal papers are from the Samuel C. Phillips Papers at the Library of Congress, from the boxes noted.

20. John H. Rubel, Deputy Director of Defense Research and Engineering, memorandum for the Secretary of Defense, subject: Management of Research and Engineering, 14 November 1961, box 19.

21. Ibid., 2.

22. It is unclear how Rubel came to the idea of "phasing." It is possible that he took it from contemporary literature. In the late 1950s, there were a number of attempts to classify the design process into phases. Some examples can be found in C. West Churchman, Russell L. Ackoff, and E. Leonard Arnoff, *Introduction to Operations Research* (New York: Wiley, 1957); and R. C. Hopkins, "A Systematic Procedure for System Development," *IRE Transactions on Engineering Management* (June 1961): 77–86. John H. Rubel, Deputy Director of Defense Research and Engineering, memorandum for the Assistant Secretary of the Air Force (R&D), subject: Standardized Agena, 4 October 1961, box 19.

23. John H. Rubel, Deputy Director of Defense R&E, memorandum for the Assistant Secretary of the Air Force (R&D), subject: Titan III Launch Vehicle Family, 13 October 1961, box 19.

24. The first description of "project management" in the aircraft industry that I encountered was William B. Bergen, "New Management Approach at Martin," *Aviation Age*, June 1954, 39–47.

25. John H. Rubel, Deputy Director of Defense R&E, memorandum for the Assistant Secretary of the Air Force (material) & the Assistant Secretary of the Air Force (R&D), subject: Planning for Titan III Phase I Efforts, 6 December 1961, box 19.

26. MMRBM was a continuation of the MRBM program mentioned previously.

27. John H. Rubel, for Harold Brown, Director of Defense Research and Engineering, memorandum for the Secretary of the Navy and the Secretary of the Air

Force, subject: Mobile Mid-Range Ballistic Missile Program Plan, 19 February 1962, box 19.

28. The contemporaneous TFX program will not be discussed here, but it went through a series of "phases" prior to its final approval. See Robert J. Art, *The TFX Decision: McNamara and the Military* (Boston: Little, Brown, 1968). For its spread to NASA, see Johnson, "Insuring the Future," chap. 6.

29. John H. Rubel, Deputy Director of Defense Research and Engineering, memorandum to the Assistant Secretaries of the Army, Navy, and Air Force (R&D), subject: Management of Research and Engineering, 9 October 1961, box 19.

30. Col. James T. Stewart, USAF/Assistant DCS/Systems, memorandum to HQ USAF (AFSSA, Col E. A. Friedlander), subject: Policies on Shortening Development Time and Reducing Development Costs and Weapon System Costs, 12 January 1962, box 19.

31. Ibid.

32. Brockway McMillan, USAF, Assistant Secretary, Research and Development, memorandum for Vice Chief of Staff, subject: Balance of Cost Versus Complexity in R&D Programs, 29 January 1962, box 19.

33. Col. Ray E. Sober, Chief, Plans & Programs Office, Ballistic Systems Division, AFSC, USAF, memorandum, subject: Rubel Management Philosophy, 25 January 1962, box 19.

34. Col. John S. Chandler, Assistant Director, Minuteman SPD, BSD/AFSC/USAF, memorandum to BSLM, subject: Rubel Management Philosophy, 31 January 1962, box 19.

35. Col. Jewell C. Maxwell, USAF, Assistant for Management Policy, AFSC, memorandum to HQ USAF (AFSDC), subject: Systems Management and Program Initiation, 9 May 1962, box 19.

36. Ibid.

37. By this, I mean decentralization away from the secretary of defense, and from the regular air force hierarchy. However, Schriever's management system was extraordinarily centralized around specific projects. Gen. B. A. Schriever, USAF/AFSC, letter to Gen. C. E. LeMay, Hq USAF, 30 April 1962, box 19.

38. Ibid., 1.

39. Ibid., 2–3.

40. Col. Otto J. Glasser, USAF, Hq, AFSC, "Current AFSC Management Environment," ca. 1962, box 19, 5-1-1.

41. Ibid., 5-1-3.

42. Department of Defense Annual Report for Fiscal Year 1961, quoted from Glasser, 5-1-5.

43. Glasser, 5-1-7, 5-1-8.

44. Schriever to LeMay, 30 April 1962, 4.

45. See Johnson, "Insuring the Future," ch. 6, 7, 8.

46. Brown argues (contrary to conventional wisdom) that concurrency did not speed up technological development. Although the case cannot be easily proven one way or the other, I believe that it did, but was extremely expensive.

47. Brown, *Flying Blind*, 26.

48. See, for example, David Packard's handling of the B-1 program. Brown, *Flying Blind*, 248–250.

49. His handling of the TFX fighter is the classic example. McNamara essentially took over project management of this fighter program, and literally pored over engineering drawings spread out over the floor of his Pentagon office. He did this without consulting with the air force project managers, with rather dismal results. See author's interview with Charles Terhune, 24, 30 September, and 14, 20 October 1998.

4

System Reshapes the Corporation: Joint Ventures in the Bay Area Rapid Transit System, 1962–1972

Glenn Bugos

The Federal Contours of Joint Venturing

Joint ventures grew increasingly popular among American corporations in each decade following 1950, and federal regulation shaped the broad contours of this growth. A wave of horizontal mergers followed World War II, as demobilized companies coped with excess capacity and led Congress to pass the Cellers-Kefauver antitrust act of 1950. This 1950 act closed a loophole in Section 7 of the Clayton Act, which had outlawed acquisition of stock in a competing company but not acquisition of productive assets. With the asset-acquisition loophole now closed, and the Eisenhower administration vigorously enforcing the Act, by the mid-1950s horizontal mergers waned—only to be offset by a boom in horizontal joint ventures.

Before 1950, joint ventures were rare. Companies thought to use them only when merger was prohibited—mostly by the complications of acquiring foreign technology or entering foreign markets. But beginning about 1950, joint ventures appeared more frequently, more often within the United States, and more clearly on the technological frontier. In the energy industries, oil and gas companies formed them to build new pipelines or pool the risks of searching for new oil. In manufacturing, joint ventures helped firms pool production expertise or build plants of minimum efficient scales. During the 1950s, the 1,000 largest manufacturing firms formed 345 domestic joint ventures. Most of these new ventures were horizontal, to circumvent the Clayton Act, though many were vertical. Virtually all, significantly, related closely to the product lines of the parent companies.[1]

Not until 1964, with the Supreme Court decision in *Penn v. Olin*, could the Federal Trade Commission prevent firms from using joint ventures to circumvent existing antitrust law. Joint venture formation should have slowed with *Penn v. Olin*, plus the regulatory uncertainty building up to it, plus uncertainty over how the FTC would apply a "rule of reason" in deciding if a proposed joint venture promoted anticompetitive incentives.[2] Yet throughout the 1960s the tempo of domestic joint venture formation never waned: 520 were formed in mining and manufacturing between 1960 and 1968, with 1,131 companies serving as parents. (The numbers of foreign joint ventures grew apace, at a ratio of about two domestic to one foreign joint venture.) More American corporations experimented with the joint venture form, and discovered new organizational advantages.[3] A slowing in horizontal ventures was offset by a rise in vertical ventures, with very few characterized as conglomerate.[4]

Of course, a more powerful movement was reshaping American industry in the 1960s—conglomeration. Eleven conglomerates—including Beatrice, W. R. Grace, ITT, G + W, Teledyne, Textron, LTV, Litton—accounted for 500 acquisitions between 1961 and 1968, mostly in completely unrelated product lines. Even nonconglomerate corporations diversified, pressed by international competition and recession in key markets, and the profit prospects of a surging Dow. In 1950, the 471 largest American corporations offered 2.55 product lines (based on four-digit SIC codes); in 1975 they offered 7.54.[5] Thus, the pace of joint venture formation and the pace of mergers ebbed and flowed in lockstep (about nine mergers to one venture) from 1950 through 1980, driven by the same economic forces, like bear and bull stock markets. However, the purposes of ventures and of mergers diverged dramatically. The ventures stayed closely tied to the parents' existing line of business, while the mergers became farther removed.

The conglomerates pioneered a new organizational form—the divisionalized corporation—since the central corporate office had so few functions beyond finance and acquisitions. Corporate officers merely set quotas for each division—called management by objective—and judged divisional performance on quantitative, financial data, rather than on the development of a product line they usually knew little about. Some conglomerate officers proclaimed their management expertise was in computerized operations research, a singular problem-solving technique that applied to any and all product lines.[6]

To see how systems thinking reshaped the corporation, look at conglomeration. Look at the corporate control centers, where managers pitched reports on their companies in a graphical and, importantly, in consistent format. Look at the constant regrouping of divisions and companies as managers look for synergy. Look at atrophy in corporate headquarters, which did nothing requiring technical expertise. And look for the language of system analysis and the gamesmanship of defining systems where none previously existed.

In 1969, a vice president for Litton Industries talked on systems thinking. A couple of years in the past, already, was the quick boost from tax shifting and stock manipulation. A couple of years ahead lay the white flag of massive divestiture. Reflecting a widespread hubris, he claimed systems thinking at the headquarters level could still make conglomeration work: "The systems approach enables a firm to enter across a wide variety of industries. From a systems standpoint, there are no industries, only problems to be solved and a methodology for solving them."[7]

Corporate indigestion followed this feeding frenzy. Overload at the corporate office, divisions that remained independent, and concern with short-term underperformance—all made it easy for corporations to sell off the companies and divisions they had just bought. Wall Street, which once championed conglomeration and diversification, turned against it. An equally intense divestiture movement followed in the early 1970s.

The pace of joint venture formation fell, too, compared to the late 1960s, yet still remained strong. The FTC observed 1,090 new joint ventures in manufacturing and mining between 1970 and 1978, with continuing close ties between parent and venture product lines.[8] Venturing remained strong as an outlet for the divisionalized corporations —ventures afforded them tremendous organizational flexibility as they maneuvered with suppliers and customers to construct complex systems along their technological frontiers. Conglomeration displayed the sizzle of systems thinking; joint venturing displayed its meat.

In the long run, the joint venturing boom will prove more significant than the conglomeration episode in mergers and acquisitions. Congress certainly thinks so. Even during the most sweeping merger movement ever, in the 1980s, politicians devoted ever more attention to joint ventures. Congress studied their anticompetitive implications as, for example, in the GM-Toyota joint venture in the NUMMI plant. But they also legalized new forms of joint ventures, like joint R&D ventures to pool American technical expertise, as in the SEMATECH chip venture, and like public-private ventures to motivate government and business to move in the same direction, as with the Joint Venture in Affordable Housing. Popular business pundits think so, too. They now talk of rediscovering a firm's core competencies, building teams with suppliers and customers, and avoiding frivolous diversification. And, I argue, historians should think so, too.

A DEFINITIONAL PROBLEM

Each joint venture is unique, forged of the historic resources and goals of the parent firms. Until 1984, there were no legal or financial templates for joint ventures as there were for corporations, partnerships, or interfirm contractual agreements. Joint ventures could operate as a close corporation, or under partnership law with the incorporated parents fully liable for its actions. Scattered case law suggested that along the spectrum of interfirm cooperation, joint ventures fell between mergers and acquisitions—the complete pooling of assets and strategy—and the contractual agreement, whereby two firms specified precisely how they would work to a common goal with no allowance for managerial flexibility or contingent decision making.

Yet all joint ventures share some salient characteristics that make them ideal vehicles for studying the business structures of technological transactions. Above all, joint ventures are project-specific. Joint venture agreements always start with the parents declaring they will work toward a common technological goal—like developing a new product, finding a new market for an existing product, building a plant, exploring an ore field, or developing real estate. Project-specific often means time limited. Most agreements specify that ventures dissolve once the project is completed, or that they will be evaluated after a few years. Experience shows, however, that few agreements accurately predict how long a venture actually will last.

Joint ventures, furthermore, are combinations of technical equity. No parent joins a venture by contributing anonymous capital; American capital markets work well enough. One parent might contribute distribution, production, a patent, or research expertise, while another contributes a completely different skill or asset. And the parents may debate valuations of their contributed skills and assets, or which parent's culture should dominate, or which parent's employees should run the venture. But always, joint ventures combine and extend the technical capabilities of the parents. Venture agreements specify which assets and expertise each parent contributes, while allowing venture officers the freedom to shape these contributed assets and expertise to project goals, using whatever management methods best apply. Often, joint venture officers champion the newfangled management methods designed to solve technical complexity—like systems engineering, program management, and network scheduling.

And joint ventures are shaped more by fear than greed. It is relatively easy for joint ventures to apportion profits—based on the ownership shares of the parents—or to return assets upon dissolution. It is vastly more difficult to apportion risk. Often firms form join ventures to pool their risk as they chase after new technology, unknown markets, or unproved mineral reserves. Often corporations form joint ventures as a liability barrier between themselves and especially capital-intensive technical systems. Often firms form joint ventures to spread the pork, and thus control the shifting political and economic environment of a new project. Each of these perceived risks, much more than any thoughts of profit, inspired the formation of nested hierarchies of joint ventures to guide the construction of the San Francisco Bay Area Rapid Transit system (BART).

About BART

BART is seventy-five miles of rapid transit track ringing the lowlands around the San Francisco Bay. The keystone of the system is a two-lane metal tube resting on alluvial mud under the Bay, spanning the four miles between downtown Oakland and San Francisco's financial district. BART planning began in 1952; detailed design began in 1962; construction began in 1966; trains began rolling in 1972; and the system was completed in 1975. BART was the first new transit system built in America since the 1930s, and the only new system—unlike those of Washington, Atlanta, Seattle, and Los Angeles—built without a financial backbone of federal funding. With an expected cost of $1.2 billion in 1962, and an actual cost of $1.6 billion (by 1974), BART stood as the largest civil construction project ever attempted to date. A project that complex, and that ambitious, could be built only by a vast network of firms working, not surprisingly, as joint ventures.[9]

What is most striking about the BART joint ventures is their diversity—both the types of organizations working jointly and the ways they built their ventures. In a conference on the spread of systems thinking, I might suggest a typology of these many ventures. Instead I will focus on two ventures—the District and PBTB (Parson-Brinckerhoff-Tudor-Bechtel), both horizontal ventures—and the three issues that best display how joint ventures helped incorporated entities adopt themselves to technological systems. First, why the incorporated entities chose the joint venture over other organizational forms. Second, how the incorporating entities reorganized themselves to deal more efficiently with their participation in joint ventures. Third, how the joint venturers manufactured technical complexity, then asserted that complexity was the issue they were best able to solve. I conclude by applying these same criteria to a related case—Rohr's work as a prime contractor in building the BART cars.

The District Tied BART to the Landscape

Special districts are special purpose governments that supply a single good or service, at a known cost, to taxpayers in a limited geographical area. Districts have all the promotional authority of general purpose government—city, county, and state—but lack authority to enforce laws. They are fully sue-able.

During the 1950s, Californians saw a profusion of districts listed on their tax bills—a doubling of new districts formed annually between 1950 and 1960.[10] This proliferation in special districts had several origins. Suburbanites settled unincorporated lands and demanded new services and infrastructure. But the number of districts grew even within the borders of incorporated cities and counties. The types of municipal services diversified, stressing the regular tax base, and cities and counties wanted to more precisely charge those who benefited. Before 1950 most districts supplied irrigation, ports, or schooling. After 1950 districts supplied mosquito and fire control, playgrounds, garbage collection, street lighting, and a variety of other public goods. And many districts supplied transportation.[11]

The Bay Area Rapid Transit District (BARTD) took the form of a special district, but that choice was not uncontested. First, private companies pushing untested monorail technology claimed they could build a transit system with just minor bond guarantees and right-of-way concessions. Second, taxpayers objected more to every new district, and instead asked local government to restructure existing buses and trolleys. Third, the state government could demand an authority rather than a district, and the state had already coopted the lead in new transportation systems. Following World War II, the U.S. Navy Department said the Bay Area would get no more defense contracts unless it solved its traffic congestion. The state traditionally controlled crossings of navigable water like the Bay, which was the biggest bottleneck. The state legislature could enable an authority, a joint venture of cities or counties that sells revenue bonds serviced by the profits earned by a new transit system. Or they could enable a district, also a joint venture of incorporated local governments, but financed through general obligation bonds paid through a fixed assessment of property taxes.

Before 1950, most transit systems were organized as authorities because they were easier to set up and manage, and most earned enough revenue to service their bonds. After 1950, that assumption no longer held. The Muni and Key bus and streetcar systems in the Bay area, like transit systems all over America, bled red ink. Bond buyers insisted BART work as a district. Every time bankers ran the numbers on BART, they saw no way that revenues earned could ever cover the capital costs needed to build the system. For any project, as BART boosters kept saying, "of such unprecedented size and complexity," a district's financial base was simply more secure than an authority's. (As proof of the obsolescence of the authority form, in the East Bay around Oakland and Berkeley, where a rail and bus system had long existed, the Alameda–Contra Costa County Transit District was formed to run it rather than an authority.)[12]

The general obligation bonds buttressing the BART District generated complex incentives. The bonds offered a public subsidy—from the pockets of property owners—for a public good—that is, less congested streets. Property owners bet that the amenity of the new technological system would make their communities better places to live, which would drive up the value of their lands. Only a bare minimum of taxpayers—60.2 percent in the November 1962 elections—made that bet, approving the bonds and moving BART into the construction stage. Then the incentives grew more complex, as the involved entities grew more adept at managing their contributions to special districts.

County government contributed political expertise, since boards of supervisors appointed BARTD board members. Voters considered districts unresponsive to democratic control. Many voters thought that they were formed only when county politicians wished to shunt technical responsibilities out of public scrutiny and into the purview of technocrats most loyal to their technical system. In 1963, state law required these counties to form "new agency formation commissions" to keep close tabs on the districts operating within their borders. These commissions helped county engineers allocate time to each district, facilitated tax collections, and prevented districts from usurping authority from county officials.

State government contributed financial expertise. In the late 1950s, the California assembly tried to rationalize the boom in special districts.[13] It drafted legislative templates that turned some special districts into standard districts, implemented uniform accounting, and established an oversight board to funnel financial data to bond-rating agencies. Furthermore, the state encouraged Bay Area regional government so that local entities had power to match the regional districts. BARTD began with a comprehensive survey of regional traffic and economic growth, and the data BARTD had collected gave it continuing power in all Bay Area construction plans. To wrest the power of the regional plan from BARTD, the state assembly funded its own Bay Area Metropolitan Transportation Survey, and placed it in the hands of an Association of Bay Area Governments (ABAG). All nine counties and every incorporated city sent representatives to ABAG, which increasingly oversaw distribution of state funds around the Bay Area based on the new regional plan. One ABAG subsidiary, the Metropolitan Transportation Commission, managed distribution of all federal funds to BART and other transportation projects.[14]

City engineers and the regulated utilities contributed technical expertise, that is, the expertise BARTD did not buy directly itself. Even before the District was funded, BARTD had hired a systems engineering and construction management contractor. A lawsuit questioned whether BARTD had contracted away responsibilities rightly reposing in a public body, so BARTD hired its own engineering staff. When costs started spiraling in 1966, BARTD hired more engineering staff. This staff helped BARTD be a better customer and oversaw work of BARTD contractors, primarily PBTB (Parsons Brinckerhoff-Tudor-Bechtel), the joint venture that actually designed BART.[15]

The JVCO Tied Together the Technology

Awarded a traffic origin-and-destination survey contract in 1952, Parsons Brinckerhoff Hall & McDonald (PBH&M) became the first consulting engineer in the new BART project. PBH&M beat out other firms mostly because it had never done engineering studies in the Bay Area. The BART Commission sought a firm that had no prior opinion to enforce, and could see without prejudice the whole regional picture. The commission also wanted a firm, respected by bond buyers, to generate numbers showing that BART could make some money. And they wanted a leader like Walter Douglas, a PBH&M partner known as a strong advocate for rail projects and the revitalization of urban America. Company founder Henry Parsons' first big project had been the Manhattan subways in the 1880s, and into the 1950s his firm had the expertise to compile a regional and financial plan that showed how urban transit was a good deal for the Bay Area.[16]

However, PBH&M had let atrophy their expertise in detailed design and construction management. The Bechtel Group was happy to provide it. From its headquarters in San Francisco's financial district, Bechtel had grown into one of the largest construction management firms in the world. And it had grown most by participating in joint ventures—for the Hoover Dam, Jubail, nuclear power plants, the Canadian mineral extraction infrastructure—and by providing turn-key plants. When Douglas asked Bechtel to work as Parsons' sub-contractor for construction management, Bechtel instead suggested a joint venture formed under partnership laws. Bechtel would take the lead by buying a 48 percent share in the partnership. It also suggested bringing in Tudor Engineering of Oakland. This firm was run by Ralph Tudor, former chief engineer for the state Division of Bay Toll Crossings, and a specialist in California transportation planning and in bridges. A Joint Venture Control Office (JVCO), located in San Francisco, housed engineers loaned from the three parents—Parsons staff did most traffic planning, Bechtel did estimating, and Tudor oversaw the trans-Bay tube. A JVCO board of control then bought staff time from the parents for detailed work on geographical sections, and issued memoranda of understanding to assure that this work integrated well. Together, the three firms of consulting engineers brought all the expertise needed for, as they routinely said, a project "of such unprecedented size and complexity."[17]

The parents reorganized themselves to deal with the demands of the JVCO. The Bechtel Group, Inc., was already a nested hierarchy of autonomous companies—grouped by countries, regions, and industries—some of which had equity partners, some of which only participated in a joint venture. This hierarchy of companies provided extraordinary liability protection to the Bechtel family, which owned most of them. Bechtel had a tradition of strong project managers and, as it did more defense and production plant work in the 1950s, actively built up the functional/departmental side of its matrix organization so that engineering specialists could better support, and bill time to, these projects. The Bechtel corporate office was kept so thin that its primary contributions to the companies were legal, purchasing, and estimating (each of which Bechtel considered the keys to its competitive advantage). Bechtel simply added a new BART division to its transportation industry group, led by John Kiely, with the authority to draw functional expertise whenever needed.[18]

Parsons, however, was less adept at dealing with joint ventures. BART was the largest project in Parsons' history, and Walter Douglas accepted Bechtel's offer to joint venture largely from concern about the enormous liability his firm would bear should BART fail. Parsons then operated as a partnership—much like a law firm—with each partner fully and personally liable for every project it did. In 1964, just before the start of BART construction, Parsons reorganized to protect itself. Parsons kept a partnership at the top of its organization to distribute profits, but gradually incorporated each of the project teams into operating companies. First was PBQ&D, Inc., incorporated under New Jersey law, which assumed Parsons' role in the JVCO. But rather than organize the companies into groups, like Bechtel, Parsons arrayed them under the aegis of a partner who became responsible for finding synergies to make each company work more efficiently.[19]

The JVCO engineers thought they might better control the entirety of BART construction by encouraging the firms that actually did construction into joint ventures. BART District staff had originally expected large construction firms to bid on each construction package and then hire smaller firms to do spot subcontracting. However, when BARTD opened the first bids—for the huge Oakland subway—only three firms had bid, and each far exceeded estimates. The partners in the JVCO expected this; they had seen the recent swell in Vietnam-era construction that kept the bigger firms busy.

The JVCO had already prepared procedures, quickly implemented, that favored bids from joint ventures of smaller construction firms primarily by shifting risk from the constructors to BARTD. First, the JVCO cut construction plans into smaller work packages so that these firms could pool their asset bases and get the legally required performance bonds. Second, the JVCO offered a new type of wrap-up insurance policy, offered by a joint venture of Bay Area insurance firms, that transferred to BARTD the cost of workmen's compensation, public liability, and safety engineering. Third, BARTD offered advances for leases of construction equipment, as well as favorable advance payments. And the JVCO imposed on the contractors a uniform time-accounting technique, specifically CPM (the critical path method of network scheduling), and prevented them from predatory hiring of foremen. Of the forty-three different construction contracts let out for bid following this repackaging of work, thirty-nine were won by joint ventures of smaller firms, with each bid close to the JVCO estimates. Some construction firms participated in several different joint ventures.[20]

The Complexity of Creating Amenity

Technical complexity was first envisioned, then manufactured, then attacked by those who created it. The joint ventures manufactured complexity socially—by encouraging diverse new groups to enter into project planning and thereby add performance and append expertise. Much of the intellectual activity in engineering in the postwar period, as evident in the work of the BART venturers, then went into issues of integration as engineers constantly reinvented ways of melding more diverse people and parts into workable systems. Two BART issues—utilities and architecture—show how combination created complexity.

One key facet of any rapid transit system was that it run on a grade-separated right-of-way. One key facet of BART was that certain tracks run straight through established downtowns. In these downtown areas BART tracks encountered many utilities—sewers, water mains, electrical lines, telephone wires, old trolley tracks, long-buried piers. The BARTD board had already decided that, in order to protect the principle of its unobstructed right of way and to buy cooperation from municipal engineers, BARTD would pay for relocating any utilities it traversed. Suddenly, every public works department in the Bay Area, plus every engineering department in a regulated utility, had some say in how and when BART intersected the Bay Area economy. JVCO engineers tried to rationalize the process but could only think of imposing CPM and standardized utility chases. The situation became especially troublesome in Berkeley and along Market Street in San Francisco, where local planners sabotaged both the schedule and standards. They held utility work ransom to force the JVCO to make design changes to local stations and track routing. By the end of construction, 1974, virtually every utility system in the Bay Area had been upgraded, at great delay, so that BART could move forward with constructing its system.[21]

Another key facet of BART was that it be a public amenity. BARTD's 1962 campaign repeatedly referred to open and safe stations, each well integrated with its host community, and linked by an airy system of tracks. However, BARTD had given little thought to how it would ensure this. The engineers that ran the JVCO were already deciding the most crucial design elements—station orientation, the support of the station box, and trestle design. The JVCO hired two renowned architects—a specialist in public buildings and a landscape architect—then prevented them from designing anything more than standard elements: signage, lighting, tiles, elevator placement, paint colors. The smaller joint venturers of constructors then each hired a local architect to mount these design elements to each station.[22]

In September 1966 the BARTD architects very publicly resigned, proclaiming that BARTD engineers were foisting ugly design on the Bay Area. When BARTD faced its first budget crunch, the JVCO engineers eliminated everything they considered an architectural nicety. Most galling to the architects, BARTD scrapped proposals by landscape architect Lawrence Halprin to protect the third rail of track by using an elegant system of grading, plantings, and retractable ladders. The engineers instead chose the ugly but cheap, encasing the entire BART right-of-way in barbed wire fencing. Finally, the JVCO flagrantly ignored the rules of collaboration carefully negotiated in the early 1960s amongst the professional societies of architects, engineers, planners, and constructors on how each profession would behave in the increasing numbers of joint ventures.[23]

The BARTD board, though confronted with countless crises, took the resignation of the architects most seriously. In a series of public meetings, the JVCO engineers argued that the highest public amenity was a transit system that functioned well without going bankrupt. The architects argued that the highest public amenity was a system that blended into each community it intersected. "We are living in an age," claimed BARTD general manager B. R. Stokes, "when architects do assume for themselves a mantle of righteousness which is hard to combat."[24] The BARTD board solved the crisis by giving the station designers more leeway to work with local city councils on the designs of stations. (Current maintenance problems in BART stations result largely from having abandoned the book of standard design elements.) The BARTD board also hired an architectural adviser who reported directly to the board and had free rein to question any JVCO designs which produced a public "disamenity." (The JVCO lost its argument that it had learned its lesson, and would integrate this architect into its team.) The consulting architect questioned, for example, the design of the automatic fare collection system, the split between bus drop-off and kiss-and-ride curbside, and the division of stations into "free" and "paid" areas.[25] And he questioned the design and comfort of BART trains.

A Prime Contractor Integrated the Trains

BART public relations hacks had promised trains that looked nothing like old-fashioned trolleys. Instead, they would look like jetliners on tracks. BART trains would accelerate smoothly, cruise over sixty miles per hour, point a sleek aerodynamic nose into Bay Area winds, and stop precisely where passengers waited in queue to board. The insides would be spacious, sunlit, and carpeted, and contain cantilevered seats. Train controls would emanate from a central room reminiscent of Houston mission control. Computers would schedule trains with ninety-second headways; electromagnetic pulses would keep tracks secure from collision, while human operators would sit back and watch from their comfortable front cabin. BART's political and economic success depended upon procuring trains unlike any others. Indeed, when it came time to accept bids, BART went out of its way to discourage those from established rail car companies—like Budd, St. Louis Car, and Pullman-Standard. BART instead found plenty of aerospace firms—like General Dynamics/Convair and Rohr—hoping to diversify away from defense work.

Key to BARTD's strategy for getting this newfangled train was building a track section near Concord, with federal funding for telemetry and computing equipment, where new components were tested. Bechtel ran the test track, claiming to know something about electricity, electronics, and systems control from its work in nuclear power plants and airports. Any manufacturer with a new component was invited to test it out. Increasingly, the track ran all night and the JVCO engineers quickly became world experts on new transit technology.

Yet the JVCO declined to assume the risks of supervising design and construction of the BART trains, as they had with the constructor joint ventures. BARTD allowed the JVCO to write performance specifications for each train component, even though its test track experience should have allowed it to write more exact hardware specifications. And the JVCO delayed extra months picking proper components, without actually integrating all the parts on a working train. In addition, BARTD let the JVCO define the contract so that the company building the car would also serve as prime contractor. (BARTD awarded only two other design/build prime contracts—to

Westinghouse for the automatic train controls, and to IBM-Almaden for the fare collection system.)[26]

Required by law to accept the lowest bidder that could post the required performance bond rather than basing its decision on a technical evaluation, BARTD awarded the contract to San Diego–based Rohr, Inc. From its start in World War II, Rohr had become the world's largest maker of aircraft engine nacelles. Rohr had plenty of cash on hand plus a new manufacturing plant, when a 1963 down-blip in earnings scared it into diversifying. Rohr experimented with building warehouses, prefab houses, boats, and big antennas, and formed divisions to manage each effort. When Congress passed the Urban Mass Transit Act in 1964, Rohr smelled easy research cash and jumped into rail transit. The firm bought rights to European technologies, invented others, and bid on the BART cars as though it were already in the business.

And with all its transit efforts, Rohr insisted on working as a prime contractor with a program manager in charge. Rohr had worked all its corporate life as a manufacturing subcontractor to airframe prime contractors. Rohr officials believed that to be master of its new diversification efforts—or to work as the equals of airframe companies as associate contractors—the firm would need to put some experience behind its advertised skills in program management and systems engineering. Westinghouse Electric Company of Pittsburgh offered to enter the project as an associate contractor. Westinghouse already had a vibrant electric transit division that built transit systems all over the world. In dollar volume, and not including final assembly costs, Westinghouse made more of the cars than Rohr. Most important, Westinghouse made the AC motors that drove the cars.

But Rohr was willing to build and integrate the BART trains at its own risk. Rohr claimed it had a stellar "first article flow chart" plus an aerospace method of "build-up testing" to keep things on track. Rohr looked forward to crossing that gray area where it transformed BART's performance specifications into the hardware specifications written into its subcontracts with Westinghouse. Nor was Rohr worried that it would work with a very diverse group of part suppliers, most new to train-building, because the JVCO wanted only products closely integrated to the BART cars. Rohr built a "corporate control center" where program managers would work their magic. To protect itself against failure, 1969 Rohr reincorporated itself under Delaware law.

Rohr did not deal well with the complexity of the cars. As it was rushing to begin delivery of BART cars in 1970, Rohr could not solve the problem of arcing and fires in the traction motors. After months of debugging, Rohr discovered that the filter Westinghouse had designed for cooling air admitted brake lining dust into the motors. Rohr was responsible for systems integration, and even though it met every manufacturing deadline, it later had to pay rent for using BART maintenance bays for making each car right.

BARTD had clearly expected these sorts of systems integration problems. BARTD appreciated that the prime contractor relationship with Rohr helped keep responsibility at a sue-able point in a single incorporated entity it could force to make good on all failures of schedule, cost, and equipment.[27] And the JVCO was relieved that it had constructed Rohr's contract as a liability between the JVCO and the complexity of the BART cars.

In the mid-1970s, after retrofitting working equipment on all the BART cars, Rohr abandoned all its transit efforts. Yet Rohr officers proclaimed success. "One of the things we learned from BART," said one Rohr engineer, "was how to be a prime contractor."[28] With such prime contractor experience and with a return to its traditional strength in nacelles, Rohr finessed more prominent positions as associate contractors on aircraft design teams. As a result, Rohr shared in both the risks and rewards from the boom in orders for jumbo jets and new jet fighters in the mid-1970s.

The Postmodern Turn

If the vertically integrated, line-and-staff corporation was how capital shaped itself for modern throughput technology, and if the multidivisional corporation was the capital shape of research-intensive and professionally managed industry, then the group-oriented corporation poised to participate in joint ventures and program teams is the shape of postwar systems technology.

Joint venturing, as manifested in the construction of BART, was part of a larger milieu of organizational innovation, as corporate leaders experimented with ways of reshaping the corporation to systems technology. The three aforementioned salient characteristics of joint ventures—project specificity, a combination of technical equity, a focus on risk and liability—appeared in many ways in many corporations as they took the postmodern turn.

Internally, corporations widely adopted the epistemological tools of systems technology—network scheduling, operations research, and systems engineering. Corporations also reorganized themselves internally, along matrix organization, in the expectation that projects would keep coming. And they made a place for experts who championed systems thinking—program managers.

Externally, new ways of managing projects across institutional borders made possible the embedded supplier relationships that characterized Silicon Valley, as well as the trialing relationships between biotechnology start-ups and established pharmaceutical firms. In monopsonistic industries, like aerospace and telecommunications, where the customer dictates technical issues and can demand a single point of responsibility within each firm, associate contracting and systems teaming became the norm. European nations used associate contracting to fix the capital of their many high-tech firms (in the 1950s), then encourage them to form national champions (in the 1960s), with the market clout to compete in a global economy (in the 1970s). As these national champions became dependent upon supranational programs, European industry invented new forms of industrial consortia—like the Airbus GIE—that fiercely protected the autonomy of the parent, national firms.

All of these—the joint ventures, the program managers, the supranational consortia—constituted a revolution in the business structures of technical transactions comparable to the impact of scientific management a half-century before. Scientific management spread via a group of experts who envisioned the modern factory, professional management, and vertically integrated firms, and then constructed the straw man of inefficiency to battle. Similarly, the prophets of systems thinking in postwar industry envisioned vast networks of diverse institutions working toward the completion of blockbuster projects. As they made complexity and risk bear upon the thinking of corporate leaders, industry began its postmodern turn.

Risk was never a fixed, zero sum. People created it by imagining larger systems, appending new performance, and allowing new types of relevant expertise. The postmodern turn continued, with creation of new intellectual methods to deal with complexity—system engineering and program management are the best examples—that allow experts to maintain complementary but conflicting goals and complementary but conflicting ways of manufacturing certainty. In the postmodern turn, groups temporarily dedicate part of themselves to make the system work, but constantly try to realign system goals to make their contributed expertise more dominant. Merger traditionally precludes the coexistence of conflicting goals—because it includes acquisition of strategy. Contractual association between firms is traditionally silent on issue of goals. Each firm cares most that associates deliver promised performance. Thus joint ventures, with their system orientation, with their partial and temporary conjoining of strategy, with their concern over where liability lies, and their concern over how to exercise effective control, are a paradigmatic structure of this postmodern turn.

NOTES

1. Daniel R. Fusfeld, "Joint Subsidiaries in the Iron and Steel Industry," *American Economic Review* 48 (1958): 578–587; "Expansion via Joint Subsidiaries," *Mergers and Acquisitions* (Chicago: American Management Association, 1957), 86–87.

2. Paul Tractenberg, "Joint Ventures on the Domestic Front: A Study in Uncertainty," *Antitrust Bulletin* 8 (1963): 797–841; Robert Pitofsky, "Joint Ventures under the Antitrust Laws: Some Reflections on the Significance of *Penn-Olin*," *Harvard Law Review* 82 (1968): 1007–1063.

3. Malcolm W. West, Jr., "The Jointly Owned Subsidiary," *Harvard Business Review* (July–August 1959): 32+; n.a., "Note: Joint Venture Corporations: Drafting the Corporate Papers," *Harvard Law Review* 78 (1964): 393–425.

4. James L. Pate, "Joint Venture Activity," *Economic Review of the Federal Reserve Bank of Cleveland* (1968): 16–23; Stanley E. Boyle, "An Estimate of the Number and Size Distribution of Domestic Joint Ventures," *Antitrust Law and Economics Review* 1 (1968): 81–92.

5. David J. Ravenscroft and F. M. Scherer, *Mergers, Sell-Offs and Economic Efficiency* (Washington, D.C.: The Brookings Institution, 1987), 28.

6. Neil H. Jacoby, "The Conglomerate Corporation," *The Center Magazine*, Spring 1969.

7. John H. Rubel, "Systems Management and Industry Behavior," in J. Fred Weston and Sam Peltzman, eds., *Public Policy toward Mergers* (Pacific Palisades, Calif.: Goodyear Publishing Co., Inc., 1969), 215.

8. Sanford V. Berg, Jerome Duncan, and Philip Friedman, *Joint Venture Strategies and Corporate Innovation* (Cambridge: Oelgeschlager, Gun & Hain, Publishers, Inc., 1982), 13, 113.

9. Richard Grefe, *A History of Key Decisions in the Development of Bay Area Rapid Transit*, DOT-OS-30176 (September 1975); Stephen Zwerling, *Mass Transit and the Politics of Technology: A Study of BART and the San Francisco Bay Area* (Praeger, 1973).

10. Institute for Local Self Government, *Special Districts or Special Dynasties? Democracy Denied* (Berkeley: The Institute, 1969), 12.

11. Stanley Scott, *Special Districts in the San Francisco Bay Area* (Berkeley: Institute of Governmental Studies, University of California, 1963).

12. Records of Legal Committee, San Francisco Bay Area Rapid Transit Commission (File A19–3, Record Group 2904, California State Archives, Sacramento); J. Allen Whitt, *Urban Elites and Mass Transportation: The Dialectics of Power* (Princeton: Princeton University Press, 1982).

13. California Legislature. Assembly Interim Committee on Municipal and County Government, *Modernization of County Government: Special Districts* (Transcript of Proceedings, 15 June 1960), 26; *Supervisorial Redistricting and Consolidation of Special Taxing Districts* (Sacramento: Assembly of the State of California, 1962).

14. Donald Chisholm, *Coordination without Hierarchy: Informal Structures in Multiorganizational Systems* (Berkeley: University of California Press, 1989); Chester W. Hartman, *The Transformation of San Francisco* (Totowa, N.J.: Rowman & Allanheld, 1984); George M. Smerk, *Urban Transportation: The Federal Role* (Bloomington: Indiana University Press, 1968).

15. *Westinghouse Electric Corporation, et al., v. The Superior Court of Alameda County, SFBARTD, et al.* Sup. 131 Cal. Rptr. 231; *Robert L. Osborne, et al. v. SFBARTD, et al.*, Civil Files 873 32, Superior Court of California, Contra Costa County (29 November 1962).

16. Minutes of Engineering Committee, San Francisco Bay Area Rapid Transit Commission (File A5–1, Record Group 2904, California State Archives, Sacramento).

17. Ralph S. Torgerson, "San Francisco's BARTD Project," *Consulting Engineer* 20 (April 1963): 67–74; David G. Hammond and Viggo C. Bertelsen, "An Owner's Viewpoint on Owner-Engineer-Contractor Relationships in Tunneling," paper presented at Conference on North American Rapid Excavation and Tunneling, 5–7 June 1972. Library of the Bay Area Rapid Transit District, Oakland, California, document number 323.008:3; hereafter abbreviated BARTD Library 323.008:3.

18. Robert Lockwood Ingram, *The Bechtel Story: Seventy Years of Accomplishment in Engineering and Construction* (Bechtel: San Francisco, 1968); Glenn E. Bugos, "Programming the American Aerospace Industry, 1954–1964: The Business Structures of Technical Transactions," *Business and Economic History* 22 (Fall 1993): 215–216.

19. On how Parsons organized its work on BART, see Henry D. Quinby papers, Library of Institute for Transportation Studies, University of California at Berkeley; Benson Bobrick, *Parsons Brinckerhoff: The First Hundred Years* (New York: Van Nostrand Reinhold, Co., 1988), 185–188.

20. Kenneth M. Graver and Jerold L. Zimmerman, "An Analysis of Competitive Bidding on BART Contracts," *Journal of Business* 50 (July 1977): 279–295; Arthur

Young & Company, "A Survey of the PB-T-B Construction Management Function," report prepared for E. A. Tillman, Assistant Director—Engineering, Department of Development Operations (December 1966) [BARTD Library 323.0188]; PBTB, *San Francisco Bay Area Rapid Transit District Insurance Specifications* (August 1964) [BARTD Library 323.0192:1].

21. "Market Street Improvements," File #297, records of the San Francisco Board of Supervisors.

22. Donn Emmons, "The Role of the Consulting Architect," in B.R. Stokes Architectural Program File, 1966–1967 [BARTD Library 323.013:10]; Wurster, Bernardi & Emmons (Public Structures, Inc.), *Manual of Architectural Standards* (San Francisco Bay Area Rapid Transit District, 1965).

23. Donn Emmons, "Reflections of the Consulting Architect," *Cry California* (Winter 1966–67): 29; Jack McGinn, "Engineer vs. Architect," *San Francisco Chronicle* (11 October 1966); The American Institute of Architects, et al., *Professional Collaboration in Environmental Design* (New York: American Institute of Architects, June 1966); David R. Dibner, *Joint Ventures for Architects and Engineers* (New York: McGraw-Hill, 1972).

24. Statement of B. R. Stokes to BARTD board meeting, 6 October 1966 [BARTD Library 323.002:4].

25. Tallie B. Maule and John E. Burckhard, "Architectural Design Procedures in the Bay Area Rapid Transit District," *Journal of Franklin Institute* 286 (November 1968): 2.

26. Robert M. Anderson, et al., *Divided Loyalties: Whistle-Blowing at BART* (West Lafayette, Ind.: Purdue University Press, 1980).

27. "Westinghouse Electric Corp. Concentrates Mass Transit Activities in One Division," *Passenger Transport* 24 (22 April 1966): 3; "BART Management Techniques Scored," *Aviation Week and Space Technology* 98 (9 April 1973): 41; Richard G. O'Lone, "Industry Influences Rail System," *Aviation Week and Space Technology* 97 (28 August 1972): 36–44.

28. Charles G. Burck, "What We Can Learn from BART's Misadventures," *Fortune* 92 (July 1975): 164.

5

Planning a Technological Nation: Systems Thinking and the Politics of National Identity in Postwar France

Gabrielle Hecht

The relationships between technology and politics have been a major theme in technology studies in the past decade or more. How, we have asked, do politics shape technology? What kind of politics shape technology? How, in turn, do technological activities or artifacts shape politics? How do these relationships vary across time and place? And no less important: how should we describe, analyze, and theorize these multiple and complex relationships?

Systems thinking in the early Cold War period provides an ideal site for deepening our investigation of these questions. This volume reveals considerable variety in the philosophies, methods, and goals of systems thinkers. But systems advocates shared at least one point in common: all engaged in what we have come to call "heterogeneous engineering." At its core, it seems to me, systems thinking consisted of deliberate attempts to fit heterogeneous elements—artifacts, institutions, people, and ideas—into a whole that was greater than the sum of its parts.

This extremely general description might give pause. How different was the systems thinking of engineers, planners, economists, and operations researchers in the 1950s and 1960s from our own attempts to describe the heterogeneous relationships that constitute technological activity? A facile answer might be "very." After all, we typically position ourselves outside, rather than inside the system. Our analytic goals differ starkly: we do not attempt to use our conceptions and theories about these relationships to shape the systems themselves (though perhaps we should?). And our analytic tools differ from theirs: we rely on qualitative rather than quantitative modes of reasoning and we do not aim to build formal predictive models. The case seems closed, the question trivial.

Yet it would be a mistake to dismiss the comparison between us and systems thinkers too quickly. Two additional questions point to the importance of more complex consideration. How did systems thinkers conceive of relationships between technology and politics? And how should we deal with their conceptions and constructions in our own explorations of these relationships? This chapter addresses the first question by examining how the systems thinking of state engineers, planners, and economists in France fit into broader debates about the relationship between technology, politics, and the national future during the 1950s and 1960s. It addresses the second question by example.

I begin by analyzing French debates between technologists, social scientists, and humanists over the meanings of the words "technocrat" and "technocracy." Ultimately, these debates were about who should hold which kinds of power in France. Who, in particular, should have the power to define and construct France's future? They thus hinged on fundamental questions about the nature of technology, politics, and the relationships between them. State engineers, planners, and other technologists thus promoted their visions of the future of France and their own role in shaping that future. In so doing, they attempted both to describe a new technological France and to define a specifically French technological style; the next part of the chapter examines these attempts. Technologists argued that they had a crucial role in shaping this new French national identity. Their conceptualizations of the means by which they would construct a new technological France relied on qualitative, informal versions of systems thinking. Finally, I explore the more formal attempts to apply systems thinking to shaping France's future and identity made by the Planning Commission. A state agency founded after the Second World War to direct the reconstruction and modernization of the nation, the Planning Commission produced multiyear Plans aimed at modeling and shaping national development. By the early 1960s these Plans constituted attempts to apply qualitative and quantitative systems thinking on a national scale; they were also self-conscious efforts by technologists to enact a certain kind of politics.

As we shall see, formal as well as informal attempts at systems thinking provided technologists with a means of articulating both their conceptions of the relationships between technology and politics and their hopes for the future of those relationships and of their nation. In the process, these attempts also gave them ways to define and legitimate their own means of doing politics.

What Is a Technocrat?

The aftermath of the Second World War led the French to rethink the role of the state in directing the economy in general, and industrial, scientific, and technological development in particular. Dominant political groups in this period agreed that by investing and guiding the modernization and expansion of French industry, the state would accomplish the dual aim of resuscitating the economy and restoring the country to its rightful place among the world's great nations. Despite considerable disagreement over procedure and form,[1] enough consensus emerged to allow the creation or reform of several state institutions and national companies.

This profusion of new and reformed institutions proved extremely congenial to a growing class of state experts. Issuing primarily from the *grandes écoles*, these men included engineers, economists, and professional administrators. They cultivated an ideology of public service, polyvalent knowledge, and meritocratic leadership that dovetailed with the prevailing view that state institutions should take primary responsibility for directing national reconstruction and modernization. Denouncing the conservative, protectionist practices of traditional French businessmen, these state experts poured into the top ranks of ministries, the Planning Commission, nationalized companies, and other state institutions. There they vigorously forged their visions of a French technological identity. Charles de Gaulle's return to power in 1958 provided political support and leadership for such efforts, which gained correspondingly in momentum, visibility, and prestige.

The important role played by state experts in running the nation did not escape the attention of their contemporaries. Quite the contrary: as the euphoria of liberation gave way to the hardships of reconstruction, private industrialists and opponents of domestic policies from both ends of the political spectrum began to accuse state experts of undemocratically imposing their will on the nation. The word "technocrat," which had been a fairly neutral term before the war, became a derogatory epithet.[2] At the same time, "technocrats" attracted the attention of social scientists and humanists as an identifiable group to be analyzed and critiqued. Debates sprang up over the meaning of "technocrat" and "technocracy." At the root of these debates lay passionate defenses of a separation of technology and politics. For social scientists as well as more virulent opponents of state experts, "technocrat" designated someone who had breached a boundary, who had moved from his area of expertise into the domain of political decision making. The dangers inherent in breaching this boundary were considerable. First and foremost among them was the capitulation of democracy to technocracy.

What was technocracy, and what relationship did it subsume between politics and technology? According to some humanists and social scientists, technocracy entailed the replacement of politicians by experts. Some, like the eminent writer André Siegfried, saw this process as the quasi-inevitable result of technological civilization:

> Indeed, is it not normal that we should be pulled toward technocracy? The primacy of our material preoccupations demands this, for the standard of living depends on technology: the quantifying spirit, the geometrical spirit in machine civilization has come to dominate. With the expert replacing the politician, it has even invaded a domain where the spirit of finesse should continue to reign. These transformations were inevitable, but perhaps they pulled the State too far down the road of a potentially oppressive technocracy. The defense of the individual, of liberalism must reorganize itself by finding new positions.[3]

Social scientists echoed these anxieties. "Techniciens"—a more polite, less loaded term than "technocrate," which I shall translate here as "technologists"—were eroding traditional political power. For some, "technologists" referred primarily to state engineers; for others, the term encompassed any expert or high-level bureaucrat involved in state administration. Either way, this erosion posed a grave danger to democracy. While in theory power remained in the hands of elected officials, in practice these officials no longer played any significant role in policy making. State planning offered a striking example: plans were devised by state experts, while elected officials, who had neither the time nor the qualifications to understand the calculations, merely approved the budget without questioning the plan. "Government by opinion is giving way to the power of initiates who have the secrets of technology, or who simply know its rules."[4] Democracy, the essence of the French republic, was threatened: crucial decisions were being made by an unelected elite.[5]

To some extent, these anxieties echoed age-old French fears about the encroaching power of the state. But they represented more than simply renewed expressions of historical themes; they also expressed fears about the apparently relentless advance of technological civilization. Without any deliberate human agency, these intellectuals feared, technological change and the lure of material goods had conspired to change the very structure of social and political relationships. Would technology go so far as to make politics irrelevant?

One might expect these anxieties to fade after General de Gaulle's return to power: he was, after all, the very incarnation of a strong and decisive politician. But they did not. True, de Gaulle made it clear that he had no intention of leaving serious decision making to experts; he himself lambasted "technocrats" on a number of occasions. But the general believed in a strong state. Powerful leadership, he felt, had to rise above party politics. He appointed several state experts with no history of professional politics to his ministerial cabinets.

Under these circumstances, debates about the nature of technocracy intensified. In 1960, for example, political scientist Jean Meynaud accused the French state of becoming increasingly technocratic. He described the technocratic mentality as "that state of mind which makes us conceive of technological achievements as the supreme evidence of humankind and that invites us to expect everything from scientific progress."[6] He located the intellectual foundation of technocracy in an unlimited faith in the value of scientific analysis. The end result of technocratic action was the "abdication (*dessaisissement*) of the politician in favor of the technologist."[7] Numerous leftist critics claimed this had occurred in 1940 with the institution of the Vichy government[8] and again in 1958 with the Gaullist overthrow of the Fourth Republic. Ultimately, Meynaud agreed. He judged technocratic ideology to be little more than an apology for the technologist and a justification of the desire to reduce politics to technology.

Meynaud was particularly suspicious of that ideology's exaltation of the technologist and his supposed ability to take a "vue d'ensemble"—a systems view. When paired with the alleged moral qualities of the technologist—such as a highly developed sense of public responsibility—this ability supposedly enabled men to make choices in the general interest. But Meynaud argued that men possessed of the technocratic mentality frequently did not have such an elevated moral sensibility. Uncertainties abounded in any kind of decision making, and technologists were no better qualified to deal with such uncertainty than politicians. Furthermore, any kind of decision, no matter how "technical," necessarily incorporated political considerations. Most ridiculous of all were those cyberneticians who proclaimed the arrival of a "governing machine" that would ultimately mechanize political decision making. For Meynaud, the advance of technocracy heralded the weakening of democracy.[9] Meynaud's analysis was representative of the writings of many social scientists. Technocrats, in this literature, were men who proclaimed the irrelevance of all ideology, called for the introduction of operations research in political decision making, argued that France had to find a middle ground between Russian rationality and American efficiency, and vehemently denied being technocrats.[10]

Social scientists and humanists found technocrats most threatening when they breached the boundary between technology and politics. Clearly, these intellectuals deemed the maintenance of this boundary crucial to the proper and decent functioning of society. When someone like Jean Meynaud acknowledged that technical decisions had political dimensions, he did not mean, as historians and social scientists of our generation might, that technology and politics could not be separated (much less that the very process of design was political). Rather, he meant to make his own political statement: namely, that elected officials should be presented with a range of technical options so that *they* could make the final decision. Democracy and justice, in other words, demanded a clear demarcation between technology and politics.

How did the men thus accused defend themselves? They adopted two strategies, contradictory on one level but eminently compatible on another. The first denounced the negative connotations of the "technocrat" and attempted to salvage a positive meaning for the word. This consisted of arguing that "technocratic" modes of action, accused of being authoritarian, in fact fit perfectly well within the democratic process. The second strategy involved articulating an identity for technologists that opposed that of politicians. On the surface, this strategy consisted of affirming separate and opposing identities for the two social groups. But in arguing that technologists were superior to politicians, this strategy ultimately militated for a kind of conflation between technological and political means of action. I will examine each strategy in turn.

Alfred Sauvy (polytechnicien, director of the Institut National d'Études Démographiques) maintained that the pejorative sense of the word "technocrat" had emerged unjustly. All technologists did was propose ideas for reforming the nation, and ideas were part of the democratic process.[11] While Sauvy admitted that the defenders of the general interest did not always communicate and negotiate as much as they should, this was largely because such men didn't have any forums where they could express themselves. Developing more and better channels of information could easily remedy this situation. It was absurd, he said, that French citizens were more suspicious of the "technocrat" who defended the public interest than they were of private interests.

Sauvy thus argued that disinterested expertise was perfectly compatible with democracy. He dismissed the word "technocrat" as nothing but a petty insult. Others, however, attempted to reappropriate the word and restore a positive meaning to it. In *Plaidoyer pour l'avenir* [Plea for the Future], Louis Armand (polytechnicien, prominent member of the Corps des Mines) and his coauthor concluded that "it is not being a 'technocrat' in the insulting sense of the term to want to base oneself on realistic data, to seek to understand [this data], and to finally attempt to synthesize it. This is being a man who loves life and wants to figure out what he can do to love it even more and to ensure that others love it."[12] Thus, far from being inhuman or mechanical, the technocrat was full of passion. Technocracy, said another enthusiast, was "exercising power inherent in scientific ... and mathematical technology in order to [ensure] the good operation of society and the success of large social entities: large companies, nations, tomorrow perhaps all of humankind."[13] Technocracy was good because it contained within it the power to transcend the petty boundaries of national politics (and indeed Armand and others like him were ardent defenders of the European Community).

Such attempts to salvage a positive meaning for "technocracy" failed. Somewhat more effective was the second, parallel strategy that technologists used to counter accusations against them: articulating an identity that explicitly opposed that of the politician—or, more accurately, the politician as seen by the technologist. In the pages of *La Jaune et la Rouge* (the alumni magazine of the Ecole Polytechnique), politicians emerged as corrupt, dishonest, and ineffective. One Polytechnique alumnus defined the contrast with particular eloquence in a speech entitled "The Cardinal Virtues of the High-Class Engineer":

> Do we want the best of our engineers to participate, through their acts, their pens, and their words, in making our economy healthy again? Well then! We must immunize them, from their very first steps, against the disease known as political thinking of which Machiavelli was the champion ... The Florentine did not hesitate to claim that the individual should sacrifice to the State not just his fortune and his life, which is very noble [*très beau*], but also his honesty, which is detestable.... In the twentieth century,... a democracy cannot accommodate long-standing deception, and I cannot conceive that economic recovery could occur without the country being fully aware of the difficulties to surmount and having full confidence in the sincerity of its guides.[14]

Democracy—a cardinal virtue of the social order about which everyone could agree—could not be served with politicians and their political values. The guiding spirit of Machiavelli would only lead the nation into further chaos. France needed different heroes: Galileo and Pasteur, for example. Such mentors could teach boys to stand up for truth and rigor under the direst of circumstances, turning them into fine, upstanding young men who would "restore our economy and give France the place it deserves in Europe. Let us pray, my dear comrades that, in the expert hands of our imperturbably devoted teaching personnel, and with the help of its cadre of officers whose enthusiasm is infectious, our old establishment can still shape men who, thanks to their talents, their virtues, and their culture of Science, will have the Glory of accomplishing the task that the Nation demands of them!"[15] In contrast to the corruption and decadence of politicians, the eternal values of the Ecole Polytechnique would thus guide the nation.

Not all technologists felt this need to attack the political class in order to articulate a distinctive identity that would naturalize their participation in running the nation. Some—like Armand—defined the difference as one of method and thought process. The problem with politicians was their inability to think synthetically and systemically. This ability constituted the great strength of technologists, and their most important difference from politicians.[16] Allowing technologists to produce and direct organized systems would not render politicians superfluous, but it would render ideology "obsolete."[17] Thanks to the systems thinking and building of technologists, politicians would no longer need ideology to demonstrate the validity of a policy. Good policies would now emerge from rational, not ideological, choices. Systems thinking, therefore, both defined and legitimated the participation of technologists in public life. I shall return to this point later.

Another strand of efforts to elucidate a distinctive technological identity crystallized in the portrayal of technologists as supremely masculine. The technologist was virile, decisive, and forward looking. He was, in a favorite expression of Pierre Massé (head of the Planning Commission), a "man of action."[18] Politicians, by contrast, were not men of action—or at least not very efficient men of action. "The political man of the Republics ... the product of chance, [is] badly prepared for the awesome task of a man of State, ignorant of international and economic problems, duped by his own ease of expression. He will be vanquished by facts."[19] It followed that only technologists, who had a true mastery of "facts," could be real "men of action." It also followed that the virtues of the man of action were honesty and directness. In Massé's experience, some politicians did possess these virtues—men who laid their cards on the table and engaged in rational discussion. With them, he could talk "man to man."[20] But such politicians were rare.

Contrary to the image that political detractors painted of the cold, hard technocrat, men of action were not without heart. Indeed, another technologist showed a passionate desire to procreate and nurture—actions, indeed, which could come directly from the male technologist, completely bypassing any feminine assistance: "in the literary mind, the love of man is merely a platonic love, [a love] which does not create human life. The technologist loves man with a more carnal love and wants to continue to nurture the being whom he loves. He will therefore try to protect [this being]."[21] Technologists were thus the virile, passionate protectors of mankind (and presumably of womankind as well). Love spurred them to action. Politicians and writers did not act—they wrote, they talked, they waffled, but they did not act. Action was what made technologists masculine and therefore powerful.

Clearly, technologists as well as humanists, social scientists, and political leaders found it vital to enact a boundary between technology and politics. But the boundary had different meanings and locations for the two groups. For nontechnologists, the boundary upheld one of the foundations of democracy. Its transgression therefore signified the collapse of the social order. The boundary was thus located in the domain of practice: technologists should not behave like politicians, nor should technological methods be applied to political decision making. For technologists, the boundary was located primarily in the domain of identity. The relevant difference was above all one between themselves and politicians. In constructing this difference, technologists adopted extremely narrow definitions of politics. Politics could mean the implementation of classic ideological stances such as communism, socialism, or liberalism. Or it could refer to the activities of politicians, caricaturized as corrupt, indecisive, irrational, manipulative, and Machiavellian.

"Politics" in the sense of classical ideology and corrupt machinations may have been "other" at one level of technologists' identity discourse. But at a deeper level lay the implication that, because of their values and knowledge, technologists were ultimately better equipped to pursue at least some of the activities in which politicians engaged than were the politicians themselves. At this level, stripped of connotations of ideology and corruption, politics took on a broader meaning, becoming part of what technologists did and should do. At this level, technologists sought to efface what they claimed was an outdated boundary between technology and politics.

THE FUTURE OF FRANCE

One important site for the effacement of the alleged boundary between technology and politics was in the discourse about the role of technology in the future of France. This discourse attempted both to describe what a future technological France would look like and to define a specifically French form of technological and industrial development. General forms of these descriptions and definitions constituted a relatively mild erosion of the alleged boundary between technology and politics: they proposed that technological achievement replace more traditional measures of national power and prestige. Discussions of how to attain such futures, however, attacked the boundary much more aggressively. Shifting from goals to means meant searching for ways to shape the future and control destiny. Systems thinking—both qualitative and quantitative—loomed large in this drive to control destiny, constituting an important means for technologists to blur the technology-politics divide and define their own role in shaping France's future and identity. In order to understand this function of systems thinking, though, we must first examine the national goals technologists sought to attain.

The fundamental premise of these discussions about a future technological France was that in the postwar world, technological achievements defined geopolitical power. A typical article stated that "the possession of industry, especially heavy industry, appears to be a necessary element for respect and independence."[22] Technologists who had sometimes vainly insisted on this equation during the early postwar years found ample political support for it once General de Gaulle returned to power: "We are in the epoch of technology," declared the general on one occasion. "A State does not count if it does not bring something to the world that contributes to the technological progress of the world."[23] Contributing to the "technological progress of the world" had several concrete meanings for both technologists and de Gaulle: among them, technological development would provide the basis for a new relationship between France and its former colonies. Writing just one year after the Algerian crisis that had brought de Gaulle to power, for example, one civil servant saw tremendous potential for this new way of conceiving geopolitical relationships: "In 1958, destiny knocked on the door. . . . In came the technocrats who would build the Franco-African industrial community. In them, the science of engineers is united with the will of captains. The new French strategy, French peace, will be brought to the world."[24] It followed from this equation of geopolitical power and technological achievement that France needed more scientists and engineers, and the 1950s and 1960s witnessed frequent appeals to this effect.[25]

But technological achievement as the new measure of power did not mean that technological pursuits all over the world were identical. Indeed, one of the great dangers of adopting this new standard lay in the loss of cultural specificity. In this respect, the American model posed the biggest danger.[26] The Groupe 1985, a collection of technologists convened by Pierre Massé in 1964 to think about the long-term future of the nation, issued a clear warning on this theme: "First and foremost it is the intellectual and cultural survival of an original and individual France which is the object of an unexpected challenge. Indeed this scientific civilization will tend increasingly to attenuate national specificities and deformities. From now on our presence in the world depends on our ability to imprint our mark on this civilization by means of significant contributions from French technology and French science."[27]

What made a technological or scientific project French? A difficult question to answer, and indeed most technologists avoided addressing it directly, concentrating instead on listing and celebrating French achievements. The pages of *La Jaune et La Rouge*, for example, were filled with praises for French technological achievements, succinctly expressed by titles such as "French Aeronautics, a Matter of Pride and Hope"; " 'The Caravelle': A National Triumph"; "The Radiance of France from the Builder's Scientific and Economic Viewpoint."[28] Such pieces described specific achievements with patriotic enthusiasm. In 1960 the explosion of France's first atomic bomb showed the "entire world the value of French technologists and considerably reinforc[ed] our country's position."[29] Two years later, the new terminal at Orly airport filled this role: "We deemed it indispensable that an undertaking of this size, destined to be seen by the entire world, should give everyone, inside as well as outside, an example of what we can do in France."[30]

Language provided another means of defining a French technological style. Many technologists worried that American terms would colonize French technical language. These terms were difficult to pronounce, they sounded ugly in French, and they threatened the precision of the French language. To guard against the wholesale invasion of American terminology, one group of technologists founded the *Comité d'Étude des Termes Techniques Français* in 1954, a committee dominated by polytechniciens, which included engineers from public and private industry, linguists, university professors, and delegates from technical professional associations. The group met monthly to elaborate or clarify the definition of specific terms and find French equivalents for those that sounded particularly horrible in French. The committee saw itself as a kind of linguistic immigration officer: "Upon entering a country there is a service that sorts immigrants in order to ensure that only the useful ones enter; similarly, we must filter foreign words as soon as they mingle with French vocabulary."[31]

There were also attempts to elucidate what was (or should be) specifically French about French technology. Mostly these appealed to a sense of tradition. Tradition, it appeared, could define or describe Frenchness fairly unproblematically. Placing modern accomplishments in a direct historical lineage with accepted traditions would therefore make them demonstrably French.

So, for example, one traditionally French quality was a refined esthetic sensibility. "The beautiful," noted one group convened to imagine the long-term future of the nation, "is a traditional export of France."[32] The time of hideous industrial landscapes had ended. Modern technology—especially in France—"engenders ... its own beauty, [the beauty] of large dams and artificial lakes..., [the beauty] of large bridges..., [the beauty] of large buildings where lines, materials, and light play with each other..., and even [the beauty] of the metal towers of high-tension power lines."[33] Beauty did not necessarily come at extra expense, but even when it did, the cost supplement remained small compared to its benefits: "Beauty brings income for tourism (it is important not to disfigure sites with inadequate equipment), it brings prestige because it represents a considerable attraction, even when there is no commercial profit. The [atomic energy commission's] installations, which receive numerous foreign visitors, would certainly gain nothing by being hideous, and the esthetic of certain nuclear reactors, whose cost is insignificant with respect to [the cost of the] equipment, does more for the radiance of France than would ten times as many millions spent on propaganda."[34]

Through beauty, tradition could legitimize modern technological achievements as being truly French. And the relationship worked both ways. France was a nation rich in tradition, but this tradition no longer sufficed to define the glory—indeed, the radiance—of the nation. Armand made a point of this: "The wealth of the setting—churches, castles, rivers and their embankments, towns which each have their own personality . . . —should mean that France would continue to be a crucible of ideas. Yet all that subsists in France in the way of tradition —in the countryside as in the army—will only have real value and will only be able to radiate if the nation as a whole is solidly of our time."[35] Hence the other side of a symbiotic relationship: just as tradition was necessary in order to make French technology truly French, modernity was necessary in order to make France truly France.

Technologists thus envisioned a future France whose power would rest on technological prowess, yet whose technological achievements would remain distinctly, identifiably French. The type and degree of nationalism in this discourse varied greatly. Some shared de Gaulle's vision of a strong, independent France, while others argued that henceforth France could be strong only by associating itself with a larger European community. As I argue elsewhere, such differences led to differences in the development strategies pursued by technologists.[36] But either way, the goal seemed clear: France had to become a technological nation. Its future depended on planning a wise route to this goal.

To attain this goal, most technologists argued that the French *mentalité* had to change. This refrain dated from the early postwar period. In those years, advocates of state planning had blamed French defeat on petty industrialists who had clung tenaciously to the status quo and refused to invest in new technologies that could have made France strong. While the most egregious material problems had been corrected by the mid-1950s, advocates of state planning still found much to complain about in the French *esprit*. For example, one former planner reminisced about an interaction between Etienne Hirsch, high commissioner for planning in the 1950s, and Monsieur de Wendel, the elderly, well-respected head of a steelworks. Hirsch had invited de Wendel to sit on the steel production commission of the Second Plan. De Wendel was puzzled, and replied: "But Mr. General Commissioner, what is this about? After the war you explained to us that we had to make a big effort to modernize the steel industry. We listened to you, we did it, we took risks . . . now we are modern! So what could you possibly want to discuss?" Exasperated, Hirsch replied: "But Mr. de Wendel, modernity is not a definitive state! You made an effort to make up for a delay and modernize certain installations, but this effort will never be exhausted once and for all!"[37] The point of this anecdote, which opposed the old-style industrialist to the modern planner, was clear: the listener was supposed to be amused that the old man did not realize that modernity was not a physical condition, but a state of mind.

This idea that modernity began with a change in attitude pervaded the discourse of technologists throughout the 1950s and 1960s. The biggest reproach of technologists, however, centered around the French conception of and approach to the future. The French could no longer stumble blindly into their future; they had to learn how to control their destiny. This involved the cultivation of *une attitude prospective.*

The notion of *la prospective* originated with the formation of the Centre International de Prospective. Though most of those involved would have responded to the label "technologist," a few were also humanists.[38] The center was intended to provide a place and a publication (the journal *Prospective*) for systematic and systemic reflection and action oriented around three related poles: "human problems" such as employment and education; the relationship between Western and other civilizations; and the consequences of new developments in science and technology. The center forbade itself from conducting "any political activity"—in the sense of corrupt ideological machinations—and made a point in its publications of lambasting "ideological" modes of reasoning.

What was the *attitude prospective*? It was one turned toward the future, especially the far future. It differed from short-term forecasting and had to be cultivated by different people. Gaston Berger, the president of the association, explained this using a military analogy:

> It would be dangerous for a combat officer to be associated with peace negotiations because his role is to fight even while peace is being discussed. But it would be unforgivable for leaders not to dream of peace while making war. In the adversary of today they must already see the colleague, the client, the friend of tomorrow . . . It even happens fairly frequently that short-term actions must be taken in a direction opposed to that revealed by a study of the long term. Those who implement such actions must pursue them with vigor, but at a higher level, responsible leaders must calculate the

importance of these accidents and situate their exact position in events as a whole.[39]

It was the "responsible leaders," in other words, who had to take an *attitude prospective*. The essence of this attitude was systemic: it involved defining a goal and figuring out how human, technical, and economic factors could be synthesized into a plan for attaining this goal. The *attitude prospective* was thus a qualitative example of what this volume of essays calls "systems thinking." Ultimately, its advocates argued, this attitude would enable men to "control their destiny," rather than "submit" to it.[40]

Many highly placed technologists—Louis Armand, Pierre Massé, and François Bloch-Lainé among others—advocated the *attitude prospective*. They took it to mean applying systems thinking to problems that were at once technological and social. Armand, for example, sought to define *la prospective* for national transportation systems, a subject he knew intimately. The issue for the future of transportation, he argued, was no longer maintenance but coordination—not just within subsystems like the railways, but also between subsystems. Airlines, railways, and roads should be coordinated to provide optimal transportation routes for travelers and goods. The current chaotic state of affairs, in which these subsystems competed with each other, merely demonstrated the "need for Governments to apply notions of political economy which are the domain of operations research."[41] He did not, however, specify *how* operations research might be applied on such a scale or to such problems.

Like modernity, *la prospective* was above all a state of mind. Taking *une attitude prospective* involved seeing life as a "continual invention."[42] It demanded intimate knowledge of "large new technologies," of which the two most important were atomic energy and cybernetics.[43] It was an activity for the "elite,"[44] an elite composed of "men of action": "Men who not only have a taste for moral or philosophical meditation, but also a concrete knowledge of men and the experience of command and responsibility."[45] The action in which men should engage involved synthesizing "all the means at the disposal of modern society in order to know and to predict, to organize and to decide."[46] Like the others involved in the center, Berger placed a heavy emphasis on real-world experience: "We do not seek a synthesis of knowledge and writings, but a synthesis of lived experience . . . only doctrinaires—[who are] inefficient but formidable—start with abstract ideas completely cut off from reality."[47] Real-world experience thus provided a nonideological foundation for action based on *la prospective*.

The central figure at the heart of *l'action prospective* was the engineer. Not just any engineer, but an engineer who could take a "vue d'ensemble," a "systems view." Such a systems thinker could master his destiny. He combated the defeatism of those who saw the human condition reflected in the myth of Sisyphus: "If the myth of Sisyphus expressed our true condition, our engineers would have already discovered the means of using the regular fall of the boulder and Sisyphus, freed from repetitions, would devote himself to other tasks."[48] Invention provided the foundation for men to build their destiny; as such, the material world had spiritual value.[49] Berger evoked the myth of Faust to demonstrate this point. In the end, he said, Faust found fulfillment not through the gifts of the Devil, but by working for other human beings. For this, God saves him from the Devil's clutches. "And what does this mean? This means that man had been a magician and became an engineer, but an engineer in the service of others. What is the magician? He is the one who uses spiritual forces for selfish goals. What is the technologist? He is the one who uses his work, his pain, his intelligence to bring men the things they need."[50] Imbued with *une attitude prospective*, the engineer, man of action and systems thinker, could shape human destiny. This was "decidedly" not technocracy. It was simply good sense.[51]

What (besides a more rational, prosperous, and powerful nation) would systems thinking produce? Armand's reply was, "Technology + Organization = Culture."[52] While technological change could certainly have disastrous effects, these could be avoided through heterogeneous, systemic organization. This, in turn, involved overstepping traditional boundaries, particularly those between technology and a certain kind of (presumably nonideological) politics. In the words of another *prospective* advocate, "If the modern world demands an increasingly large number of specialized technologists and researchers, alongside [such people] a certain number of young men must be able to dominate their technologies in order to define and implement general industrial policy. For this, they must be or become more than just technologists."[53]

La prospective essentially consisted of a qualitative approach to systems thinking: by taking into account human and cultural factors in their inventive efforts, engineers and other "men of action" had the means to set goals for the future and trace out trajectories through which those goals could be attained. Though some advocated specific methods for tracing those plans—particularly the techniques of operational research—even they kept their contributions to the journal *Prospective* quite general. This was not the case, however, for the Plan.

The Plan

The Plans elaborated by the Planning Commission were the ultimate instruments for shaping the future and destiny of the nation, the ultimate effort to constitute and define a large-scale system.[54] In the Plan and its planners, the various themes examined thus far come together. The planner epitomized the technocrat—or the broad and forward-thinking "technologist," depending on one's perspective. Architects and advocates of the Plan often held it up as a quintessentially French achievement, the ultimate marriage of (certain select) traditions and modernity. Making the inevitable reference to Descartes, one enthusiast noted that "the effort towards increased rationality that the French Plan represents conforms to one of our best national traditions."[55] Planners themselves promoted the Plan as an instrument not only of national cohesion and internal economic development, but also of national power. The introduction to the Fourth Plan began thus:

> . . . [G]oing beyond individual destinies, [the national goals of the Fourth Plan] define themselves as survival, progress, solidarity, [and] radiance. They consist of ensuring our defense by combining the modernization of the military with a reduction in its personnel, of giving research the material power necessary to ensure the full participation of the French spirit in the great scientific and technological enterprise of this century, of giving regions and less-favored groups —be they the aged, repatriated soldiers, employees or low-income farmers—concrete proof of a solidarity indispensable to national cohesion, and finally of pursuing our aid to the less-developed nations of the third world, especially those French-speaking African States which decided to keep special ties with our nation.[56]

Given such lofty goals, serving the Plan was not only an "ardent obligation" but potentially a quasi-religious experience: "Inasmuch as the Plan has become a necessity, all reticence is abnormal, even stupid. The only logical and efficient recipe is to make planification a psychological force of progress and solidarity. Serving the Plan, participating in it at different levels, can feed the transcendence of each one [of us], provide satisfaction for the civic-minded."[57] Thus the Plan could furnish a means for enacting the spiritual dimension of the material world, of invention. A quintessential manifestation of the *attitude prospective*,[58] the Plan, by its very existence, would turn the nation into a system.

One way in which the Plan would systematize the nation was by providing information. Left to their own devices, private industries or even larger, organized sectors of the economy would, in the disastrous manner of the prewar period, pursue independent policies conceived from the sole point of view of the industry or sector in question. They would have no way of knowing how their actions affected other industries or sectors; nor would they know how the actions of others affected them. This ignorance might well lead them to make faulty decisions, not just in terms of the national interest, but potentially also in terms of their own interests. By providing decision makers with increasingly detailed maps of the infinite interconnections that bound the nation's economy together, successive Plans would enable a new kind of decision making: one that was not only more rational and efficient, but also more systemic. Indeed, it would ultimately be in everyone's best interest to work toward the national goals set by the Plans (for the Plans were not merely descriptive, but also prescriptive), even if in the short term these goals asked people to make decisions that appeared to go against their immediate interests. The whole would be bigger than the sum of the parts, and the parts would benefit in consequence. Thus both *describing* and *prescribing* the system were also, in a sense, supposed to *create* the system—assuming, of course, that the actors operating within it paid attention to the maps and prescriptions provided by the planners. Assuming, in other words, that they accepted the notion of themselves as actors in a system.

More complex and sustained systems thinking in the Planning Commission began with the Fourth Plan. The methods for elaborating the first three Plans had been primarily qualitative, and even the most fervent admirers of state planning conceded that the Plans themselves had been important primarily for psycho-cultural reasons, serving to change the *mentalité* of the French by articulating a dynamic, modern future. In contrast, the Fourth and Fifth Plans—under the impulse of their new commissioner, economist Pierre Massé, one of the primary French architects of linear programming—were developed using a combination of qualitative and quantitative methods. As such, they pursued systems thinking on a national scale far more intensely.

The first three plans had used material indices such as industrial equipment and economic indices such as productivity and efficiency in order to set goals for national economic and industrial performance. In contrast, the Fourth Plan sought to develop dynamic models that would describe the economy, and the social and material relations that drove it, as a whole. The economy was composed of diverse and overlapping heterogeneous "sub-systems" ("sous-ensembles"). These included diverse sectors such as agriculture, industry, commerce, and transportation; geographic regions of the nation; economic constructs such as balance of trade, consumption, savings, and investment; and socio-economic relationships such as employment and the labor market. These sub-systems interacted in complex ways. With the help of the mathematical services of the Ministry of Finance and the INSEE (the National Institute for Economic and Statistical Studies), planners aimed to model these interactions as closely as possible.

Though they fell short of building a single model for the entire nation, they nonetheless produced a set of models that described the national economy as a heterogeneous system composed of technological and economic artifacts, people, and social relationships, and driven by the interactions between these various components. Using these models, they defined an "optimum" growth rate. They then turned the growth rate and parts of the models over to the modernization commissions. Convened separately for each Plan and not composed of expert planners, these commissions studied specific industrial, economic, geographic, or cultural sectors. Based on the information they received from the expert planners, the commissions drew up tentative plans for their sectors. Finally, the expert planners collected these sub-plans and modified them in order to fit them into a system both described and constituted by the final plan.

In passing over the details of these models and the processes by which they were built, I in no way mean to imply that these were unproblematic or uncontested. Planners and experts from the INSEE and the Ministry of Finance negotiated nearly every step. At no point was the planning process seen (by participants or expert observers) as a "black box."[59]

Intended to cover the period 1962 to 1965, the Fourth Plan outlined policy directions for a broad swath of French economic, industrial, social, and even cultural life. First and foremost came the need to develop scientific and technical research. "The fate of a people is increasingly determined by the energy it deploys in opening new routes to knowledge, which is the very source of its radiance and the indispensable condition of the [continuous] renovation of its technologies."[60] Beyond this, the Plan set growth and development objectives for the sectors of the economy defined by its models, recommending for example that the nuclear industry seek to develop several different types of power plant design. Pointing to increasing international competition, it recommended that large industrial establishments pursue efforts to both merge and specialize. It set goals for urban development throughout the nation, recommending the destruction of dilapidated buildings, the construction of wide roads, and the creation of parks and sporting facilities in city centers. It introduced the idea of regional planning. In prosperous regions, it recommended policies for developing infrastructural facilities to keep pace with local growth, while it recommended that the state offer more substantial aid to poor regions. For all regions, it outlined schemes for the modernization of rural areas, including the widespread development of public utilities such as electricity and running water.

In addition to making recommendations for material development, the Plan promoted cultural harmony through technological and institutional means. A second television station would contribute to "increasing the information and cultural development of the population and to spreading the radiance of France beyond its frontiers."[61] Meanwhile, the construction of "culture houses" throughout the nation would enable "culture" (which remained undefined) to "remedy that which often seems discordant and inhuman about technological civilization, ... to penetrate the daily life of men and especially to become, through the exercise of diverse professions, as immediate a concern as hygiene and stable employment."[62]

The Fourth Plan thus represented an attempt to chart the future of the nation on every possible front. In theory, the national system that it conceived was open and unbounded. Even if some parts of the system could not be described quantitatively, ultimately no aspect of national life lay outside the whole. In greater or lesser detail, labor policy, industrial growth, urban development, technological change, investment, and cultural enrichment could all be related to each other and planned for the greater good of a new, modern France. The modernity of the nation could be described, in some instances measured, and in all cases enhanced. And sweeping as the Fourth Plan was, the Fifth Plan covered even more ground and used even more elaborate quantitative methods.

Clearly, these Plans were hybrids of technology and politics in the broadest sense of both terms. Indeed, the architects of the Plan saw matters in precisely these terms. In an INSEE report written to promote the diffusion of the programming methods of the Fifth Plan, one technologist wrote of the "simultaneously technical and political nature" of the Plan's "elaboration process."[63] Throughout its entire course, the planning process wove together technical and political methods: "The kind of variables involved in the technical work [and] the determination of their values or of the relationships between them are themselves tied to explicit, or more often implicit, political choices. This web of political choices and technical work is woven during the preparation of the Plan."[64] While no one went so far as to argue, as we might, that the technical and political aspects of the process were indistinguishable, it seemed clear to all those involved that they were at least closely related, and furthermore that a close relationship was necessary for the success of the Plan.

In a sense, taking a systems approach constituted an attempt to naturalize the erosion of the alleged boundary between technology and politics. Conceiving of the nation as a system and arguing that all its heterogeneous components were interrelated implied that all these components could be planned. In other words, all components—technological, regional, cultural, and economic—fell within the purview of the planners.

The enrollment of an extremely heterogeneous group of people into the planning process provided a key means of enacting this erosion. While planners themselves remained a small, select group of experts, the modernization commissions were populated by a great variety of people: industrialists, labor union representatives, bankers, regional administrators, architects, urban planners, even a few artists and writers. Their participation was meant to ensure the democratic character of the plan. It was also intended to increase the investment of different social groups in the plan, thereby increasing compliance.

In a sense, the Plan as a whole formed an attempt to enroll the entire nation in a broad program of sociotechnical development. Because the Plan was not coercive, it could not impose its agenda on industrialists, workers, regions, banks, or flows of money. But it could and did attempt to persuade the nation to follow its lead. Its very existence was one form of persuasion, and the High Commissioners for Planning engaged in other forms by visiting politicians, regions, and industries to promote the Plan. Witness, for example, how Massé described his job two decades after retiring:

> My profession was to send a message that would not falsify the truth, but that would be accessible to labor union members, politicians, [and] public opinion. I had to convince the Government to adopt my plan project, to convince the Economic and Social Council to emit a favorable overall opinion, [and] to convince Parliament to vote it [into effect]. I repeat the word "convince" three times because this was an essential part of my job, carried out for seven years with a respectable measure of success. Voting the plan [into effect] is an important act—the Nation's commitment to itself . . . In sum, I had a responsibility of a political nature that went beyond the mission of the experts.[65]

Here, politics was not ideological but persuasive, an essential and evidently enjoyable (given the relish with which Massé described his many voyages and conversations throughout his autobiography) part of building a modern nation.

Conclusion

The relationship between technology and politics was a vital issue in debates over running French society and constructing a modern future and identity for France during the 1950s and 1960s. Much was at stake, from the professional identities of political scientists and engineers to the principles underlying the proper functioning of a democratic society. For many social scientists and humanists, keeping technology and politics separate appeared crucial not just for their own professional survival, but for the survival of democracy itself. Technologists also sought to maintain a boundary, but they thought the relevant line of demarcation lay between their identity and that of politicians. This boundary served not to defend an existing order but to valorize a new one, in which technologists, thanks to their expertise and morality, would forge a better and more modern democracy.

Technologists envisaged a modern France whose strength and "radiance" would rest on technological achievement. In articulating the mechanisms through which the nation would attain this future, and their own role in shaping those mechanisms, technologists sought to erode rather than maintain a boundary between technology and politics. Systems thinking provided an ideal means for describing and legitimating this goal. If heterogeneous elements were related to each other in identifiable, describable, controllable ways, and if technologists could predict and control these systems, then they could not only engage in politics but could do so better and more reasonably than could politicians.

Donald MacKenzie has described how the Americans who designed nuclear missile guidance systems sought to maintain a discursive boundary between technology and politics while at the same time eroding that alleged boundary through their practices.[66] The case of French systems thinkers presents a striking contrast: these men explicitly and joyfully advocated the erosion of any such boundary. But the comparison with MacKenzie's engineers reveals a similarity as well as a difference. In both cases, articulating conceptions of the relationship between technology and politics served a strategic purpose. In both cases, issues of legitimation and authority lay at the heart of these conceptions. What is the significance of this similarity—and of this difference? How did systems thinkers in other nations conceive of the relationship(s) between technology and politics, and what role did these conceptions play in shaping the systems models or schemes that they developed? This questions remain open for discussion.

French planners used a systems approach to naturalize the erosion of the boundary between technology and politics, to defend it and make it normal. In arguing that technological systems are socially, culturally, and politically constructed, we in technology studies have shown that the boundary between technology and politics (and society and culture) is at the very least fuzzy, in some cases not even meaningful. The ineffability of the boundary, we have said, is business-as-usual in the technological world; it reflects the normal state of affairs. Indeed, our modes of analysis appear rather congenial to (at least some) French state engineers. I presented "Political Designs"[67]—an article in which I analyze nuclear reactors as political constructs and as instruments of political action—to two former members of the French nuclear industry. Their only comments were of a technical nature: both appeared to agree completely with my argument. I still do not know how to interpret this acquiescence. Does it simply mean that I "got it right"? Or that I was merely echoing their vision of the world?

Of course a big difference between us and the systems thinkers I have described in this chapter is that we do not argue that technological expertise constitutes a privileged form of knowledge; nor, therefore, do we argue that the systems approach constitutes a privileged form of political action. Indeed, our analyses frequently contain an implicit critique of this sort of privileging, this means of doing politics that can sometimes appear downright underhanded. And we can also offer a different perspective by looking beyond the systems conceived by systems thinkers. We do not, therefore, merely echo technologists' vision of the world.

Nonetheless, we should not get too comfortable. For social scientists as well as technologists in 1950s and 1960s France, conceptions or models of the relationship between technology and politics were themselves political statements in the broadest sense. In what ways is this true for us and our conceptions? Many of us have argued for the inseparability of technology and politics in an analytic sense. Are we thereby also arguing for their inseparability in a political sense? And, if we do not wish to ally ourselves with technologists, what does such an argument imply? These may not be new questions for historians and sociologists of technology, but studying the systems approach gives us a particularly fruitful field in which to explore them.

NOTES

1. The politics of the immediate postwar period have been well covered by historians. For example Jean-Pierre Rioux, *La France de la Quatrième République*, vol. 1: *L'ardeur et la nécessité* (Paris: Seuil, 1980); Maurice Larkin, *France since the Popular Front: Government and People, 1936–1986* (Oxford: Clarendon Press, 1988); Richard F. Kuisel, *Capitalism and the State in Modern France: Renovation and Economic Management in the Twentieth Century* (Cambridge: Cambridge University Press, 1981).

2. For analyses of technocratic movements before World War II, see Kuisel, *Capitalism and the State in Modern France*; Gérard Brun, *Technocrates et Technocratie en France* (Paris: Albatros, 1985); Philippe Bauchard, *Les Technocrates au pouvoir* (Paris: Arthaud, 1966). This last also functions as a primary source and example of the turns taken by debate about technocracy in the 1960s.

3. André Siegfried, "Le Problème de l'Etat au XXe siècle en fonction des transformations de la Production," 13–25 in Gaston Berger et al., *Politique et Technique* (Paris: Presses Universitaires de France, 1958).

4. Marcel Merle, "L'Influence de la technique sur les institutions politiques," in Berger et al., *Politique et Technique.*

5. Marcel Waline, "Les Résistances techniques de l'administration au pouvoir politique," in Berger et al., *Politique et Technique.*

6. Jean Meynaud, *Technocratie et politique. Etudes de Science Politique* (Paris, 1960), 7.

7. Meynaud, *Technocratie et politique*, 39.

8. See Kuisel, *Capitalism and the State in Modern France*, for an analysis of this.

9. Meynaud wrote several other critiques of technocracy, including: "Les Mathématiciens et le Pouvoir," *Revue Française de Science Politique* IX, 2 (1959): 340–367

(this article actually came out in favor of more moderate approaches to cybernetics, in which cybernetic techniques might be used as thinking tools, but still argued strongly against mathematicians having any sort of decision-making power); Jean Meynaud, "A propos de la Technocratie," *Revue française de science politique* XI, 4 (1961): 671–683; Jean Meynaud, "A Propos des Spéculations sur l'Avenir. Esquisse bibliographique," *Revue française de la science politique* 13 (1963): 666–688; Jean Meynaud, *Technocracy (Technocratie, Mythe ou Réalité?)* (New York: Free Press, 1964, 1968). All make similar arguments to the book discussed here.

10. See, for example, Jean Touchard and Jacques Solé, "Planification et Technocratie: Esquisse d'une analyse idéologique," 23–34 in Fondation Nationale des Sciences Politiques et Institut d'Etudes Politiques de l'Université de Grenoble, *La Planification comme processus de décision* (Paris: Armand Colin, 1965). See also articles by the sociologist Nora Mitrani: "Reflexions sur l'opération technique, les techniciens et les technocrates," *Cahiers internationaux de sociologie* XIX, 1955; "Les mythes de l'énergie nucl. et la bureaucratie internationale," *Cahiers internationaux de sociologie* XXI, 138–148; "Attitudes et symboles techno-bureaucratiques: refléxions sur une enquête," *Cahiers internationaux de sociologie* XXIV, 148–166; Bernard Chenot, *Les enterprises nationalisées* (*Que sais-je?* n. 695, Paris: Presses Universitaires de France, 1956); Georges Lescuyer, *Le contrôle de l'Etat sur les entreprises nationalisées* (Paris, 1959); Henri Migeon, *Le Monde après 150 ans de technique* (nouv. ed., Paris, 1958); Gabriel Veraldi, *L'humanisme technique* (Paris, 1958).

11. Alfred Sauvy, "Lobbys et Groupes de Pression," in Berger et al., *Politique et Technique*.

12. Louis Armand and Michel Drancourt, *Plaidoyer pour l'avenir* (Paris: Calmann-Lévy, 1961), 230.

13. Dominique Dubarle, preface to Jean-Louis Cottier, *La Technocratie, Nouveau Pouvoir* (Paris: Edition du Cerf, 1959), 26–27. A Catholic priest, Dubarle also wrote his own book on the subject: Dominique Dubarle, *La civilisation de l'atome* (Paris: Éditions du Cerf, 1962).

14. André Léauté, "Les Vertus Cardinales de l'Ingénieur de Grande Classe," *La Jaune et la Rouge*, 120 (1958): 36–42.

15. Ibid., 42.

16. Armand and Drancourt, *Plaidoyer pour l'avenir*, p. 107 passim.

17. Ibid., 17.

18. Pierre Massé, "Propos incertains," *Revue française de la recherche opérationelle* 2e trim., n. 11 (1959): 59–71. Quote p. 60 and passim.

19. Jean Barets, *La fin des politiques* (Paris: Calmann-Lévy, 1962), 113.

20. Pierre Massé, *Aléas et Progrès: Entre Candide et Cassandre* (Paris: Economica, 1984). Throughout these memoirs, Massé uses this expression to confer high praise on the reasonableness of certain interlocutors.

21. Jean Barets, *Revue de Défense Nationale,* mai (1963).

22. Georges Villiers, "Industrie, Technique et Culture," *Prospective* 4 (1959): 21–32.

23. Quoted pp. 4–5 in Olivier Wieviorka, *Charles de Gaulle, la technique et les masses* (Paris: 1990), G0004: 1–13. For a more sustained discussion of Gaullist discourse on technology and power, see Gabrielle Hecht, *The Radiance of France: Nuclear Power and National Identity after World War II* (Cambridge: MIT Press, 1998).

24. Cottier, *La Technocratie, Nouveau Pouvoir,* 41.

25. A few examples: Henri Longchambon, "Prenons Garde!" *La Jaune et La Rouge* 93 (1956): 26–31; Alfred Sauvy, "Lobbys et Groupes de Pression"; *Refléxions pour 1985* (Paris: La Documentation Française, 1964); Louis Armand, "Technocrates et Techniciens," *La Jaune et La Rouge* 216 (1967): 4–8.

26. Richard Kuisel, *Seducing the French: The Dilemma of Americanization* (Berkeley: University of California Press, 1993).

27. *Refléxions pour 1985,* 13–14.

28. Albert Caquot, "Le Rayonnement de la France aux points de vue scientifique et economique du constructeur," *La Jaune et La Rouge* 112 (1958): 23–28; Ingénieur général de l'Air Dumanois, "L'Aéronautique Française, sujet de fierté et d'espoir," *La Jaune et La Rouge* 106 (1957): 27–28; Jacques Lecarme, "Un triomphe nationale: 'la Caravelle,' " *La Jaune et La Rouge* 118 (1958): 25–42.

29. Pierre Couture, "Explosion de la première bombe A Française," *La Jaune et La Rouge* 136 (1960): 32–35.

30. Pierre Cot, "La Nouvelle Aérogare de l'Aéroport d'Orly," *La Jaune et La Rouge* 158 (1962): 10.

31. Anonymous, "Défense de la langue française: Le Langage Technique," *La Jaune et La Rouge* 107 (1957): 27–36. Quote p. 29.

32. *Refléxions pour 1985,* 88: "Le beau est une exportation traditionelle de la France."

33. *Refléxions pour 1985,* 85.

34. *Refléxions pour 1985,* 88.

35. Armand and Drancourt, *Plaidoyer pour l'avenir,* 238.

36. See Hecht, *The Radiance of France.*

37. Quoted in François Fourquet, *Les Comptes de la Puissance: histoire de la comptabilité nationale et du plan* (Paris: Recherches, 1980), 237.

38. In particular, Gaston Berger, the founder and president of the Centre Internationale de Prospective, was a philosopher—though a rather special kind of phi-

losopher, since he also sat on the scientific advisory board of the French atomic energy commission. But the six vice-presidents of the Centre all fit under the label "technicien" in its broadest sense: Louis Armand, François Bloch-Lainé, Pierre Chouard, Jacques Parisot, Georges Villiers, Arnaud de Voguë.

39. Gaston Berger, "L'attitude prospective," *Prospective* 1 (1958): 1–10. Quote p. 4.

40. Berger, "L'attitude prospective," 6.

41. Louis Armand, "Vues prospectives sur les transports," *Prospective* 1 (1958): 37–44.

42. Gaston Berger, "Culture, qualité, liberté," *Prospective* 4 (1959): 1–12. Quote p. 5.

43. See the second issue of *Prospective* on "conséquences générales des grandes techniques nouvelles," focused exclusively on these two technologies.

44. See for example André Gors, "Avant Propos," *Prospective* 2 (1959): 2.

45. Berger, "Culture, qualité, liberté," 6.

46. Gors, "Avant Propos," 7.

47. Berger, "Culture, qualité, liberté," 7.

48. Ibid., 5.

49. The spiritual dimension of the "attitude prospective" was mainly propounded by Berger, a devotee of the Catholic thinker Teilhard de Chardin. I will not elaborate on this aspect of his thought here; see Gaston Berger, "L'idée d'avenir et la pensée de Teilhard de Chardin," *Prospective* 7 (1961): 131–153.

50. Berger, "L'idée d'avenir et la pensée de Teilhard de Chardin," 151.

51. Berger, "Culture, qualité, liberté," 7: "La prospective s'oppose décidément à toute technocratie."

52. Armand and Drancourt, *Plaidoyer pour l'avenir*, 153.

53. Georges Villiers, "Industrie, Technique et Culture," *Prospective* 4 (1959): 21–32. Quote p. 21.

54. Unless otherwise noted, my description of the Plans and of the planning process is based on the following sources (not necessarily in order of importance): "Quatrième Plan de Développement Economique et Social (1962–1965)," Commissariat Général au Plan, 1961; Pierre Bauchet, *La Planification Française: du premier au sixième plan* (Paris: Editions du Seuil, 1966); Fourquet, *Les Comptes de la Puissance*; Fondation Nationale des Sciences Politiques et Institut d'Etudes Politiques de l'Université de Grenoble, *La Planification comme processus de décision*; INSEE, "Méthodes de Programmation dans le Ve Plan," *Études et Conjonctures* 21, 12 (1966); Etienne Hirsch, "Les méthodes françaises de planification," *Mémoires de la Société des Ingénieurs Civils de France* 112, II (mars–avril 1959): 81–94; Massé, *Aléas*

et Progrès; François Perroux, *Le IVe Plan Français (1962–1965)* (2e édition; 1e éd. 1962) ed. *Que Sais-Je? ("Le Point des Connaissances Actuelles") no. 1021* (Paris: PUF, 1963); Bernard Cazes, "Un Demi-Siècle de Planification Indicative," in M. Lévy-Leboyer and J.-C. Casanova, eds., *Entre l'Etat et le marché: L'économie française des années 1880 à nos jours* (Paris: Editions Gallimard, 1991): 473–506; Kuisel, *Capitalism and the State in Modern France*; N. J. D. Lucas, *Energy in France: Planning, Politics and Policy* (London: Europa Publications Limited, 1979); John H. McArthur and Bruce R. Scott, *Industrial Planning in France* (Boston: Division of Research, Graduate School of Business Administration, Harvard University, 1969); Henry Rousso, ed., *De Monnet à Massé: Enjeux politiques et objectifs économiques dans le cadre des quatre premiers Plans* (Paris: CNRS, 1986).

55. Perroux, *Le IVe Plan Français (1962–1965)*, 126.

56. "Quatrième Plan de Développement Economique et Social (1962–1965)," Commissariat Général au Plan, 1961, 4–5.

57. Armand and Drancourt, *Plaidoyer pour l'avenir*, 209.

58. Indeed, Pierre Massé, the third high commissioner of planning, was an ardent supporter of this attitude. See for example Pierre Massé, "Prévision et Prospective," *Prospective*, 4 (1959): 91–120.

59. Perroux, *Le IVe Plan Français (1962–1965)*.

60. "Quatrième Plan de Développement Economique et Social (1962–1965)," 153.

61. Ibid., 274.

62. Ibid., 265.

63. INSEE, "Méthodes de Programmation dans le Ve Plan," *Études et Conjonctures* 21, 12 (1966), 11.

64. Ibid., 12.

65. Massé, *Aléas et Progrès*, 183.

66. Donald MacKenzie, *Inventing Accuracy: A Historical Sociology of Nuclear Missile Guidance* (Cambridge: MIT Press, 1990).

67. Gabrielle Hecht, "Political Designs: Nuclear Reactors and National Policy in Postwar France," *Technology & Culture* (October 1994): 657–685.

6

A Worm in the Bud? Computers, Systems, and the Safety-Case Problem

Donald MacKenzie

Introduction

At the heart of nearly all the large technical systems of the late twentieth century are computers. The year 1958 saw the initial operation of the first large-scale, computerized, real-time system: the multibillion dollar SAGE system, controlling North American air defenses.[1] Over the following decade, computers became central to the systems of nuclear attack and of spaceflight, and they began to be used on board civil aircraft. The experience of SAGE helped make possible the first truly large-scale commercial real-time network: the SABRE computerized airline reservations system, delivered by IBM to American Airlines in 1962.[2] The role of computers in management and in the handling of large numbers of financial transactions steadily grew until they became indispensable to large organizations. Computers became central to telephone systems, gradually replacing human operators and electromechanical switches, and the combination of computer and communication technologies turned many other large organizations, previously coordinated by labor-intensive flows of paper, into explicitly technical systems. By the 1980s, it had become difficult to identify a large technical system in the advanced industrial countries in which computers were not central. By the early 1990s, microprocessor control was becoming pervasive even in familiar, mundane technologies such as washing machines and motor cars; in the late 1990s, the Internet was spreading at a pace probably unparalleled in the history of technology.

Computerization has given technical systems the capacity to perform tasks that previously were difficult or impossible: imagine, for example, the difficulty of trying to design a reservation system for a modern mass-transit airline that did not use a computer, or trying to construct a noncomputerized defense against ballistic missiles. Computerization has also greatly increased the efficiency with which many tasks are performed. However, in apparent paradox, the computer has also been a source of unease: if a crucial computerized system went wrong, might catastrophe not ensue? Almost at the beginning of the age of the computerized system, there were glimpses of potential risk. On 5 October 1960, the newly operational, computerized Ballistic Missile Early Warning Station (BMEWS) at Thule, Greenland, interpreted radar signals reflected from the rising moon as incoming missiles, and a message warning of nuclear attack was sent to the war room at North American Air Defense Command (NORAD) headquarters in Colorado. Fortunately, cool heads prevailed at NORAD, and the warning, unsubstantiated by other evidence of an attack, was diagnosed as a false alarm; confidence was helped by the conviction that the Soviets were unlikely to attack while premier Khrushchev was in New York, as he was on 5 October.[3] However, two years later disaster did strike, though fortunately not in a system where there was a threat to life: on 22 July 1962, the first U.S. interplanetary space probe, Mariner 1, was lost because of the interaction between a hardware fault and a simple error (an omitted bar) in the equations embodied in the software controlling the trajectory of its Atlas booster.[4]

By the end of the 1960s, fears of out-of-control computers found expression in popular culture. Few who have seen Stanley Kubrick's now classic 1968 film, *2001: A Space Odyssey*, can have forgotten the role played in it by the spacecraft's computer system, Hal.[5] Of course, there was little resemblance between Hal and the real-world computers of the late 1960s, nor have the decades since then closed the gap to any great extent. However, fear of the computer should not be seen simply as the product of a combination of science-fiction imaginings and popular ignorance. It is to be found amongst computer specialists as well. One of the oldest of the Internet's many news groups is the RISKS bulletin board, set up in August 1985. RISKS is very widely read by specialists, those contributing to it include some of the famous names of computer science, and it is almost certainly the most influential of all the news groups. The view of computerization presented there is seldom reassuring.[6]

This, then, is the paradox I want to explore here. The computerization of technical systems is one of the greatest technological success stories of the late twentieth century. Yet it is also a matter of deep concern, even to many of those whose work has helped make it possible. It is hard to think of another sphere of technology in which a leading figure would put her or his name to a sentence analogous to the following: "Nobody trusts a computer; and this lack of faith is amply justified."[7] Yet that statement is to be found in a 1986 U.K. Cabinet Office report, signed by C. A. R. Hoare, one of the world's most influential computer scientists; and the sentiment is far from unique. For example, doubts about software reliability raised by computer scientists, notably critical-systems specialist David Parnas, were a major feature of the 1980s' debate about the most ambitious system yet planned, the "Star Wars" strategic defense system.[8]

So the relevance of computerization to the spread of the systems approach goes beyond its role in enhancing the capacities and efficiency of technical systems, and therefore encouraging their dissemination. An incompatibility has emerged between computerization and a key aspect of the systems *approach* as manifest in other spheres of systems engineering. Ways of showing the noncomputerized aspects of systems to be safe have broken down when applied to their computerized aspects. No *generally* applicable, *consensually* credible, way of demonstrating the safety of the latter has yet emerged. Modern technical systems need computerization: it greatly enhances their capacity and efficiency. But there may be a worm in this technological bud: the problem of showing such computerization to be safe, of constructing "safety cases."

Note that the problem of computer safety cases is not necessarily the same as the problem of computer safety. The actual safety record of the application of the computers to the systems of the 1960s, 1970s, and 1980s is, I shall argue below, reasonably good: indeed, surprisingly good, given the depth of the "software crisis" diagnosed at the end of the 1960s. The trouble is that in the absence of robust safety cases it would be deeply imprudent to assume that this reasonably good record will continue, especially given the ever-increasing extent, ambition, and safety-criticality of computerization.

The Computer and the System

The first large-scale system in which it was intended that computers should play a crucial real-time role was SAGE (Semi-Automatic Ground Environment), intended to integrate the previous ad hoc, separate, manually controlled air defenses of North America into a single integrated system. SAGE's origins go back to the late 1940s, to worries about air defense, to MIT's pioneering "Whirlwind" digital computer, and to the intimate links between MIT and the armed services, which gave key figures such as Jay Forrester, Perry Crawford, and Robert

Everett the capacity to turn into technological reality their conviction that "practical real-time operations lay within the power of electronic information systems."[9]

Whirlwind provided the development context for one of the two technologies that permitted computers to grow greatly in power and robustness during the 1950s and beyond: Jay Forrester's invention of random-access magnetic core memory.[10] The air defense problem was also the testing ground for the capacity to link digital computers into real-time, geographically spread out networks. The Digital Radar Relay, which permitted radar data to be transmitted down an ordinary telephone line, can be seen as the predecessor of the now ubiquitous modem. Another harbinger of the future was the method of human-computer interaction that was developed in the air defense work: data were displayed on cathode ray tubes, and operators were provided with a "light pen" they could use to interact with the system by "writing" on the display. As early as April 1951, a combination of Whirlwind and the Missile Early Warning Radar at Hanscom Air Force Base, Massachusetts, demonstrated the capacity to track an incoming aircraft and to guide an interceptor toward interception. An expanded system linking several radars in the Cape Cod area to Whirlwind was providing experimental data by March 1953.[11]

Yet turning these promising technologies and successful early experiments into an operational, continentwide air defense system proved a daunting task: SAGE was a harbinger in another, less positive, respect. The Cape Cod system "worked—surprisingly well—less than a year after it was undertaken," wrote J. C. R. Licklider, professor of electrical engineering at MIT and former director of the Advanced Research Projects Agency's celebrated Information Processing Techniques Office:

> The SAGE system was to be essentially a scaled-up and replicated Cape Cod system, hence easy to estimate and schedule. Yet the number of man-years of programing required was underestimated by six thousand at a time when there were only about a thousand programers in the world.[12]

In the early 1950s, a computer program involving twenty-five instructions could be seen as "large"; SAGE was eventually to require programs totaling over a million instructions. At the start of SAGE development it was thought that there might be a need to employ 100 programmers. Since, in 1955, there were perhaps only around 200 people in the United States with the programming skills necessary for a system of SAGE's complexity, it was already clear that recruitment and training were going to be problems. By 1959, the System Development Corporation had 800 programmers working on SAGE.[13]

Two particular difficulties—later found to be pervasive in other contexts[14]—caused the SAGE software task to grow far beyond initial expectations. First, the system specification was not stable. The bomber threat evolved, and the weapons and procedures to meet it changed too. Half of the 800 programmers "were continuously modifying SAGE and attempting to control the orderly implementation of myriad changes."[15] Second, it turned out to be impossible to use a standard "turn-key" software solution, and extensive new programming was needed to fit the particularities of different air defense sites and the demands of the commanders in charge of them. Even after an installation became operational, the System Development Corporation had to leave a team of eight programmers at each site to correct errors and update programs.[16] By 1961, all the sites were declared "operationally ready," but, even in the late 1960s, the operators in some SAGE installations were reportedly to be found using "makeshift plastic overlays on the cathode-ray displays" and "the 'scope watchers were bypassing the elaborate electronics—operating more or less in the same 'manual mode' used in World War II."[17]

Such problems were not universal. Where the tasks to be performed by computer systems were kept very simple—as, in the 1960s, they still had to be onboard missiles or spacecraft—the programming problems were less daunting. Missiles and space provided the crucial early market for the most important of all the component technologies of modern computers, the integrated circuit, launching that technology on its "Moore's law" trajectory of annual or eighteen-month doubling of the number of components on a state-of-the-art microchip.[18] Despite occasional troubles (Minuteman II, the first intercontinental ballistic missile (ICBM) to employ integrated circuits, suffered physical cracking of the silicon, temporarily disabling large sections of the force; much later, a 1980 integrated circuit failure led to another false warning of nuclear attack),[19] computer hardware development seemed to soar ahead. In comparison, by the late 1960s software was beginning to be seen as the key "reverse salient" (in Thomas Hughes's terminology)[20] in the "two or three dozen complex electronic systems for command, control and/or intelligence operation . . . being planned or developed by the military" in the 1950s and 1960s United States. In Licklider's words:

> ... [W]e have learned that the brash confidence of the "systems salesman" usually fades into the background soon after the development contract is let—and that schedules slip, the costs mount, and the delivered product falls short of the promise. We have learned that the misestimation of time, cost, and performance are usually worst for the most complex subsystems. And we have learned that in many systems the most complex subsystem is the computer software, that is, programing ... [W]hen software gets very complex, you can pour more and more men and money into it without causing it to be completed.... You begin to understand the possibility that [programs] may literally never be "debugged" and integrated.... All the large software systems that exist contain "bugs."[21]

The Software Crisis

Licklider's comments were occasioned by Strategic Defense Initiative's predecessor: the schemes current, in the late 1960s, to build computer-controlled antiballistic missile defenses (including, let us recall, a plan to station nuclear-armed defensive missiles around major American cities). He feared that, because of software error, such schemes were "potentially hideous folly."[22] Yet his observations were far from unique, as increasingly serious software development problems were experienced by the wider computer industry: particularly traumatic were the difficulties encountered developing the operating system for IBM's famous System/360 series of computers.[23] These problems were first systematically aired at what has become perhaps the most famous of all computer science meetings, the conference on "Software Engineering" held at Garmisch in the Bavarian Alps, under the auspices of the NATO Science Committee, in October 1968.

The small but select Garmisch meeting deliberately brought together leading academic, industry, and government figures in computing. Much of its discussion was framed by the diagnosis of a "software gap," or, more famously, "software crisis."[24] As one participant noted: "It seems almost automatic that software is never produced on time, never meets specification, and always exceeds its estimated cost." Others pointed out that the consequence could be more serious than just management and financial problems:

> Particularly alarming is the seemingly unavoidable fallibility of large software, since a malfunction in an advanced hardware-software system can be a matter of life and death, not only for individuals, but

also for vehicles carrying hundreds of people and ultimately for nations as well.[25]

Not all the participants at Garmisch were persuaded that there was indeed a crisis of the wider software industry, but the majority of speakers seemed deeply concerned about software systems on which life depended. The following extract from the transcript of the proceedings gives a flavor of the debate:

> *R. C. Hastings* [IBM Corporation]: I am very disturbed that an aura of gloom has fallen over this assembly. I work in an environment of many large installations using [Operating System] 360 . . . People are doing what they need to, at a much lower cost than ever before, and they seem to be reasonably satisfied. Areas like [air] traffic control, hospital patient monitoring, etc., are very explosive, but are very distinct from general purpose computing.
>
> *A. Perlis* [Carnegie-Mellon University]: . . . Is bad software that important to society? Are we too worried that society will lose its confidence in us?
>
> *B. Randell* [IBM Corporation]: Most of my concern stems from a perhaps over-pessimistic view of what might happen directly as a result of failure in an automated aircraft control system, for example. I am worried that our abilities as software designers and producers have been oversold.
>
> *A. Opler* [IBM]: As someone who flies in airplanes and banks in a bank I'm concerned personally about the possibility of a calamity, but I'm more concerned about the effects of software fiascos on the overall health of the industry.
>
> *K. Kolence* [Boole and Babbage, Inc.]: I do not like the use of the word "crisis." It's a very emotional word . . . There are many areas where there is no such thing as a crisis—sort routines, payroll applications, for example. It is large systems that are encountering great difficulties. We should not expect the production of such systems to be easy.
>
> *D. T. Ross* [MIT]: It makes no difference if my legs, arms, brain, and digestive tract are in fine working condition if I am at the moment suffering from a heart attack. I am still very much in a crisis.[26]

No clear, consensual way forward out of the "crisis" emerged from the Garmisch meeting, or from the follow-on conference held in 1969 in Rome. A proposed International Software Institute was never actually founded: "It got lost in international politics, in quarrels over whether it should be in Munich or Luxembourg or Geneva."[27] More fundamentally, the widely accepted slogan "software engineering"—"implying the need for software manufacture to be based on the types of theoretical foundations and practical disciplines, that are traditional in the established branches of engineering"[28]—was just that, a slogan. Giving it concrete meaning was much more difficult.

The very term "engineering" carried diverse connotations. For some, the emphasis lay on providing the requisite "theoretical foundations," while for others the key was the "practical disciplines": " 'modular' software development, in which separate functions are implemented in separate, clearly articulated sections of program code ... milestones, test plans ... early prototypes ... thorough documentation."[29] In part, the theoretical/practical divide was a divide between academics and those in industry. For one key participant at Garmisch, the Dutch computer scientist Edsger Dijkstra, the divide also seemed to map onto a transatlantic difference. Dijkstra told the Garmisch meeting:

> We, in the Netherlands, have the title Mathematical Engineer. Software engineering seems to be the activity for the Mathematical Engineer par excellence ... our basic tools are mathematical in nature.[30]

Dijkstra, however, recalls that when he introduced the phrase "Mathematical Engineer":

> ... my American colleagues started to laugh, because the mathematician was unpractical and the engineer was practical, and never the twain shall meet.... As soon as you cross the Atlantic, all important words slightly change their meaning....[31]

At Garmisch, and in the years that followed, a variety of different approaches—mathematical, technological, and organizational—to software engineering were proposed with great enthusiasm and met with some success, but none proved to be a panacea. Reviewing the history of software engineering since 1970, Stuart Shapiro concludes:

> No single approach in any single aspect of software technology could fully satisfy. Precise dogma finding its expression in a single programming language, design technique, metric type, or management methods is no doubt ... emotionally satisfying, but nevertheless impractical. Effective technological practice demands technical pluralism operating in the context of local knowledge and within a

> framework for choice ... The pragmatic response was combination and accommodation ... splitting the difference, synthesizing complementary approaches, or accommodating inescapable trade-offs.[32]

Nor did all these approaches filter down into industry, especially in the key safety-critical field of real-time software. Robert L. Glass of Boeing wrote in 1980:

> The literature may picture modern methodologies, but the real-time practitioners are still coaxing along their antique products and using machine language and patching to hold them together. The result is a product that usually works, but not always; a product that may be very economical, if the developers are skilled, but may on the other hand cost a fortune; a product that is virtually unschedulable; and a product whose quality is intimately tied to the skill of the software mechanic. The result can be satisfying and satisfactory; but when it is bad, it is very bad.[33]

By the 1980s, it was clear that, in the famous words of Frederick P. Brooks, Jr., project manager for both hardware and operating system for the IBM System/360, there was "no silver bullet," no single technical or organizational solution to the problems of software development.[34] Between the middle 1950s and the start of the 1980s, "hardware prices have dropped by a factor of a thousand, while software—still regarded as more an art than a science—costs little less than it did when SDC built SAGE."[35]

SINCE THE SOFTWARE CRISIS

Nevertheless, despite the absence of a silver bullet, and despite the continuing high cost of software development, the situation diagnosed at the end of the 1960s by Licklider and at Garmisch did improve somewhat. "Moore's law" meant that software developers could, increasingly, be profligate with hardware resources, a situation that has many benefits from the viewpoint of system reliability. For example, rigorously structured programming could be pursued even though it could be less efficient by two orders of magnitude, in terms of efficiency of use of hardware resources, than its more ad hoc equivalent.[36] Both individuals and organizations grew more experienced; a number of technical systems to support aspects of software development emerged; and standard "packaged" solutions to key tasks, such as database management, became available. Developing an airline reservations system, for example, was not as difficult a task by the 1980s as it had been at the time of SABRE. Building complex systems from scratch remained expensive and risky—even in the 1990s it could still lead to organizational catastrophe—but much of software development came to involve modifying, and adding features to, existing systems, rather than producing entirely new systems. In consequence, by the 1980s the programming team on a typical project was much smaller than the 800 people working on SAGE, thus reducing the managerial demands.[37] In short, "the software crisis was not 'solved,' but several aspects of it eased." However, as computer systems became more and more integral to organizations, issues of user relations, and eventually interorganizational issues, became more problematic. The problems of software development focused on at Garmisch turned out to be merely one aspect of deeper *system* development issues, and those cannot be said to have eased.[38]

Nevertheless, in the area that was the cause of greatest concern at Garmisch—computerized systems upon which human life depended—calamity was largely avoided. Using as my main source the RISKS reports in the Association for Computing Machinery's *Software Engineering Notes*, I have attempted to construct a list of all the computer-related fatal accidents that had taken place, worldwide, up until the end of 1992. I accumulated evidence on several dozen incidents,[39] resulting in a total of around 1,100 such fatalities. Two things are striking about these data. The first is that the number of deaths involved is modest: I suspect the Garmisch participants would have been relieved that the total, accumulated over almost a quarter century since their meeting, was not much larger than this. The second is that their main concern, software error, has not been the leading cause of computer-related accidental deaths. Only 3 percent of total deaths appear to have been the result of software error, and nearly all of these occurred in a single incident. Interestingly, in the light of Licklider's strictures, that incident was a failure of antiballistic missile defenses, which took place in Dhahran, Saudi Arabia, in February 1991, leading to the deaths of twenty-eight American soldiers in an Iraqi Scud missile attack. An omitted call in the system software to a subroutine improving the accuracy of the conversion from decimal to binary representations of time led to loss of tracking by the antimissile defenses.[40]

My data source is indirectly dependent upon media coverage of accidents (fatal accident reports in the RISKS forum are predominantly derived from the media), and it is perfectly possible that this coverage is incomplete, and fatal accidents are therefore underreported. Indeed, in a limited sphere—computer-related industrial accidents—this is definitely the case: independent sources (the industrial safety authorities in the United Kingdom and France) reveal a number of deaths that did not reach RISKS.[41] There is, furthermore, a particular problem in the diagnosis of software error as the cause of an accident. Unlike, say, a bolt that has sheared, software error tends to leave no physical traces.

Nevertheless, major accidents to computerized systems, such as crashes of "fly-by-wire" airliners,[42] are the subject of intense scrutiny (and much discussion in the RISKS forum), so I think it is unlikely that the figure of 1,100 deaths is too gross an underestimate. Furthermore, there are several highly computerized systems where there is now an extensive record of usage and no serious computer-related accidents. Perhaps the most impressive case is full-authority digital engine controls (FADECs) for aircraft, which have now accumulated around a billion hours of operational use without a serious accident.[43] But there are other cases too. Although the U.S. space shuttle has been afflicted by a number of serious software errors, these have not led to an accident: the 1986 Challenger disaster was not (at least in any direct sense) computer related.[44] Since 1985, a growing proportion of the rail network in Britain has been controlled by a computerized system, Solid State Interlocking, which has replaced the older electromechanical interlockings of signaling and points system, without being implicated in any accident.

System Safety and Risk Assessment

That computerized systems have, to date, turned out to be safer in practice than was feared at the end of the 1960s does not, however, resolve the problem of safety-critical computing. Many safety-critical computer systems have to be introduced into contexts where there are powerful "audiences," such as regulatory bodies, that need to be convinced—in advance—that systems are safe. Since the early 1960s, a systematic "safety engineering" way of tackling the problem of safety has evolved, in part to meet the needs of such audiences, and the resultant analysis of risk has become a vital part of the modern "systems approach." However—as I shall try to show—deep problems have been encountered in extending this way of dealing with risk from the noncomputerized to the computerized aspects of systems.

At the heart of conventional system safety engineering are "fault trees" and "hazard mode and effects analyses."[45] A fault tree is a geometric notation for rendering explicit the combinations of events that can lead to system failure. The technique was first applied, around 1961, in safety evaluation of the Launch Control System of the U.S. Minuteman ICBM. The goal was to prevent accidental warhead detonation or inadvertent missile launch, and those involved worked "backwards" from those undesired outcomes in an attempt to discover the full range of events (including events such as a squirrel gnawing through the insulation on a cable) that could lead to them. Hazard mode and effects analysis, in contrast, works "forwards" in an attempt to discover the consequences of "all conceivable failures [of] every component in [a] system."[46]

If one has sufficient confidence in the completeness of a fault tree or hazard mode and effects analysis, and if, furthermore, one can assign probabilities to the events or component failures involved, then it becomes conceivable to discover—in advance of use—the overall probability of system failure. This crucial further step in the system safety engineering is known as *probabilistic risk assessment*. Again, it has its origins in analysis of the safety of the U.S. defense and space systems of the 1960s. NASA subsequently abandoned quantitative risk analysis for a number of years, following "bad experience with some probability estimates during the Apollo program,"[47] but probabilistic risk assessment became a crucial weapon in the nuclear power industry's attempts to fend off criticism of nuclear safety, which was increasingly vociferous from the late 1960s onward. While controversy over nuclear power was a uniquely strong motivator, probabilistic risk analysis spread to other fields, and, in the 1980s, returned to at least a degree of favor in the space program. As one of the technique's proponents put it in 1988:

> The era of quantitative risk assessment and risk management is here. Technologists are being held accountable to the public for answering the basic question of, "What is the risk?" of their engineered systems. The [probabilistic risk assessment] thought process is the most effective means for answering this question in a comprehensive and quantitative manner.[48]

Not everyone was convinced as that by probabilistic risk analysis—critics saw it as "an inherently dubious process of compounded speculation"[49]—and, furthermore, use of the technique seems to have done little to modify public perception of the dangers of nuclear power. Nevertheless, amongst key, specialized, powerful, audiences, notably the nuclear regulatory bodies in the United States and Europe, the demand for probabilistic risk assessment has become entrenched to a significant degree. Increasingly, such bodies have been demanding explicit "safety cases": arguments and bodies of evidence to show that a proposed new system poses acceptably low risk. Fault trees, failure modes and effects analysis, and probabilistic risk assessment have, therefore, become key aspects of modern systems engineering.

THE COMPUTER-SYSTEM SAFETY-CASE PROBLEM

If there was a guaranteed way of producing bug-free software, or of designing computer systems that could be guaranteed to be free of design errors, then computerization would pose no particular problems for risk assessment. Computers could be treated like other system components are, with attention focused on their physical failure. If the probability of the latter had unacceptable consequences for the overall probability of system failure, then additional, redundant computers could be installed, with a system of "voting" to isolate faults. In probabilistic risk assessment, it is normal to treat physical failures of components (even physically identical components) as independent events. Since the probabilities of independent events combine multiplicatively, modest degrees of redundancy of only moderately reliable components can yield reassuringly low probabilities of system failure. (Suppose, for example, system failure requires failure of three independent components, each of which has a probability of failure of one in a thousand, or 10^{-3}, during, say, a mission of typical duration. Then the probability of system failure is $10^{-3} \times 10^{-3} \times 10^{-3} = 10^{-9}$, or one in a billion—the kind of level that is demanded in safety-critical avionics systems.) So physical failure of computer hardware posed no special problems for the construction of safety cases, even with the modestly reliable hardware available in the 1960s. Software error, however, was a different matter. The public diagnosis, at the end of the 1960s, of a software crisis, the attendant widely circulated fears of "bugs" in crucial systems, and the failure to find a "silver bullet" to guarantee error-free software, meant that it was hard to assign a probability of zero to software error.

But, if not zero, what? Commentators such as Licklider were claiming that the probability of error, at least in software of any complexity, should, in effect, be treated as being one (i.e., certainty of error). That would imply the unacceptability of any system in which software of any complexity played a safety-critical role. This conclusion actually seems to have been reasonably widespread amongst engineers in a variety of contexts, from the 1960s—when safety-critical software first became a real, widespread possibility—at least until the 1980s, when the practical and economic advantages of computerization began to overwhelm resistance. Even by 1995, there were, to my knowledge, only two software systems in British nuclear power stations that are deemed safety-critical. One is the primary protection system of the Sizewell B light-water reactor; the other is a reactor trip system retrofitted to the gas-cooled Dungeness B reactors. In other reactor protection systems in the United Kingdom, the necessary logic is "hard-wired" by permanently, physically configuring arrays of magnetic cores so that they implement the necessary logic. That these technical judgments are situation-dependent is revealed by the fact that nuclear engineers in France, typically working upon more recent systems, seem to have been much readier than their counterparts in Britain to adopt safety-critical software. However, a reasonably widespread conservatism, amongst engineers, about safety-critical uses of software may well be one reason why software error has led to far fewer fatal accidents than the Garmisch participants would have expected: the move to safety-critical computing has been significantly slower than would have been "technically possible."

Testing and the Construction of Safety Cases

If an a priori zero probability of software error lacks credibility, and a probability of one paralyzes safety-critical computerization, then a way must be found either to assign an intermediate probability or plausibly to justify a value of zero. The obvious route is empirical testing. Computer system testing was not originally seen as part of the construction of an explicit safety-case, and from a critic's point of view the inferences ordinarily made on the basis of testing could seem sloppy. One computer scientist phrases the dominant criticism memorably:

> ... [T]hose who today use defect-detection methods claim a connection between those methods and confidence in the tested software, but the argument seems to be the following:
>
> I've searched hard for defects in this program, found a lot of them, and repaired them. I can't find any more, so I'm confident there aren't any.

> Consider the fishing analogy:
>
> I've caught a lot of fish in this lake, but I fished all day today without a bite, so there aren't any more.
>
> Quantitatively, the fishing argument is much the better one: a day's fishing probes far more of a large lake that a year's testing does of the input domain of even a trivial program.[50]

Garmisch participant Edsger Dijkstra phrased the point pithily in 1972, when being awarded the Association for Computing Machinery's Turing Prize. Using words that were quoted and requoted many times in the years to follow, he said: "Program testing can be a very effective way of showing the presence of bugs, but it is hopelessly inadequate for showing their absence."[51]

System developers had always devoted a large part of their time and budget to testing. Considerable effort was put into the development of tools to assist testing. For real-time systems (and most safety-critical systems are real-time), the most important of these was the environmental simulator, "an automated replication of the external system world built for computer hardware/software test." By 1980, environmental simulators were "almost universally used" in real-time software development.[52] However, even when such simulators were used, testing was for many years seen primarily as part of the informal practice of "debugging," not as a means of producing systematic knowledge of the properties of the system being tested, nor as a subject for computer-science attention in its own right. It emerged in the latter role only during the 1970s. In 1972, opening what was, as far as I am aware, the first conference devoted to program testing, its organizer, William C. Hetzel, commented: "At the base of our testing problems is that there is currently no established discipline to act as a foundation."[53] Another contributor, Fred Gruenberger, complained that program testing "is all art, and . . . demands a form of low cunning on the part of the programmer."[54]

During the 1970s, several important papers helped to clarify ways of making testing more systematic, for example, by striving to ensure that testing exercises all the control-flow paths through a program.[55] This work was drawn upon in the first textbook devoted exclusively to program testing, which appeared in 1979. However, the book's title, *The Art of Software Testing*, together with much of its content, implied a continuing vital role for the testers' experience and skill. A seemingly inevitable part of real-world testing was what the textbook admitted to be the "intuitive" process of "error guessing."[56] Certainly, it was emphatic that exhaustive testing—in the sense of a complete search of a program's input domain—was in general impossible in practice.

Beyond the particular testing strategies formulated in the 1970s, the two most important general attempts to increase the credibility of conclusions based on testing, and to move the latter from the status of an "art" to that of a "science," have been reliability growth modeling and random testing. The former is the less radical challenge to established testing practice: it is an attempt to systematize, and to give an empirical justification to, the practitioner's intuitive sense that the reliability of a program improves as errors are detected and corrected. The key early work on reliability growth modeling was done at TRW, the large corporation that developed out of the collaboration, in the 1950s and beyond, between Simon Ramo and Dean Wooldridge in the management of the U.S. ICBM program.[57] In reliability growth modeling (as in conventional computer systems development) a program is tested until it fails. The error causing the failure is found and corrected, and the program is tested until it again fails; and the process is continued through many iterations. The difference from conventional development is that the times between failures, or some other measure of the extent of testing, are fitted into a mathematical model (the reliability growth model) derived from previous experience of similar development processes, until the model indicates a sufficiently low failure probability (or, equivalently, a sufficiently long expected mean time between failures).

It was widely agreed that reliability growth modeling was a useful tool in the management of software development, but there were worries about whether there would always be strong enough similarities between the past development processes, on the basis of which the model had been constructed, and the current one. There were also concerns about the potential of error correction processes to reduce reliability (by introducing new errors) rather than to increase it. Critics argued that

> . . . for safety-critical applications one must treat each modification of the program as a new program. Because even small changes can have major effects, we should consider data obtained from previous versions of the program to be irrelevant.[58]

Furthermore, in practice extremely large amounts of error-free testing were found to be needed to use reliability growth modeling to justify claims of very low error probabilities.[59]

The second main approach to making testing a "science" was random testing. This was a more radical departure from existing practice, in that the "commonsense" approach to testing had been a highly focused, purposive search for the test cases most likely to reveal errors. In random testing, test cases are selected from the domain of possible inputs to a system by an impersonal mechanism with known statistical properties, such as a random number generator. One advantage of random over purposive testing was that the former was easier to automate, and its advocates suggested that it was not inferior as a means of finding errors.[60] The other, dominant, advantage was that randomness permitted the application of the theory of probability to the results of testing. Instead of what critics saw as the mere empirical curve-fitting of reliability growth modeling, random testing could draw upon the credibility of well-established theories of statistical inference.

However, while it was straightforward to produce an algorithm to generate random test cases from a distribution of input cases, there was no algorithm for ensuring that the latter distribution was realistic (i.e., representative of what would be met in use). Instead:

> Deep knowledge and experience with the application area will be needed to determine the distribution from which the test cases should be drawn.[61]

Furthermore, the theory of random testing had a feature that seemed suspicious to some. The theory treated programs as "black boxes," whose internal structure (below the level at which testing was being conducted) was irrelevant. In particular, the size of the program being tested did not enter into the equations, derived from classical statistical theory, governing the inferences to be made on the basis of testing. These formulae therefore predict:

> ... exactly the same reliability for a 10-line program computing roots of a quadratic equation as for a 100,000-line accounting system, if each is tested with the same number of points. The result seems intuitively wrong. ...[62]

Even if this aspect of the theory of random testing was set aside, there was again the problem that very large amounts of error-free testing are needed to justify confidence in very low error probabilities. For example, two NASA computer scientists calculated that under realistic conditions (with, for example, only a modest number of systems available for testing in parallel), the sort of failure rate demanded for life-critical avionic systems required infeasible quantities of testing: "to quantify 10^{-8}/h[our] failure rate requires more than 10^{8} h[ours] of testing."[63] Theirs was an application of standard, "frequentist" statistical theory, but the alternative Bayesian approach (in which statistical inference is seen as the modification, by experience, of prior beliefs) yields scarcely more optimistic results:

> [U]nless the Bayesian brings extremely strong prior beliefs to the problem ... both approaches will generally give results that agree to within an order of magnitude.

For example, "if we require a median time to failure of 10^{6}, and are only prepared to test for 10^{3}, we essentially need to start with the belief in the 10^{6} median."[64]

These problems are in practice exacerbated by the fact that wholly automated testing—and, clearly, automation is necessary if very large amounts of testing are contemplated—is, in some cases, difficult or impossible. One fundamental issue is what those involved sometimes call the "oracle" problem: the need for an independent criterion against which to judge the correctness of the responses of the system being tested.[65] The difficulty is, of course, that if a reliable, efficient way of calculating correct response already exists, then a new system would not be needed, while if it does not exist, one may have to rely on slow, unreliable "pseudo-oracles," such as expert human judgment of the correctness of response. In addition, the construction of suitable test inputs may also be hard to automate, and may therefore be slow.

The consequence of this sort of problem, when combined with the above statistical analyses, is that practicable amounts of testing may fall far short of what is needed to justify claims of very high reliability. Two observers noted, in 1991, that "we are not aware of anyone yet having achieved an effective test demonstration at the 10^{-4} level [1 failure per 10,000 demands]."[66] In 1995, working parties of the International Electrotechnical Commission, drafting a possible future generic standard for safety-critical systems, judged the limits of permissible claims for the failure probabilities of programmable systems of any complexity to lie in the region of 10^{-4} to 10^{-5} per demand or, in the case of continuously operating systems, 10^{-4} to 10^{-5} per year.[67]

N-Version Programming and the Knight-Leveson Experiment

Component failure probabilities of 10^{-4} or 10^{-5} are encountered routinely in probabilistic risk assessment, and, as noted above, are no insuperable barrier to very high reliability if independent, redundant components are available. This, for example, seems to have been the form of argument that persuaded the U.K. Nuclear Installations Inspectorate to permit Sizewell B to go on stream. Once the Inspectorate was convinced that a failure probability of the primary software system of 10^{-4} was justifiable (justifying even that relatively modest figure turned out to be time consuming and difficult), the existence of a different, hard-wired secondary protection system made it possible to claim an overall failure probability of 10^{-7}.[68]

To require computerized systems always to be accompanied by an independent, noncomputerized backup would, however, remove many of the practical and economic advantages of computerization. An attractive option, therefore, is to install redundant, different computer systems, comprising different computer hardware and/or different software. The systems have to be *different*: while redundant identical systems can be, as noted above, a good insurance against physical failure, they are clearly no insurance against design flaws or software errors (which would be expected to manifest themselves simultaneously in all of the redundant systems). This approach was applied to several important systems, especially in Europe, in the late 1970s and 1980s: to a Swedish railway control system; to the computerized slat and flap control system of the Airbus A310 airliner; and in the flight control system of the first fly-by-wire civil airliner, the Airbus A320. For example, each of the latter's safety-critical computers continually runs, and compares the results of, two programs, each written in different languages by different programming teams.[69]

There is, however, a potential difficulty in using software redundancy, or "N-version programming," to justify, via the multiplicative law of independent probabilities, very low overall failure probabilities. The difficulty is whether the different programs will be sufficiently different to justify the assumption that their failures will be independent. A leading practitioner of the N-version approach suggests that it will be necessary to aim for both technical diversity (e.g., use of different programming languages) and social diversity (even use of programmers from "diverse ethnic backgrounds").[70]

Sharp controversy over the independence assumption was sparked by an experiment by computer scientists John Knight and Nancy Leveson, in which they compared the errors made by twenty-seven graduate and senior-level students who were asked to write programs to satisfy the same requirements' specification (for part of a simple antiballistic missile system). Knight and Leveson found that in certain cases, different programmers made effectively the same error, and concluded that "the assumption of independence of errors that is fundamental to some analyses of N-version programming does not hold."[71] While this conclusion was fiercely contested by the proponents of N-version programming, the widely circulated doubts cast on the independence assumption by the Knight-Leveson experiment are a barrier to the use of the multiplicative law of independent probabilities to justify very high reliability figures.[72] It is, for example, of interest that Boeing abandoned plans to apply N-version programming in its first fly-by-wire airliner, the 777: it seems to have be judged that the resultant decrease in failure probability did not justify the increased complexity involved.[73]

Mathematical Proof and the Construction of Safety Cases

So there are problems with the escape route, via the multiplicative law, from the limitations of inductive, test-based, knowledge of the properties of computer systems. Another potential escape route—a very widely advocated one—is to shift the basis of inference altogether, from induction to deduction. After all, our knowledge that $1 + 1 = 2$ is not ordinarily thought of as constrained by the finite number of times we have placed one apple, then another apple, in a fruit bowl, and found that it contains two apples; nor is that knowledge usually considered to be undermined by the occasions when we see one raindrop merge with another raindrop to form only one raindrop.[74]

Since computer systems can be seen as implementing basic mathematical and logical operations, and many computer scientists come from backgrounds in mathematics and logic (rather than in engineering), it is not surprising that deductive proof should figure large as a possible basis for computer-system safety cases. The Garmisch conference convened at a point in time, the late 1960s, when there was an efflorescence of interest in ways of showing, mathematically and deductively, that a program was a correct implementation of its specification.[75] Dijkstra's remark at Garmisch on the need for the "Mathematical Engineer" was, in part,[76] reference to this growing interest. By the 1970s, program proof, or formal verification, was a "hot topic" for computer science researchers, especially in the United States. Indeed, the 1970s' literature on program testing is characterized by a certain defensiveness vis-à-vis its more prestigious deductive rival.[77] The latter received extensive research and development funding, especially from national security interests in the United States, which saw in formal verification a potential solution to the analogue of the computer-system safety-case problem: the problem of showing that systems are secure, in the sense of preserving the confidentiality of the information they contain.

I have written elsewhere on the many fascinating issues raised by the application of deductive proof to computer systems.[78] Here, it is sufficient to note that the effort has been dogged by its sheer practical difficulty. Unless programs are developed in such a way that the development process and the construction of proof are one and the same—a procedure advocated, above all, by Edsger Dijkstra, but very rare in industrial practice—proof has tended to be a slow, expensive addendum to the development process, and one that can be applied, in practice, only to relatively short, simple programs. Although many computerized systems to assist such proofs have been developed, they are nearly all automated theorem provers, not automatic ones.[79] Considerable skill is required to use them effectively, and the work is not to everyone's taste: the proofs involved are often extremely intricate, but repetitive and lacking in "depth," not the simple, elegant, deductive reasoning prized by mathematicians. In consequence, the industrial uptake of formal proof is, in the 1990s, still at best patchy, with uptake concentrated in sectors supplying systems critical to national security (but in retreat there, as governments increasingly seek "off-the-shelf" solutions to their computer security needs), and in a few areas of safety-critical computing (the U.K. Ministry of Defence has played a leading role in demanding it).[80] Interestingly, formal proof seems to be making faster inroads into verification of the design of computer hardware than into program proof: the relevant proofs seem more susceptible to automation.[81]

Even if the technical, organizational and economic barriers to the industrial uptake of proof-based methods could be overcome,[82] they would not resolve *all* the problems of the construction of safety cases. As those involved emphasize, formal proof can show only that a mathematical model of a program or hardware design is a correct implementation of its specification. Deductive reasoning alone cannot demonstrate that the specification is "correct," or that a program as run on an actual physical computer, and the actual silicon implementation of a hardware design, will behave as the mathematical models of them suggest they should.[83] These problems have to do with the relationships between mathematical models, the intentions of designers, and the real world, and perhaps also with the behavior of human users. If nothing else, these considerations suggest that empirical testing will always have a part to play in assuring the safety of computer systems, and that deductive knowledge of the latter's properties will always have to be complemented by inductive knowledge.

Conclusion

The culture of computer science is in many ways a deeply egalitarian, anarchic one. Its very vibrancy, and its proneness to explicit, public disagreement (in many ways more akin to the social sciences than to the natural sciences or much of engineering) have perhaps been reasons why the construction of computer system safety cases has been hard. The tradition of critical public comment, by computer scientists, on the safety of computer systems—a tradition begun at Garmisch, and stretching through to the RISKS newsgroup and beyond—was, arguably, part of a professionalizing project: showing the need for improved standards of rigor by pointing to the deficiencies of current practice. However, its very success has meant that computer-system safety-case construction begins against a background of skepticism, and takes place in a context in which bad publicity is an ever-present possibility. That background and that context make computer-system safety-case construction harder. RISKS may not be entirely unique, but in few other professions, in engineering or elsewhere, has there been such a long-standing, determined, public effort to discover system risks and unmask failings.

The cynical conclusion might be that computer scientists should keep their mouths shut, their disagreements and failures private, and thereby build their professional authority and ease the construction of computer-system safety cases by lessening the skepticism they encounter. From a wider, less cynical, viewpoint, however, that conclusion would be quite wrong. One of the unusual characteristics of beliefs about safety is that they tend to have a self-negating aspect. Beliefs that "bugs" are pervasive, that computer systems can be dangerous, have been one of the factors that have kept the death toll in computer-related accidents modest. There has been no headlong rush into safety-critical computing: on the whole, the steps involved have been careful, measured ones. The fear of calamity—expressed, for example, by Licklider and at Garmisch—has been one reason why calamity has largely been averted. Distrust of the computer—the fact, for example, that those in command at NORAD in 1960 knew that BMEWS "was still being run-in"[84]—has helped keep computerized systems safe.

The persistence of distrust of the computer is fortunate because the problem of ensuring safe human-computer interaction is at least as intractable as the problem of computer-system safety cases. Completely automatic safety-critical systems have been plausible only in situations where there is a clear-cut "safe state" that can be reached straightforwardly. Trains can safely be stopped, and chemical processes or nuclear reactors can be shut down, but an aircraft in flight, for example, has no equivalent safe state.[85] Designers have been reluctant to make computerized systems without safe states entirely automatic, so such systems have depended—and for the foreseeable future will continue to depend—on the safety of human-computer interaction. The driverless train is a reality (at least in certain limited situations); the pilotless passenger aircraft is still some way off.

There is no known, a priori, way of ensuring safe human-computer interaction.[86] Problems in this sphere swamp all other causes of computer-related accidental fatalities.[87] Complacency and misplaced trust by human operators in computer system output are prominent among the causes of these problems. Increasing the safety and reliability of the "technical" aspects of systems might conceivably worsen these dangers: a computer system that errs frequently (and is therefore distrusted) is, under some circumstances, less dangerous than one that *almost* never errs. A solution to the safety-case problem that reduced or eliminated distrust of the computer might well make computerized systems more, not less, dangerous. Frustrating though it obviously is for both regulators and practitioners, there are advantages to a situation in which the latter "are capable of constructing systems that are safe," but do not know how to demonstrate that they are safe.

Because of its self-negating aspects, the fear of computer-generated calamity is a healthy fear. The tradition in computer science that has emphasized its possibility, and has highlighted deficiencies in current systems and in current development practices, has been a healthy tradition. The existence of that fear and that tradition have made the safety-case problem harder, but they have made systems themselves safer.

Acknowledgments

This research was supported financially by U.K. Economic and Social Research Council (grant R00029008), originally via its Joint Committee with the Science and Engineering Research Council (grant H74452). It also forms part of a larger project on "Communication in Safety Cases: A Semantic Analysis," funded by the U.K. Safety-Critical Systems Research Programme (Engineering and Physical Sciences Research Council grant J58619). I am grateful to Graham Spinardi and Maggie Tierney for help with the research drawn on here, and to Stuart Anderson, Alan Bundy, and Arne Kaijser for useful comments on an earlier draft.

Notes

1. See Paul N. Edwards, *The Closed World: Computers and the Politics of Discourse in Cold War America* (Cambridge: MIT Press, 1996), chap. 3; Thomas P. Hughes, *Rescuing Prometheus* (New York: Pantheon, 1998), chap. 2; and Claude Baum, *The System Builders: The Story of SDC* (Santa Monica, Calif.: System Development Corporation, 1981). SDC, formerly the Rand Corporation's System Development Division, was set up largely for the task of writing the SAGE software; in 1981, it was taken over by the Burroughs Corporation.

2. Charles J. Bashe et al., *IBM's Early Computers* (Cambridge: MIT Press, 1986), 310 and 516–522; and Paul E. Ceruzzi, *Beyond the Limits: Flight Enters the Computer Age* (Cambridge: MIT Press, 1989), 76.

3. Anon., "Moon Stirs Scare of Missile Attack," *New York Times*, 8 December 1960, 71; John Hubbell, "You Are under Attack!" *Readers Digest* 78 (May 1961): 47–51.

4. Ceruzzi, *Beyond the Limits*, 202–203.

5. Arthur C. Clarke, *2001: A Space Odyssey* (London: Hutchinson, 1968).

6. RISKS reports have been the basis of a book by the forum's moderator, Peter Neumann, *Computer-Related Risks* (New York: ACM Press and Reading, Mass.: Addison-Wesley, 1995). See also popular accounts such as Leonard Lee, *The Day the Phones Stopped* (New York: Fine, 1992); Lauren Ruth Wiener, *Digital Woes: Why We Should Not Depend on Software* (Reading, Mass.: Addison-Wesley, 1993); and Ivars Peterson, *Fatal Defect: Chasing Killer Computer Bugs* (New York: Random House, 1995).

7. Cabinet Office, *Advisory Council for Applied Research and Development, Software: A Vital Key to UK Competitiveness* (London, HMSO, 1986), 78.

8. David L. Parnas, "Software Aspects of Strategic Defense Systems," *Communications of the ACM* 28 (1985): 1326–1335.

9. Kent C. Redmond and Thomas M. Smith, *Project Whirlwind: The History of a Pioneer Computer* (Bedford, Mass.: Digital Press, 1980), 193.

10. The other technology was, of course, the transistor.

11. Redmond and Smith, *Project Whirlwind*, 192–193, 203, 216.

12. J. C. R. Licklider, "Underestimates and Overexpectations," in Abram Chayes and Jerome B. Wiesner, eds., *ABM: An Evaluation of the Decision to Deploy an Antiballistic Missile System* (New York: Harper and Row, 1969), 118–129, at 121.

13. Baum, *System Builders*, 23 and 34–35.

14. Andrew L. Friedman with Dominic S. Cornford, *Computer Systems Development: History, Organization and Implementation* (Chichester, West Sussex: Wiley, 1989).

15. Baum, *System Builders*, 38.

16. Ibid., 36.

17. Ibid., 36; Licklider, "Underestimates and Overexpectations," 121.

18. "Moore's law" was formulated in 1964 by Gordon E. Moore, then director of research at Fairchild Semiconductor: see Robert N. Noyce, "Microelectronics," *Scientific American* 237, no. 3 (September 1977), reprinted in Tom Forester, ed., *The Microelectronics Revolution* (Oxford: Blackwell, 1980), 29–41.

19. John S. Gasper, interviewed by author, Anaheim, Calif., 9 September 1986; Alan Borning, "Computer System Reliability and Nuclear War," *Communications of the ACM* 30 (1987): 112–131.

20. See, e.g., Thomas P. Hughes, "The Evolution of Large Technological Systems," in Wiebe E. Bijker, Thomas P. Hughes, and Trevor Pinch, *The Social Construction of Technological Systems: New Directions in the Sociology and History of Technology* (Cambridge: MIT Press, 1987), 51–82, at 73.

21. Licklider, "Underestimates and Overexpectations," 125, 123.

22. Ibid., 123.

23. Eloína Peláez, "A Gift from Pandora's Box: The Software Crisis" (Ph.D. diss., University of Edinburgh, 1988), chap. 3. I owe much of my initial interest in this topic to Peláez's work.

24. Peter Naur and Brian Randell, eds., *Software Engineering: Report of a Conference Sponsored by the NATO Science Committee*, Garmisch, Germany, 7–11 October 1968 (Brussels: NATO Scientific Affairs Division, 1969), 120.

25. Ibid., 120, 122.

26. Ibid., 120–121. Authors' initials and affiliations have been added to this extract.

27. Peláez, "Pandora's Box," 189–190.

28. Naur and Randell, *Software Engineering*, 13.

29. Baum, *System Builders*, 36.

30. Naur and Randell, *Software Engineering*, 127.

31. Edsger Dijkstra, interviewed by E. Peláez, as quoted in Peláez, "Pandora's Box," 193.

32. Stuart Shapiro, "Splitting the Difference: The Historical Necessity of Accommodation and Compromise in Software Engineering, 1970–1993" (Uxbridge, Middlesex, and Edinburgh: Programme in Information and Communication Technologies, 1993), 52.

33. Robert L. Glass, "Real Time: The 'Lost World' of Software Debugging and Testing," *Communications of the ACM* 23 (1980): 264–271, 266.

34. Frederick P. Brooks, Jr., "No Silver Bullet: Essence and Accidents of Software Engineering," *Computer* 20 (April 1987): 10–19.

35. Baum, *System Builders*, 8.

36. C. A. R. Hoare, "How Did Software Get So Reliable without Proof," typescript, University of Oxford Computing Laboratory, December 1995.

37. For example, in the 1981 official history of the System Development Corporation, the SAGE programming manager talks of "today's much smaller staffs" (Baum, *System Builders*, 38).

38. Friedman and Cornford, *Computer Systems Development*, 174.

39. I do not attempt to be precise, because this figure is affected greatly by whether the same mistake, repeated many times (as in cases of errors in radiation therapy), is accounted as one incident or as many. For details, see D. MacKenzie, "Computer-Related Accidental Death: An Empirical Exploration," *Science and Public Policy* 21 (1994): 233–248.

40. For details, see ibid., 240–241. The U.S. Army claims only a 70 percent success rate, in Saudi Arabia, for antimissile defenses, and critics (such as Professor Theodore A. Postol of MIT) have suggested that the true rate may have been much less, so it is possible that defenses may have failed even in the absence of this software error.

41. These deaths are included in the above total: see ibid., 237–238.

42. In a fly-by-wire aircraft, pilots do not directly control engine thrust and control surfaces; instead actions by pilots are inputs to a computerized control system.

43. They are known as "full-authority" because pilots have no way of changing engine state other than via the FADEC system. See John Rushby, *Formal Methods and the Certification of Critical Systems* (Menlo Park, Calif.: SRI International, 1993, SRI-CSL-93-07), 127.

44. See, e.g., National Research Council, Aeronautics and Space Engineering Board, *An Assessment of Space Shuttle Flight Software Development Processes* (Washington, D.C.: National Academy Press, 1993).

45. The detailed techniques, terminology, and degree of quantification vary somewhat from field to field. For example, in the chemical industry, qualitative hazard and operability studies (HAZOPs) predominate. For a brief account of some salient differences as they stood in the late 1980s, see B. John Garrick, "The Approach to Risk Analysis in Three Industries: Nuclear Power, Space Systems, and Chemical Process," *Reliability Engineering and System Safety* 23 (1988): 195–205.

46. Sol W. Malasky, *System Safety: Technology and Application* (New York: Garland, 1982), 155–157.

47. Garrick, "Approach to Risk Analysis," 201–202. Note, for example, the absence of explicit probabilistic analysis in the case of the Space Shuttle solid rocket booster O-rings, as described by Diane Vaughan, *The Challenger Launch Decision: Risky Technology, Culture, and Deviance at NASA* (Chicago: University of Chicago Press, 1996).

48. Ibid., 203.

49. Dale Hattis and John A. Smith, Jr., "What's Wrong with Quantitative Risk Assessment," in Robert F. Almeder and James M. Humber, eds., *Biomedical Ethics Reviews 1986* (Clifton, N.J.: Humana Press, 1986), 57–105, at 57.

50. Dick Hamlet, "Foundations of Software Testing: Dependability Theory," *Software Engineering Notes* 19, no. 5 (December 1994): 128–139, at 130.

51. E. W. Dijkstra, "The Humble Programmer," *Communications of the ACM* 10 (1972): 859–866, at 864.

52. Robert L. Glass, "Real Time: The 'Lost World' of Software Debugging and Testing," *Communications of the ACM* 23 (1980): 264–271, at 266.

53. William C. Hetzel, "A Definitional Framework," in Hetzel, ed., *Program Test Methods: Computer Program Test Methods Symposium, 1972, University of North Carolina* (Englewood Cliffs, N.J.: Prentice-Hall, 1973), 7–10, at 7. I am grateful to Susan Gerhart for helpful discussions of the history of testing in the 1970s.

54. Fred Gruenberger, "Program Testing: The Historical Perspective," in Hetzel, *Program Test Methods*, 11–14, at 11.

55. A strategy advocated by J. C. Huang, "An Approach to Program Testing," *Computing Surveys* 7, no. 3 (September 1975): 113–128. Among influential early testing papers are John B. Goodenough and Susan L. Gerhart, "Toward a Theory of Test Data Selection," *IEEE Transactions on Software Engineering*, SE-1 (1975): 156–173; and Thomas J. McCabe, "A Software Complexity Measure," *IEEE Transactions on Software Engineering* SE-2 (1976): 308–320.

56. Glenford J. Myers, *The Art of Software Testing* (New York: Wiley, 1979), 73.

57. See T. A. Thayer, M. Lipow, and E. Nelson, *Software Reliability: A Study of Large Project Activity* (Amsterdam: North-Holland, 1978).

58. David L. Parnas, A. John van Schouwen, and Shu Po Kwan, "Evaluation of Safety-Critical Software," *Communications of the ACM* 33 (1990): 636–648, at 644.

59. Bev Littlewood and Lorenzo Strignini, "Validation of Ultrahigh Dependability for Software-Based Systems," *Communications of the ACM* 36 (1993): 69–80.

60. Joe W. Duran and Simeon C. Ntafos, "An Evaluation of Random Testing," *IEEE Transactions on Software Engineering* SE-10 (1984): 438–444.

61. Parnas, van Schouwen, and Kwan, "Safetey-Critical Software," 647.

62. Richard Hamlet, "Random Testing," in John J. Marciniak, ed., *Encyclopedia of Software Testing* (Chichester: Wiley, 1994), vol. 2, 970–978, at 977. The objection is less of a problem for a Bayesian approach to random testing (see below), in that beliefs, prior to testing, about the reliability of the two programs would be expected to be quite different.

63. Ricky W. Butler and George B. Finelli, "The Infeasibility of Quantifying the Reliability of Life-Critical Real-Time Software," *IEEE Transactions on Software Engineering* 19 (1993): 3–12, at 10.

64. Littlewood and Strignini, "Ultrahigh Dependability," 73.

65. See, e.g., Elaine J. Weyuker, "On Testing Non-testable Programs," *Computer Journal* 25 (1982): 465–470.

66. D. M. Hunns and N. Wainwright, "Software-Based Protection for Sizewell B: The Regulator's Perspective," *Nuclear Engineering International* 36, no. 446 (September 1991): 38–40, at 39.

67. International Electrotechnical Commission, Sub-Committee 65A, Working Groups 9 and 10, *Draft IEC 1508—Functional Safety: Safety-Related Systems* (Geneva: International Electrotechnical Commission, June 1995), part 1, 28.

68. Hunns and Wainwright, "Software-Based Protection." The situation was rather more complicated than simple application of the multiplicative law to two systems with failure probabilities of 10^{-3} and 10^{-4}: the safety analysis considered different types of demand on the reactor protection systems, the probabilities of such demands, and the reliability of both the software and hard-wired systems in responding to these demands.

69. J. R. Taylor, Letter from the Editor, *ACM Software Engineering Notes* 6, 1 (Jan. 1981): 1–2; D. J. Martin, "Dissimilar Software in High Integrity Applications in Flight Controls," in *Software for Avionics* (AGARD Conf. Proc. 330, January 1983), 36–1 to 39–9; J. C. Rouquet and P. J. Traverse, "Safe and Reliable Computing on Board the Airbus and ATR Aircraft," in W. J. Quirk, ed., *Safety of Computer Control Systems 1986: Trends in Safe Real-Time Computing Systems* (Oxford: Pergamon, 1986), 93–97.

70. Algirdas Avizienis, "The N-Version Approach to Fault-Tolerant Software," *IEEE Transactions on Software Engineering* SE-11 (1985): 1491–1501, 1495.

71. John C. Knight and Nancy G. Leveson, "An Experimental Evaluation of the Assumption of Independence in Multiversion Programming," *IEEE Transactions on Software Engineering* SE-12 (1986): 96–109, at 103.

72. See, e.g., the comments on the issue in Parnas, van Schouwen, and Kwan, "Safety-Critical Software," 638.

73. This decision was in its turn to be criticized: see Charles Walker, "Boeing Slated over 777 Software Set-Up," *Computer Weekly*, 25 May 1995, 1. I do not know whether doubts over the independence assumption played any part in Boeing's decision.

74. For some important underlying issues here, see David Bloor, "What Can the Sociologist of Knowledge Say about $2 + 2 = 4$?" in P. Ernest, ed., *Mathematics, Education and Philosophy* (London: Falmer, 1994), 21–32.

75. Key early papers were J. McCarthy, "Towards a Mathematical Science of Computation," in C. M. Popplewell, ed., *Information Processing 1962: Proceedings of the IFIP Congress 62* (Amsterdam: North Holland, 1963): 21–28; P. Naur, "Proof of Algorithms by General Snapshots," *BIT* 6 (1966): 310–316; R. W. Floyd, "Assigning Meanings to Programs," *Mathematical Aspects of Computer Science: Proceedings of Symposia in Applied Mathematics* 19 (1967): 19–32; C. A. R. Hoare, "An Axiomatic Basis for Computer Programming," *Communications of the ACM* 12 (1969): 576–580 and 583.

76. For a fuller account of the mathematicization of computer science in the late 1950s and 1960s, see Michael S. Mahoney, "Computers and Mathematics: The Search for a Discipline of Computer Science," in Javier Echeverria, Andoni Ibarra, and Thomas Mormann, eds., *The Space of Mathematics: Philosophical, Epistemological, and Historical Explorations* (Berlin: de Gruyter, 1992), 349–363.

77. This is most explicit in Andrew S. Tanenbaum, "In Defense of Program Testing, or Correctness Proofs Considered Harmful," *SIGPLAN Notices*, May 1976, 64–68.

78. See, e.g., D. MacKenzie, "The Fangs of the VIPER," *Nature* 352 (1991): 467–468, and "Negotiating Arithmetic, Constructing Proof: The Sociology of Mathematics and Information Technology," *Social Studies of Science* 23 (1993): 37–65.

79. D. MacKenzie, "The Automation of Proof: A Historical and Sociological Exploration," *IEEE Annals of the History of Computing* 17 (1995): 7–29.

80. See Ministry of Defence, *Interim Defence Standard 00–55: The Procurement of Safety Critical Software in Defence Equipment* (Glasgow: Ministry of Defence, Directorate of Standardization, 1991).

81. Because much of hardware design can be modeled in terms of the connectives in propositional logic (AND, OR, NOT, etc.), the fast, efficient techniques for

automated propositional logic based on Binary Decision Diagrams have been remarkably effective here. The "stem" paper for much of this work is R. E. Bryant, "Graph-Based Algorithms for Boolean Function Manipulation," *IEEE Transactions on Computers* C-35 (1986): 677–691.

82. For the shape of a possible government program to help overcome them, see G. Cleland and D. MacKenzie, 'The Industrial Uptake of Formal Methods in Computer Science: An Analysis and a Policy Proposal," *Science and Public Policy*, 22 (1995): 369–382.

83. See, e.g., Avra Cohn, "The Notion of Proof in Hardware Verification," *Journal of Automated Reasoning* 5 (1989): 127–139.

84. Hubbell, "You Are under Attack!" 50.

85. Fly-by-wire civil airliners are still aerodynamically stable (unlike some advanced military aircraft), so the possibility exists for reversion to manual control in the event of some categories of computer failure: in that sense, a partial "safe state" still exists.

86. That is not to deny that there are useful guidelines to be found in the literature of this field, and, for example, more generally in the more sociological work of the "Berkeley" high-reliability theorists. For the latter, see, e.g., Gene I. Rochlin, "Defining 'High Reliability' Organizations in Practice: A Taxonomic Prologue," in Karlene H. Roberts, ed., *New Challenges to Understanding Organizations* (New York: Macmillan, 1993), 11–32.

87. MacKenzie, "Computer-Related Accidental Death."

7

Engineers or Managers? The Systems Analysis of Electronic Data Processing in the Federal Bureaucracy

Atsushi Akera

In *The Engineers and the Price System* (1921) Thorstein Veblen called upon engineers to bring about a social revolution. Writing in the midst of the radical sentiment that was spreading across the globe, Veblen weighed the possibility of revolution in industrialized countries and saw the engineer as a new agent of change. After all it was the engineer, he reasoned, who owned the knowledge and experience needed to operate modern industry. Suggesting that it ought to be natural for engineers to equate fairness with a sense of efficiency, Veblen called upon the engineer to end the exploitations of the corporate elite.[1]

Yet if historian Alfred Chandler is right, it was the accountants rather than the engineers who won this war. A new rank of business administrators, marketing managers, and personnel officers learned to employ new techniques of accounting to measure the efficiency of modern enterprise. This use of quantitative tools of measurement allowed them to govern modern enterprise without direct technical expertise. But if Veblen's vision was untenable in the 1920s during management's stellar ascent, it emerged as a salient issue in the U.S. economy of the 1950s. Postwar diplomatic and military situations prompted the nation to build up its research and development capabilities, placing engineers in charge of operations far more vast than they had overseen in the past. Emboldened by their new status, engineers ventured into broader areas of expertise such as marketing, operations research, and project management.[2]

Against this backdrop, I want to address the question of who controlled the introduction of electronic data processing (EDP) systems into the federal bureaucracy. This was a case that lay at the outer boundaries of dispute. Here the technicians were clearly treading upon managerial terrain, employing their knowledge of computers to justify the study of government bureaucracy. But while data processing may sound like a mundane topic requiring little research or technical expertise, early computers had to be reengineered to meet the needs of administrators as opposed to scientists. Moreover, bureaucrats welcomed the technicians' input. Amid a climate of government reform, EDP systems offered a technological solution to the mounting scale of federal paperwork and the associated politics of accountability. What remained to be worked out was the relationship between experts of a technical and managerial bent.

For reasons I will explore below, the early responsibility for advising federal agencies on EDP procurement fell on the shoulders of the National Bureau of Standards (NBS), a technical organization. The word "fell" here is a misnomer. Samuel Alexander, the head of NBS's Data Processing Systems Division, actively sought to establish an EDP advisory service for government bureaucracies. NBS was requested to conduct or participate in the EDP requirements studies of major agencies, including the Census Bureau, the Social Security Administration, and an array of military bureaus and offices. This work coincided with national efforts, such as those of the Hoover Commission, to bring techniques of professional management into the framework of public administration. NBS engineers soon ran up against an official of the federal Bureau of the Budget, who then sought to appropriate the engineer's method of systems analysis. The eventual fate of NBS's effort would also reflect broader outcomes with respect to postwar engineering management as a whole.

Computers and the Politics of Government Organization

Efforts to reform the federal government have a longstanding history. The origins of a U.S. professional civil service date back to the Woodrow Wilson administration, and Herbert Hoover is regarded as one of the early champions of government reorganization and efficiency. A former secretary of commerce, Hoover sought to bring the new tools of the corporation into the government. The New Deal brought a temporary hiatus to this corporatist model of administrative reform. However, the postwar expansion of the government, fueled in part by Cold War escalation, brought renewed efforts to rationalize and professionalize the federal bureaucracy. This was not without opposition. In fact, it was by no accident that the professionalization of public administration remained in its infancy even in the 1940s. The corporate model of a strong executive who oversaw professional bureaucrats threatened legislative authority. Congress was long wary of extending presidential powers, and blocking efforts to strengthen executive authority represented one way to ensure that federal agencies would remain responsive to legislative interests. Nevertheless, as the federal budget grew, there emerged new pressures to demonstrate efficiency and public accountability, playing into the hands of the proponents of administrative reform.[3]

One of the most visible postwar reform efforts was that of the Hoover Commission. Hoover was swept out of office in 1932, and had no opportunity to complete the extensive changes he had sought. Thus, when the problem of government administration assumed renewed urgency, Hoover again made it his personal mission to introduce modern management techniques into the postwar bureaucracy. Hoover's interest coincided with Eisenhower's election in 1952. By then, the scale of the federal bureaucracy was attracting considerable criticism. Bureaucratic reform was thus an item in Eisenhower's election campaign, as it played into the new Republican administration's general preferences for small government. Eisenhower was successful in curbing government expenditures until Sputnik (1957). But he took steps to temper the impact of the Hoover Commission. While Hoover was a fellow Republican and any president was sympathetic to a commission calling for greater executive authority, an investigation into government operations was still a critique of the current administration. Given his political latitude, Eisenhower chose not to give Hoover free rein to proceed with unbridled authority.[4]

Yet although administrative reform was constrained under Eisenhower, electronic data processing had a special place in this debate. This is best seen in the context of the actual paperwork operations of a federal bureaucracy. The Social Security Administration was established in 1935 through one of the New Deal's more permanent legislative acts. Its Bureau of Old Age and Survivors Insurance—what people normally associate with Social Security checks—emerged as a federal office with a massive burden for administrative paperwork, especially after a legislative amendment changed the system of benefits from annuities to monthly checks. The Bureau maintained 48 million active records, which was more than that maintained by the largest insurance company.[5]

From the outset, the Bureau set out to mechanize its operations through the use of IBM punched card machinery. For example, to post the quarterly wage records submitted by employers onto individual workers' ledgers, work began by punching the employers' records of employee earnings onto standard IBM punch cards. These cards were sorted and tabulated to produce a card summarizing each worker's annual earnings, brought to another station, and then paired with and recorded onto each worker's master ledger. Yet despite the partial mechanization, this was still a cumbersome process. Receiving and checking the employers' wage records took nearly 120,000 hours of work. Punching and verifying individual wage records took an additional 160,000 hours. Variations and errors were integral to the process so that "preliminary rejects," which made it necessary to list, reinstate, and balance the record, constituted some 11 percent of the total. All in all, it took some 451,000 hours, or roughly 900 clerks working a forty-hour week, to process all the quarterly records. This constituted the "new" method of recording employee earnings as reported in 1947.[6]

A photograph depicting one part of the Central Planning Unit depicts a room containing some 100 women operating key-punch machines. Such was the situation of federal clerical work that made EDP technology so attractive to bureaucrats. Viewed in the light of fiscal conservatism, the paperwork operations of government bureaucracy had become embarrassingly large. EDP systems promised to reduce or limit further growth in the agency's clerical ranks. This was particularly appealing to agencies like the Social Security Administration whose liberal roots lay in the New Deal programs of the 1930s. More generally, it was about maintaining administrative autonomy. Just as the President sought to gain independence from Congress, senior bureaucrats sought to keep presidential oversight at bay. A smooth-running bureaucracy provided some measure of protection against political critique. Carried forward by their own enthusiasm, bureaucrats sought to obtain their own EDP system.[7]

Meanwhile, NBS became involved with computers through two separate origins. Research mathematicians, by basically riding on the coattails of the nuclear physicists they aided during the war, convinced the military to sponsor a new National Applied Mathematics Laboratory (NAML) at NBS. The second effort, Samuel Alexander's Electronic Computers Group, grew out of the interests of a small group of electronics engineers who worked on proximity fuzes during the war. At first, NAML was given broad authority to oversee a computer procurement program on behalf of other federal agencies. However, this obligation of service bound NAML to a range of expertise, and the knowledge it required came to reside increasingly within the domain of the electronics engineer. Alexander slowly usurped most of NAML's authority over machine development work, and at one point secured the funds with which to build NBS's own stored program computer.[8]

The Standards Eastern Automatic Computer (SEAC), publicly dedicated in June of 1950, was the first modern stored program computer to be put into regular operation in the United States. This was a technical milestone, but diplomatic events—the Soviet detonation of an atomic bomb—launched SEAC into further prominence. There was no other U.S. machine capable of doing important design calculations for the hydrogen bomb.[9]

But while public attention earned the interest of other agencies, machine development work proved to be a dead end within the confines of a national laboratory. The most significant reason was the success of NBS's own computer procurement program, since its very mission was to foster commercial sources of supply. However, political events accelerated the demise. Although NBS was a civilian agency and a unit of the U.S. Department of Commerce, it conducted most of its postwar work through the use of military funds transfers. This stood uneasily among certain factions of the military establishment. NBS won one political skirmish in 1950, but a deadly blow came in the wake of the "AD-X2 battery additive controversy." During 1953, NBS was accused of neglecting its basic standards and product-testing responsibilities by virtue of its excessive involvement with military research. It was instructed to return to its basic mission in standards and testing, and divest itself of all military development work. Military funds transfers to NBS did not cease after 1953. Nevertheless, NBS lost 40 percent of its staff and 60 percent of its budget between 1952 and 1955, mostly as a result of divesting its major military development projects.[10]

The crisis, however, was not a uniform disaster for Alexander. He rode on the wave of the SEAC's success. Also, given the popularity of computers among federal agencies, this was one area the NBS director, A. V. Astin, was unwilling to let go. In fact, Alexander benefited organizationally, if not financially, from the turmoil. In the course of NBS's divestitures and reorganization, Alexander was asked to head up a new computer division at NBS. But it was no accident that Alexander chose to name his new division the *Data Processing* Systems Division. Strongly discouraged from seeking military contracts for machine development work, data processing seemed a safer avenue. He realized that most of the agencies that made use of NBS's procurement service used their computers for data processing. Alexander proceeded to recast his own history, hoping that the computer would then shed some of its military colors. This was important, given that most of the funds for the new division would continue to come from military sponsors. The strong association among computers, bureaucracy, and the politics of fiscal restraint was something Alexander actively sought.[11]

ESTABLISHING AN ADVISORY SERVICE

Alexander did not regard the turn to data processing as an end to hardware research. While circumstances made it difficult to secure the funds to build an entire computer, novel design work and component research were still real possibilities. Tasks such as characterizing germanium crystal diodes or evaluating vacuum tube reliability had been the mainstay of the group's early work, and there were many specialized studies that were not yet within the interests of commercial manufacturers. Alexander first saw the data processing advisory service as a natural extension of the old procurement service. NBS could provide an independent source of advice on what equipment government agencies should purchase from the manufacturer, and there were many agencies that continued to ask for NBS's advice. Moreover, an advisory service created opportunities for securing component and design work. A technical evaluation might lead not only to a recommendation for commercial purchases, but to specialized work on components and to larger configurations and designs that would facilitate an agency's particular needs for data processing.[12]

However, the advisory service turned out to be its own kind of work. Clients first asked for a paper study based on an analysis of the current paperwork operations of the agency. Alexander's engineers were not always suited for this kind of work. It required neither experiments nor engineering design. Instead it called for conducting interviews, sitting through management meetings, and writing extensive reports. In recognizing the vastly very different work requirements, Alexander assembled a small staff under his own direction. When requests continued to pour in from other organizations, Alexander gave the group a formal identity and encouraged it to define new billable projects.[13]

It was not Alexander but his assistant Mary Elizabeth Stevens who came to head up NBS's advisory service. Although she herself had a technical background, Stevens was hired as administrative staff in conjunction with the new data processing studies. While she acquired the title of "administrative officer," Stevens worked alongside a male engineer who was placed in charge of the more technical aspects of the work. In other words, Stevens was relegated to paperwork largely by virtue of her gender. But EDP studies *were* about doing paperwork. Thus, when Alexander became preoccupied with the work of a division head, it was Stevens to whom he began relinquishing his responsibilities. As the significance of NBS's advisory services grew, Stevens's responsibilities grew accordingly.[14]

Both circumstance and effort propelled Stevens forward. One of the early studies in which Stevens played a major role was the U.S. Patent Office mechanization study. The number of patents filed with the Patent Office expanded rapidly after World War II, and substantial lags began to appear in the patent examination process. The increased volume and specialization of patents also made it more difficult for patent attorneys to survey the state of the art. With lawyers complaining in one ear, and Eisenhower touting the advantages of technology in the other, the commissioner of patents and the secretary of commerce became interested in automated patent retrieval.[15]

The secretary of commerce, Sinclair Weeks, asked Vannevar Bush to direct a study of the Patent Office. Bush, a senior statesman of science, assembled a committee of respected experts in which Samuel Alexander was asked to participate. The committee met three times in late 1954. They collected testimony from various organizations, including those that hoped to conduct a systems study on behalf of the Patent Office. However, rather than recommending a study contract, Bush placed his own committee in the role of producing the study. The hearings served as the principal means for collecting the facts by which to draw up recommendations.

Alexander was called on to testify on electronic data processing *equipment*. Bush then proceeded to assign the primary responsibility for drafting the committee's final report to a Patent Office group that was already working on mechanizing the patent retrieval process. However, when this group issued a report recommending complete reliance on older punched card technology, Bush responded negatively to the group's technical conservatism. Bush reconvened the full committee, this time asking Alexander to provide a survey of electronic data processing *systems*.[16]

This was one task Alexander handed down to Stevens, perhaps because his own background was in the technical aspects of computer design rather than how computers were installed and used. Stevens compiled this information based on the substantial experience NBS had already gained, including the experiences from its earlier computer procurement service. She provided a survey of some of the largest data processing installations, complementing this with an account of the latest systems offered by manufacturers. As a result of her efforts, the Patent Office signed an agreement to have NBS study the computerized approaches to patent retrieval. Dubbed the PILOT project, Alexander also placed this task under Stevens's authority.[17]

Toward the end of 1955, Astin, the NBS director, was called upon to describe NBS's research before a congressional subcommittee. Astin asked each division to report on its current work. Alexander took this rather literally, submitting a considerable volume of technical material describing the division's activities. After being called in to a meeting, Alexander had to admit that "he wanted something different from what I expected."[18]

Alexander was scheduled to depart for Europe and could not revise his report in time. Upon mentioning this to Astin, Astin recalled the "yeoman" work Stevens had done during the Patent Office study. Asked by the NBS director to produce a report, Stevens spent the next eight days on this task.

In her report Stevens chose to deemphasize NBS's early machine development work, and proceeded instead to discuss the broader data processing needs of the federal government. She spent some time describing NBS's early role in supporting the interests of the Census Bureau and the Office of Air Comptroller. Borrowing from the plans and procurement language of the comptroller's office, Stevens made it evident that one benefit of data processing systems was to establish better fiscal accountability in government organizations. Stevens found other justifications for data processing research. Reviewing her own work for the Patent Office, Stevens suggested that mechanization was essential, "if the patent system is to continue to make its contribution to our economy." Quoting an assistant commerce secretary, Stevens also spoke of the troublesome increase in clerical workers. But being aware that computers could not always be justified on the grounds of costs alone, she emphasized how "the increased use of automatic data processing systems for paperwork handling should enable us to do things we cannot now do in any other way. . . ." In assembling what were quickly becoming familiar justifications for computers, Stevens was translating the division's efforts into terms that the director and Congress could understand. Stevens found herself promoted to a section chief of a new section on applications engineering.[19]

Although Stevens's career advanced, her advisory service turned out to be difficult to expand. While many agencies sought NBS's advice, few were willing to pay for it. The formal request for NBS's assistance usually came from a senior official, and usually this came with a request for NBS to volunteer its expertise. NBS was obligated, by its charter, to advise other federal agencies, and neither Alexander nor Stevens could refuse the request. Moreover, during the Cold War, science advising fell into a pattern of volunteerism that had specific historical origins. Part of this had to do with the broader strategies of objectivity rearticulated during the period of progressive reforms. But it was also a strategy picked by the postwar scientific community because of its fears of federal influence. The voluntary participation in science advising even by the most senior statesmen of science was about preserving the higher objectivity of technical knowledge. Operating an advisory service under such a circumstance would prove a fiscally challenging task.

From Data Processing to Systems Analysis

The limits of volunteerism and technical expertise are made apparent in a study Stevens conducted for the Department of Health, Education, and Welfare (HEW). This 1958 study was the first time Stevens was exposed to the more formal method of systems analysis. The HEW study also reveals the delicate nature of Stevens's authority when confronted with a formalized engineering management technique.

The HEW study was set up to evaluate the feasibility of automating the paperwork operations of HEW's central administrative offices. As with other studies, the request came from the top, namely the HEW secretary. The subsequent arrangements were then handled by the HEW's Office of Management Policy (OMP). It was probably this office that initiated HEW's interest in EDP automation. The HEW was a relatively new department, having been born of the postwar reforms. In 1953, the decision was made to consolidate federal welfare programs under a single administrative department. And if any group within HEW subscribed to the larger program of rationalization, it was the OMP. Because OMP had no experience with computers, it assembled a panel of voluntary "consultants" from other agencies. Recognizing NBS's experience, OMP asked Stevens to be its chair.[20]

Despite Stevens's title, it was the director of a new computing unit at the U.S. Geological Survey, Frank Reilly, who became the panel's most vocal member. While the details of Reilly's background are not known, Reilly had absorbed the methods of systems analysis that grew mostly out of wartime operations research. By the early 1950s, systems analysis entered into computer engineering circles, both through a technical lineage of seeing computers as computer *systems*, and through military contract practices that saw computers as components within larger weapons systems. Reilly employed his knowledge of systems analysis to design the Geological Survey's new computing facility, using its principles of careful planning to ensure that the facility simultaneously met the needs of the agency's scientific and administrative staff. Confident that this planning technique would work equally well for other agencies, Reilly was one of the panel's most enthusiastic participants.

However, the initial meetings were somewhat of a disappointment. Again, the panel was asked only to evaluate the prospects of automating the existing operations of HEW's central offices. However, given this constraint, the answer had to be that all of the tasks could, at most, be improved through the increased use of tabulating machinery. But this conclusion was frustrating for the panel. After all, in composing a panel of computer experts, OMP also assembled a panel of computer enthusiasts. Thus, even though the existing tasks of the administrative offices could not justify an electronic computer, each panel member saw the more extensive range of data processing activities within the HEW. Properly integrated, these might justify a computer system.

It was at this point that someone, probably Reilly, suggested that the difficulty was one of problem formulation. This was a common strategy in systems analysis. The panel converged upon this idea.

> In the course of these considerations, notwithstanding the initial limitations to the requirements now handled in departmental headquarters, the committee unanimously concluded that other major questions of thorough-going management analysis and systematic exploration of data processing requirements, both administrative and technical, throughout the Department, should, in its opinion, take precedence over the specific questions it was asked to answer.[21]

This did require that the panel seek proper authority to expand its task. This was true, in no small part, because the panel was now asking to perform what was clearly a broad management review. OMP probably recognized the limits of its own authority in defining the initial scope of the study. If there was a tension between the president and the departmental secretaries, every secretary also had to contend with the diverse objectives of his own agency heads. This was especially true for a department like HEW, which was built out of established organizations including the Social Security Administration that were accustomed to their autonomy. However, Stevens obtained the OMP's approval. If OMP had yet to plan a general management review, its disposition was sympathetic to it. Moreover, by suggesting that the complex features of the computer made it necessary to evaluate costs in the context of a general review, Stevens and Reilly rendered the issue of management reorganization into a matter of technical expertise—both to the panel and OMP's benefit. OMP gave the panel authority to "explore these broader questions," with the caveat that this would only be a preliminary analysis.[22]

Given this authority, the panel proceeded to visit various agencies within HEW, complementing a broad survey with a close study of the most computing intensive operations. This survey confirmed some suspicions. Existing work routines could not justify EDP automation. But following the logic of their own approach, the panel turned to the idea that an extensive systems analysis would reveal underlying inefficiencies within HEW bureaucracy.

> The most pressing needs and problems of the data processing users cannot be solved by a mere recourse to ADPS equipment. What is required with urgency and in some cases desperation is a thorough going systems approach to all data processing applications both administrative and scientific.[23]

These were Reilly's words. To support his position, Reilly drew upon evidence he uncovered in HEW's computing activities. After questioning the users of HEW's Statistical Processing Branch (SPB)—a facility that provided statistical services for other HEW offices—Reilly was troubled by the fact that few clients "had awareness of any basic systems planning concepts which should be applied at their work." He then criticized the "very large amounts" of manual preparation done both before and after machine tabulation, and how so many tasks were submitted to the SPB and a similar Medical Research Tabulating Unit before problems were sufficiently thought through. "This is a most inefficient way to use equipment," he observed.[24]

Reilly also came to usurp much of Stevens's authority during the course of the deliberations. Following one meeting, Stevens asked each member to submit his or her thoughts in writing. But Reilly submitted his analysis not as a statement of his opinions, but as the panel's final report. He sent this to Stevens with the somewhat dismissive remark,

> I enjoyed our little excursion into the HEW data processing function and congratulate you on your excellent performance as chairman. You got the show on the road and kept it on schedule. This, we all appreciated.[25]

He then suggested that no further meetings were necessary.

But further work was necessary. More seasoned in producing these reports than Reilly, Stevens recognized that the report's tenor would alienate HEW officials. Beginning with the Patent Office study, Stevens had grown increasingly aware that any analysis of data processing operations could become an implicit critique of the existing bureaucracy. Whatever the net efficiencies, any recommendation to alter procedures could challenge existing managers, and Stevens knew the dangers of stepping on the terrain of entrenched bureaucrats. Reilly, being the head of his own computing unit, had yet to gain such sensitivities.

But while Stevens moved to soften the tone of the report, she remained aligned with Reilly's commitment to a full-scale systems study. In articulating what this would entail, her report suggested that HEW hire a team of ten full-time analysts for six to nine months' time. This would cost $50,000 to $75,000. Systems analysis did not come cheaply. But the cost could be recovered through substantial savings generated from the improved operations. Being aware that systems analysis could transcend organizational jurisdictions, Stevens emphasized that it was critical for the study to have the support of top management, namely the HEW secretary. The report also warned that HEW should eschew short-term gains, and include future requirements in its planning process.[26]

There was one curious omission. As much as the report's recommendations were about automating vast amounts of work, it made little mention of workers. This was a substantial omission, given that EDP automation would, in due course, eliminate substantial categories of clerical work. Moreover, there was nothing inherently incompatible between systems analysis and a study of the labor process. Indeed, a complete systems analysis had to include the detailed study of paperwork, since EDP systems could be installed only by articulating, transforming, and automating those procedures that could be encoded into a machine. While not all clerical tasks were eliminated by the computer, a wholesale transformation of a labor process within the federal government might have drawn the panel's consideration.

Being a group of volunteers, the panel had neither the time nor resources by which to conduct a detailed study of clerical work. But there was more to it than that. The panel had a broader objective, namely to present systems analysis as a new scientific method applicable to clerical work. In its report, the panel compared systems analysis against an older tradition of methods analysis. This was a technique, born of an earlier movement in systematic management, that was used to design the kind of IBM tabulating facilities established at the Social Security Administration. In comparing the two, the panel suggested that systems analysis led to systemwide efficiencies, whereas methods analysis provided only local gains. Methods analysts would have probably disagreed. But more important, the new claims for systems analysis ran exactly parallel to those offered by earlier proponents of scientific management.

Indeed, the parallels between scientific management and systems analysis are inescapable. In both cases, engineers were brought in as consultants on matters of managerial reform. Although technical knowledge was a wedge, proponents of scientific management used this entry to advocate the systematic evaluation of factories and organizations as the first step toward efficiency. Historical studies of Taylorism reveal that time study was only one component of the overall reforms, and that systemwide planning and efficiency was, in principle if not practice, the foremost aim of scientific management. Recent historiography also suggests that the success of scientific management hinged as much on a consultant's ability to sell the idea to top management as on their ability to avert the concerted opposition of labor. As consistent with this particular point, Stevens's report did not address the tension between management and labor, but the one between the engineering and management professions. Her claims for systems analysis was a technical expert's appeal to legitimacy in matters of managerial review.[27]

Still, Taylor did not ignore the workers. But the panel was dealing with a very different labor force. Taylor confronted the labor aristocracy. In contrast, Stevens and Reilly were dealing with a largely feminized, white-collar workforce that was often presumed to be unskilled and unspecialized. As pointed out by Richard Edwards and others, female clerical workers occupied a secondary or even tertiary labor market that bore the brunt of distinctions and discrimination. Moreover, even Taylor first presumed a docile response on the part of the skilled machinist. Similarly, early EDP systems analysts took their control of the labor process to be a matter of fact.[28]

This did not matter, since the panel failed to receive the audience it sought. Stevens tried to make the report more palatable to HEW officials, especially by suggesting the possibility of a more limited initial effort. However, there is no indication that the HEW secretary or the OMP heeded any of the panel's recommendations. HEW established the panel with the idea that it would perform the analysis necessary to justify an HEW computing facility. When the panel returned with a recommendation for yet another study, its recommendation was probably just dropped. NBS, in any case, was not asked to perform or participate in a larger systems analysis.

Enter the Accountants

The same reform movement that created an interest in EDP systems also created an impulse to limit their procurement. The Hoover Commission, with its tentative support from Eisenhower, established a Paperwork Task Force to address the perennial issue of paperwork reduction. This task force spawned the Business Machines Group, which in turn assembled a Subgroup on Electronic Data-Processing Machines. Despite extensive deliberations, nothing much came of this subcommittee except for its recommendation to establish a permanent Interagency Committee on Automatic Data Processing.[29]

This Interagency Committee came to rival NBS as the main unit in the government responsible for promoting and overseeing the introduction of EDP machinery. Because this was an interagency committee, it had broad representation. It included many organizations enthusiastic about EDP equipment, including military agencies. But there was now a critical voice in their midst. Beginning with the record of delays and cost overruns that accompanied early computer development work, officials of the Bureau of the Budget (BoB) and the General Accounting Office (GAO) were familiar with the inflated budgets that were typical of the field. This was beginning to raise alarms, and both BoB and GAO chose to have representatives on the Interagency Committee. This committee also reported to the BoB.[30]

The Interagency Committee served a number of functions, not all of them addressed toward oversight. It served as a forum for the exchange of technical knowledge. It provided a place to conduct negotiations for sharing and borrowing equipment, which was itself a concession to increasing concerns about the duplication of research. Much of the detailed work of the committee was conducted through task forces assembled to produce specific studies. One early study was on the coordination of scientific literature on EDP research, while later work dealt with such topics as establishing a government software library and the effects of automation. But another early task force was set up to study the problem of EDP *procurement*. This task force, established in late 1958, was assigned to produce a set of procurement guidelines for those interested in acquiring a computer.[31]

But the chair of the Interagency Committee, the BoB management analyst William Gill, had a separate agenda. Immediately following World War II, BoB consolidated its fiscal authority over other government agencies. Having secured the authority to review and modify each agency's budgetary requests, BoB proceeded to push for much broader authority over managerial and organizational reform. The early 1950s brought mixed results. BoB retained much of its fiscal authority, but it lost its early gains over administrative reform. This occurred at the hands of the Hoover Commission. BoB was criticized for its own labyrinthian management structure, and its powerful Division of Administrative Management was dismantled in 1952, leaving a smaller Office of Management Organization (OMO) in its place.[32]

Gill was a member of this office. OMO was trying to forge new strategies for carrying out administrative reforms, and they had begun to articulate a new program in bureaucratic training, leadership, and coordination. It was no accident that Gill presided over an interagency committee that reported to the BoB. This too was one of OMO's new strategies. But Gill soon realized what Stevens and Reilly had seen. Namely, an EDP procurement study, and especially a systems study, could itself be a managerial review of agency operations. Gill became interested in refashioning systems analysis into an instrument by which BoB could regain its leadership over administrative reforms. With the enthusiasm of Reilly, as opposed to the restraint of Stevens, Gill assumed the chair of his own task force.

Gill also chose to exceed the task force's initial authority. The task force was originally instructed to establish procurement guidelines only for the central agencies of the government. This basically meant the executive office of the president, along with some of the major civilian agencies. But in sensing the gross inefficiency that might result from a piecemeal approach, Gill chose—in this case by his own authority—to produce a procurement guideline applicable to the entire federal government. This would include all agencies of the Department of Defense.[33]

Gill attempted to legitimize his study by designing a wide and balanced investigation. He distributed the investigative work among the task force members, each one being asked to interview an operating official and a technical staff member at several different agencies. Altogether, the task force assembled information from eighty units of the government at the bureau and agency level. The task force completed its interviews before the year's end, and then reconvened to assemble their collective intelligence.

Gill then drafted the final report. He invoked what, by then, was the typical formula for systems analysis. Each agency interested in procuring a computer was instructed to set up a systems study with multiple study phases. It would begin with a preliminary study whose recommendation, he emphasized, could never be to procure a computer system. It could recommend, at most, a full-scale systems study of the organization's EDP requirements. Gill also recognized, as Stevens had done, that a systems study would transcend organizational jurisdictions, and thus require top management support. However, given his managerial training, Gill explained this using the principles of line and staff management.

Gill assembled an array of other, familiar recommendations. A systems study was to aim for "vastly different systems" for handling data rather than merely replacing existing punched card procedures. Each agency was to hire a permanent staff of systems analysts to conduct such studies, and to be "given the strength and stature it deserves." Because there was a dearth of trained systems analysts, the government also had to establish an interim program for training systems analysts. For the sake of efficiency, the government also had to have an agency with the authority to coordinate EDP research, and to establish a central index of current EDP research. The report's frequent references to systems studies and systems analysis belies an enthusiasm found only in a novice to the field.[34]

The tensions between the management and technical professions could not have been spelled out more clearly in Gill's report. Thanking the "technicians" for their early service, Gill called on the management profession to reassume direct authority over the systems analysis of EDP. Early EDP experiments were technically valuable, but they produced numerous and expensive mistakes due to their repeated failure to conduct proper systems studies. Gill blamed senior government officials rather than the technicians for this mistake. But in exonerating the technicians, Gill was simultaneously appropriating systems analysis as the rightful domain of management.

Gill did not go so far as to suggest that BoB conduct these systems studies. This would have been a clear violation of the well-coveted separation between the different units of the government. But after consulting with the Civil Service Commission, whose charter included that of training federal employees, Gill won the concession that BoB could serve as the interim office for training new systems analysts.

Gill may not have realized the full implications of his report. Placed in perspective, Gill had mandated the review of the entire federal government's bureaucratic operations. He laid out the procedures through which this would be done, and took steps to install management analysts sympathetic to BoB's point of view. EDP procurement was no longer just about acquiring a computer system. BoB would withhold computer procurement as a means of compelling a general management review. This was a tail wagging some very big dogs.

Most agencies rejected Gill's recommendations. One Department of Defense official wrote that the office could not turn over its information concerning EDP procurement decisions on the basis of national security alone. A Census Bureau official, while politely endorsing the report's intent to coordinate federal EDP activities, criticized the report's general tone and its tendency to "police" rather than advise the agencies. Both letters failed to comment on the legitimacy of systems analysis as a technique for managerial review. This was probably an explicit choice, since shifting the report's evaluation onto familiar political terrain was a rhetorical strategy for dismissing the study as a whole.[35]

Gill simply failed to take into account the politics of the situation and the limits of BoB's authority. However, systems methodologies, upon which Gill's own study was based, were misleading in this respect. By employing the systems method, Gill had grasped the entire federal bureaucracy on a piece of paper and envisioned an orderly world in which the BoB would oversee the process of bureaucratic rationalization. Gill tried to make some concessions to politics. But the promise of efficiency gained through systems analysis produced a self-legitimating image, which obscured the transgressions Gill lacked the power to overcome.

Yet amid this hostile reception, the National Bureau of Standards, which more than any other group was the "technicians" Gill referred to in his text, responded with cautious optimism. While NBS questioned the timing by which a transfer of duties would take place, it basically conceded that the systems analysis of the federal bureaucracy was ultimately the responsibility of the management profession. Why this concession?[36]

What survives in the historical record is a copy of Gill's report that Alexander circulated among his staff, accompanied by handwritten comments from several NBS staff members who read the report. From these comments and the markings on the report itself, what is clear is that Alexander and his staff were drawn to one other recommendation that appeared in Gill's report. Namely, Gill had proposed an experimental service bureau, which would provide data processing services to agencies interested in setting up an EDP facility. This experimental bureau could take on actual data processing jobs on an interim basis, thereby providing realistic measurements of the cost and procedural difficulties associated with EDP automation. NBS's PILOT project, set up for the Patent Office, was already modeled on such principles, and NBS intended to extend this service to other agencies.[37] As one of his staff members suggested, this was an "opportunity for NBS to be of unusual service during this experimental service bureau period. [I] think that the benefits of BoB attention, IF NBS DOES A GOOD IMAGINATIVE JOB, will be of lasting benefit."[38]

NBS was no passive victim. Moreover, NBS's enthusiasm can be better understood from the point of view of its finances. The budgetary constraints at NBS forced Alexander and his Data Processing Systems Division to continue to use the SEAC at a time when IBM was introducing its third high-end model of scientific computers. NBS needed to have a larger computer to expand the PILOT project. But it also needed a new computer to do the broader work of the Data Processing Systems Division. While reminding themselves that they should avoid becoming a "rent-a-truck" operation, NBS was willing to schedule routine data processing operations on behalf of other agencies in exchange for a state-of-the-art computer. The value of the Bureau of Budget's approval did not escape Alexander or his staff.[39]

The apparent consent to managerial authority was strictly a matter of negotiated exchange. NBS was ready to cede its leadership in EDP advisory services because of what it could gain in the deal. But in taking this stance, Alexander also saw to it that his group was reverting to and reinforcing traditional distinctions between engineering and managerial work.

What is missing from the comments is Stevens's voice. Although EDP advisory services were her responsibility, she remained silent in the written exchange. It is possible that Alexander acted pragmatically, killing Stevens's project as readily as he created it based on the larger interests of his division. But the two had a close working relationship. More likely, the two had a verbal conference that simply left no written records.

There is reason to believe that Stevens also consented to the decision. After all, the advisory services, despite all their popularity, turned out to be difficult to expand. And although a service bureau would transfer overall responsibility for systems analysis to BoB, it would allow NBS to expand its EDP services on a fully cost-reimbursed basis. Albeit, this work would run more along technical lines. But Stevens's staff had already made this turn. In addition to the work on developing an indexing system for patent retrieval, Stevens's staff was studying statistical techniques and foreign language representation in an effort to extend the system's capabilities. Stevens came from a technical background herself, and was not wedded either to management or systems analysis by her early training.[40]

In any case, Gill's report came to naught. Facing criticism on almost every front, his task force was disbanded and a new group established in its place. The final guidelines that did emerge endorsed the principle of systems analysis as it was spelled out in Gill's report. But there were no teeth to the recommendations. The report emerged as a set of guidelines to be followed at the discretion of each agency. BoB was given no statutory authority to oversee agency reviews or to engage in staff training. The guidelines suggested only that each agency set up its own EDP automation task force, and,

> Such reviewing authorities may contact the ADP staffs of the Bureau of the Budget or the General Accounting Office for consultation or suggestions as to methods of reviewing studies. However, it should be understood that in all cases the decision to acquire and utilize ADP equipment rests solely with the agency involved, subject to its internal departmental or other controls.[41]

NBS did eventually get its computer and service center. But as the capabilities and limitations of new data processing equipment became more familiar to engineers and bureaucrats alike, NBS continued to cede its authority over management and organizational studies in favor of more technical lines of work. With support from the National Science Foundation, Stevens and her staff turned to such areas as machine translation, optical character readers, and other topics in information processing. With this shift, Stevens's section entered a line of work that fell more squarely within the emerging boundaries of the new discipline of computer science. In fact, the decisions made at NBS contributed, in small part, to the constitution of these very boundaries.[42]

Analysis and Conclusion

If events pertaining to electronic data processing developments at the National Bureau of Standards are indicative of the conduct of postwar science and engineering as a whole, then they reveal the many complexities to be found in this particular crucible of history. Within the course of events just described were shifts in notions of computing expertise, the reconstruction of professional boundaries, and redefinitions of the role of a federal laboratory in the development of a new technology. National politics, technical enthusiasm, and the new method of systems analysis all had their place in this unfolding history. In the end, it was less about success or failure than about the particular roles different experts carved out for themselves during postwar advances in computer technology.

The very flexibility of this history, and particularly of how people composed themselves in the midst of their circumstance, is certainly suggestive of symbolic interactionist literature. Especially early works in the field, such as Howard Becker et al.'s *Boys in White* (1961), with its notions about careers and occupational identity, seem pertinent on many counts.[43] Still, even without this layer of theory, Mary Stevens's efforts to carve out a successful career within a government laboratory, and the manner in which she composed herself, present a complex yet credible account of the kinds of choices people face in the course of their work lives.

The danger, nevertheless, remains in placing too much emphasis on flexibility. After all, despite the many opportunities Stevens found in her career, both gender and professional identities influenced the course of her actions. Dominant practices, such as the volunteerist approach to postwar science advising, delimited the ground upon which Stevens's EDP advisory services could take shape. Even something as simple as the formal organization of NBS—its hierarchy of division, sections, and groups—served to reinforce disciplinary boundaries while at the same time permitting managers like Samuel Alexander to promote or demote specific programs subordinate to his level of authority. If history has its many turns, they are not without their direction.

In fact, symbolic interactionism has been frequently criticized as an astructuralist theory. Part of this critique comes from the field's origins, since the theory gained its momentum by rejecting structural functionalism as typified by Talcott Parsons. Yet interaction and social structure are not necessarily antagonistic notions. There is no reason why prevailing beliefs and practices, along with more fleeting circumstance, cannot be seen to influence an individual's actions and hence the course of historical events. Yet despite recent advances in the field, symbolic interactionism remains a field where no extensive theorization of social structure has taken place.[44]

Without aiming to resolve this at the level of theory, I suggest that a detailed historical narrative, such as the one provided here, provides important clues toward this reintegration. The important thing to note is that many of these so-called structural influences are quite familiar things. It is about gender prejudices, traditions, and professional identities, and even the material culture surrounding old tabulating machines. What does need to be noted is that these influences do not always sit easily within the categories of social structure normally found in social theory. Distinctions between macro, mezzo, and micro sociology do have their heuristic value, but ultimately they are the product of academic distinctions with no direct grounding in history. The sedimentary metaphor they invoke instead poses the danger of the illusion that there are fixed and immutable structures at the foundations of each society. It is hardly surprising that, when pressed to a corner, it always proves difficult to point to exactly what the most fundamental structures of a society are.

This is not to say that there aren't different kinds of social knowledge, some more persistent or more pervasive than others. NBS, as a bureau of the U.S. Department of Commerce and as an organization built within a larger industrial capitalist society, experienced recurring pressures to define a line of research compatible with industrial interests. Free market ideologies, despite their lack of precise articulation, pervade

U.S. institutions. Likewise, pervasive notions of class, which specifically have no clear articulation among the U.S. middle class, fueled the tension between the engineering and management professions.

Other influences are more ephemeral, or are grounded in a specific history. Eisenhower's 1952 election, with its emphasis on fiscal restraint, had a series of repercussions for administrative reforms. Yet these reforms also referred to a longer history of government reform, and to a specific reform movement initiated by Herbert Hoover during the 1920s. Hoover himself embodied this history, carrying its principles and strategies into the postwar years. Ultimately, it is necessary to also reject the traditional distinction between structures and events, focusing instead on the specific histories of ideas and actions that both define and reshape society.

Symbolic interactionism remains particularly useful in reflecting upon the complexity of historical change because, as a body of theory, it begins with a close study of the specific sites of interaction where social reproduction and transformation occurs. Knowledge and beliefs provide a sense of direction to a narrative not by forcing any particular outcome, but through their recurrent and intersubjective incorporation into individual actions. It is therefore not that notions of class, labor, or a free market determine an event. It is rather how such ideas, combined with more local knowledge, are brought to bear at a particular point of action. History, in the end, is underdetermined not because of a lack of influences, but because its very profusion generates choice.[45]

Consider for example Alexander's decision regarding Gill's report. Alexander decided to cede authority over EDP *advisory services* to the *management profession*, judging the *political* value of BoB's endorsement of a *state-of-the-art* computer to be more important to the overall *research* program of his *Data Processing* Systems Division, as compared to the entry they made into the *systems analysis* of federal *bureaucracies*. Still, as a *technical organization*, he consulted other members of *his staff*, seeking to establish *consensus* in support of his decision as the division *head*. The italics here are somewhat arbitrary. Yet in looking more closely at this instance, Alexander's understanding of such concepts, presumably with their resonance to issues of professional interest, institutional stature, advanced research, and staff relations, were all invoked in rendering his final decision. There is ample evidence in the historical records to show that all such concerns were brought to bear.

Yet, at the same time, invoking a decision—here to cede EDP advisory services—meant reinforcing certain associations at the expense of others. Notably, action is both constructive and selective in the very same instance. Charles Sanders Peirce, along with others who more recently have taken the linguistic turn in studies of culture, is correct to point out that a sign can carry multiple meanings. Symbolic interactionism adds to this by getting at how these constructions, and deconstructions, constantly take place at the site of everyday acts. Moreover—and unlike those poststructuralist analyses that describe society at successive moments to unveil the possibility of change—the historian approaching a series of actions will see, for better or for worse, those ideas and distinctions which not only perish, but persevere.[46]

What ideas or values were constructed, sustained, or chipped away during the course of NBS's involvement with EDP advisory services? The answer is necessarily as complicated as the history itself. But the framework of theory is useful in understanding the central question as framed in this chapter. Namely, given that the final outcome sustained the distinctions between technical and managerial work, why did tensions emerge over this issue? Cast more generally, why were encounters over disciplinary authority so often an issue in early computing developments?

I suggest first that all new technologies open up new epistemological possibilities that require the renegotiation of existing meanings and responsibilities. Neither the traditional definition of an engineer nor manager sufficed to provide the kind of organizational analysis pertaining to the installation of an electronic data processing system. If a computer was going to be installed in an office, it had to be defined what role technical people would play in its installation. The eventual plethora of computing specialists—computer scientists, database specialists, network technicians, professional indexers, data processing directors, and data entry clerks, just to name a few—reveals the extent to which roles had to be defined and reworked during the field's relatively brief history.

But I also suggest that the computer was a special kind of device. Once the technical community coalesced around the notion of a *general purpose* computer, it had in fact opened up a flexible epistemological space that was subject to recurring questions of authority. General purpose computers contained a paradox in that by standardizing the central devices that performed computation, it was necessary to always reconfigure and encode the machine to satisfy the needs of each application. Who would assume what aspects of this act of translation were always open to dispute? Moreover, given the flexibility of the computer, such disputes necessarily recurred within each site and discipline where the modern computer was applied.

It was precisely the size of this epistemic space that made systems methodologies so popular in the field of computing. So long as role definitions remained fluid, members of a given profession or field had ample opportunities to test the boundaries of their own authority. In this context, systems methodologies, including systems analysis, provided a particularly attractive means by which members of a technical profession could lay claims for their ability to address issues that were broader than their present domain of expertise. Both Stevens and Gill employed systems concepts in this manner, much as systems concepts were employed in fields such as operations research, cybernetics, and even ecology.

But the use of a systems methodology did not guarantee its success. Whatever the technical merits of an evaluation, the success of a systems study hinged on its ability to understand and incorporate the politics of a situation. In this account, Alexander understood this better than any other individual. Stevens successfully reduced it to a body of practice. The Bureau of the Budget's William Gill, a novice to the field, was most clueless in this respect. But as much as politics determined immediate outcomes, systems studies, and decisions rendered during the course of their evaluations, could nevertheless realign professional boundaries and responsibilities.

Not surprisingly, as roles stabilized, systems analysis and its allied techniques assumed an increasingly technical pallor within the engineering disciplines. For example, while systems engineering and systems analysis did emerge as specialized techniques within computer and software engineering work, its textbooks now place far less emphasis on problem reformulation and the potential for social reorganization.[47] The requirements of a particular application—an automated teller machine network, for example—is taken more or less as given, and the work directed more toward reconfiguring artifacts to meet a predefined concept of service. Devoid of historical context, this might appear as the myopia of the technical professions. But what the case of EDP systems analysis and NBS's negotiations with BoB suggests is that the limited vision of the technical professions was itself a product of historical choices that were already played out in the past.

The power of systems analysis, however, was that its use was not confined to the technical professions. As indicated by Gill's enthusiasm, and despite his specific failure, systems methodologies found other, more powerful enthusiasts within the management profession. Thus, in contrast to the limited vision of computer engineers, systems methods found greater extension in the defense procurement studies and the social programs of the Kennedy and Johnson administrations. These were very different spaces where the boundaries of professional authority remained yet to be fixed.[48]

Notes

1. Thorstein Veblen, *The Engineers and the Price System* (New York: Viking, 1936; first published in 1921).

2. Alfred Chandler, *The Visible Hand: The Managerial Revolution in American Business* (Cambridge: Harvard University Press, 1977), especially 444–448. For a direct study of corporate paperwork and accounting, see Sharon Strom, *Beyond the Typewriter: Gender, Class, and the Origins of Modern American Office Work, 1900–1930* (Urbana: University of Illinois Press, 1992). For an entry into the subject of Cold War science and engineering, see Stuart Leslie, *Cold War and American Science: The Military-Industrial-Academic Complex at MIT and Stanford* (New York: Columbia University Press, 1993); and Roger Geiger, *Research and Relevant Knowledge: American Research Universities since World War II* (New York: Oxford University Press, 1993). While there are a number of canonical works that describe U.S. industrial research, including Leonard Reich, *The Making of American Industrial Research: Science and Business at GE and Bell, 1876–1926* (Cambridge: Cambridge University Press, 1985), few works describe the postwar industrial research laboratory. On project management and operations research, see Thomas Hughes, *Rescuing Prometheus* (New York: Pantheon, 1998); and Stephen Johnson, "Three Approaches to Big Technology: Operations Research, Systems Engineering, and Project Management," *Technology and Culture* 38 (1997): 891–919.

3. For a general account of the politics of government organization, see Frederick Mosher, *A Tale of Two Agencies: A Comparative Analysis of the General Accounting Office and the Office of Management and Budget* (Baton Rouge: Louisiana State University Press, 1984). See also Louis Galambos, ed., *The New American State: Bureaucracies and Politics since World War II* (Baltimore: Johns Hopkins University Press, 1987); Brian Balough, "Reorganizing the Organizational Synthesis: Federal-Professional Relations in Modern America," *Studies in American Political Development* 5 (Spring 1991): 119–172, especially 141–167.

Public administration historians correctly describe how this tension is rooted in the notion of the separation of powers, and how this trend toward a stronger administrative state had a difficult political origin. Frequently mentioned in this context is the Brownlow report of 1937, which advocated a strong administrative presidency. Franklin Roosevelt's Executive Order 8248 (1939) established the Executive Office of the President, and much of the expanded executive authority was placed in the Bureau of the Budget (BoB). BoB is central to the latter part of this article. See Mosher, *A Tale of Two Agencies,* 63–69. On Hoover, see Ellis Hawley, ed., *Herbert Hoover as Secretary of Commerce: Studies in New Era Thought and*

Practice (Iowa City: University of Iowa Press, 1981). Hoover explicitly brought the tenets of scientific management into public administration. See Charles Walcott and Karen Hult, "Management Science and the Great Engineer: Governing the White House during the Hoover Administration," *Presidential Studies Quarterly* 20 (Summer 1990): 557–579.

4. There were actually two, not one "Hoover Commissions," both of which operated under the formal title of the Commission on Organization of the Executive Branch of the Government. For an account of the Hoover Commissions, see Ronald Moe, *The Hoover Commissions Revisited* (Boulder, Colo.: Westview Press, 1982). The larger context for administrative reorganization can be seen in Frederick Mosher, *A Tale of Two Agencies*; and Gary Dean Best, *Herbert Hoover: The Post-presidential Years, 1933–1964* (Stanford, Calif.: Hoover Institution Press, 1983).

5. On the Social Security Administration, see Edward D. Berkowitz, *Mr. Social Security: The Life of Wilbur J. Cohen* (Lawrence: University Press of Kansas, 1995).

6. "Operations" and other handwritten documents; and Federal Security Agency, Social Security Board, Bureau of Old Age and Survivors Insurance, Accounting Operations Division, "The Establishment and Maintenance of Wage Records" (Baltimore, Md.: Bureau publication, n.d.). Both in Record Group 167, Records of the National Institute of Standards and Technology, Applied Mathematics Division, Electronic Computers 1939–54, Records (Hereafter AMD-ECR), box 3, folder: The Heat Problem. National Archives and Records Administration, National Archives II, College Park, Md. (hereafter, NA-II). These items were apparently filed in the wrong folder. Federal Security Agency, Social Security Administration, Accounting Operations Division, "The New Method of Recording Employee Earnings," 1 July 1947. NA-II, AMD-ECR, box 3, folder: Federal Security Old Age Insurance.

7. The central political debate over Social Security under Eisenhower had to do with the size and scope of entitlement, and not the efficiency of its bureaucracy or of its paperwork operations. As discussed by Berkowitz in *Mr. Social Security*, postwar economic prosperity created an environment favoring the expansion of Social Security, and even Eisenhower conceded to popular demand. EDP equipment could have been instrumental in producing statistical data conducive to the kind of "mechanical objectivity" described in Theodore Porter's *Trust in Numbers: The Pursuit of Objectivity in Science and Public Life* (Princeton: Princeton University Press, 1995), but I did not discern such discussions within the sources I examined.

8. For an account of Samuel Alexander and his Electronic Computers Laboratory, see chap. 4 of my Ph.D. dissertation, "Calculating a Natural World: Scientists, Engineers, and Computers in the United States, 1937–1968" (University of Pennsylvania, 1998). For a general history of NBS, see Rexmond Cochrane, *Measures for Progress: A History of the National Bureau of Standards* (New York: Arno Press, 1976).

9. On the public dedication of the SEAC, see "SEAC National Bureau of Standards Eastern Automatic Computer," 20 June 1950; and E. U. Condon, "Address

at the Dedication of the National Bureau of Standards...," 20 June 1950. Both CBI 50, Edmund Berkeley Papers, box 8, folder 26A, University of Minnesota, Charles Babbage Institute for the History of Information Processing, Minneapolis, Minn. (hereafter, CBI). On the AEC's use of the SEAC, see NBS, Applied Mathematics Advisory Council, Minutes, 11 May 1951. CBI 45, box 7, folder 34. Although the work was classified, the veil of secrecy was itself a mechanism that drew attention to the machine.

10. On NAML's procurement program, see NBS, "The National Applied Mathematics Laboratory of the National Bureau of Standards: A Progress Report Covering the First Five Years of Its Existence," written by J. H. Curtiss, 1 April 1953, 10–12, CBI 45, Margaret Fox Papers, box 1, folder 2. While these developments are also described, in part, in my dissertation, this is also the subject of current study by historian Pap Ndiaye. His observations, especially with regard to the political economy of NBS's postwar machine procurement program, precede my own. For primary sources on the 1953 controversy, see Churchill Eisenhart, "Chronology of the Battery Additive Controversy and Its Aftermath," which appears in the finding aid for NA-II, RG 167; Committee for the Evaluation..., "Minutes of First Meeting, 29 April 1953"; and Ad Hoc Committee for Evaluation..., "A Report to the Secretary of Commerce," 15 October 1953. Both NA-II, RG 167, Records of the National Institute of Standards and Technology, Records of Allen Astin (hereafter, Astin), box 5, folder: Kelly Committee, Correspondence and Reports. NBS's expenditures fell from $52.5M to $20.0M between 1952 and 1955 (−62 percent), while staff fell from 4,781 to 2,800 (−41 percent). Rexmond Cochrane, *Measures for Progress*, Appendix F. The details of NBS's response to the cutbacks can be seen in its annual reports to the Visiting Committee, in NA-II, Astin, boxes 40 and 41.

11. Ad Hoc Committee on Evaluation, "A Report to the Secretary," 18; also S. N. Alexander, "Division 12—Data Processing Systems" (1955); and E. W. Cannon, "Division 11—Applied Mathematics" (1955). Both in NA-II, Astin, box 41, folder: A—NBS Visiting Committee, 1955.

12. S. N. Alexander, "Division 12—Data Processing Systems" (1955).

13. Alexander was reporting the "Analysis of Computer Applications" as a billable project within the Electronic Computers Laboratory (Section 12.3) as early as 1951. NBS, "Scientific Program, Fiscal Year 1952," NBS Report (1951), 342; NBS, Division 12, Data Processing Systems, "Projects, Fiscal Year 1956," NBS Report (1955). Both CBI 45, box 1, folder 4.

14. M. E. Stevens, "First Annual Progress Report on Applications of Electronic Data Processing Techniques to Supply Management Problems, July 1 1953 through June 30 1954," NBS 3786, September 1954. NA-II, AMD-ECR, box 1, loose item.

15. Correspondence pertaining to the Patent Office study appears in three folders, CBI 45, box 3, folders 19–21. The formal committee was named the Advisory Committee on the Application of Machines to Patent Office Applications. The minutes of its meetings appear in folder 19.

16. Advisory Committee. . . , "Third Meeting, November 8, 1954," Minutes. n.d. CBI 45, box 3, folder 19.

17. Vannevar Bush to Norman Ball (NSF; Executive Secretary to Advisory Committee), 19 November 1954; Advisory Committee. . . , "Fourth Meeting, November 22, 1954," Minutes; Norman Ball to Members, Advisory Committee. . . , "Draft Text of Final Report," 2 December 1954; Joint Patent Office–NBS Committee, "News of Subcommittee on Recommendation 1, Meeting of December 29, 1954," n.d.; all in CBI 45, box 3, folder 20.

18. S. N. Alexander to Mary Stevens, Memo, 28 September 1955. CBI 45, box 8, folder 40.

19. Mary Stevens to A. V. Astin, Memo, 6 October 1955; "Testimony of Dr. A. V. Astin. . . ," draft, 27 October 1955. Both in CBI 45, box 8, folder 40. NBS, Division 12, Data Processing Systems Division, "Projects, Fiscal Year 1957," NBS Report (1956), 32–41. CBI 45, box 1, folder 4.

20. Richard Simonson to Advisors on Automatic Data Processing System Analysis, 27 June 1958, and other documents in CBI 45, box 2, folder 19.

21. Ad Hoc Committee of Consultants on Data Processing to Rufus Miles (Director of Administration, HEW), with attached report, "Report of the Ad Hoc Committee on Automatic Data Processing Analysis," 5 November 1958. CBI 45, box 2, folder 19.

22. Ibid.; and Frank Reilly, "Summary of Preliminary Analysis of ADPS Requirement in HEW," draft, 15 July 1958, CBI 45, box 2, folder 19.

23. Frank Reilly, "Summary of Preliminary Analysis," 15 July 1958.

24. Ibid.

25. Frank Reilly to Mary Stevens, 15 July 1958. CBI 45, box 2, folder 19.

26. "Report of the Ad Hoc Committee. . . ," 5 November 1958.

27. Daniel Nelson, *Frederick W. Taylor and the Rise of Scientific Management* (Madison: University of Wisconsin Press, 1980). Brian Price, "Frank and Lilian Gilbreth and the Motion Study Controversy, 1907–1930," in Daniel Nelson, ed., *A Mental Revolution: Scientific Management since Taylor* (Columbus: Ohio State University Press, 1992). See also Kathy Burgess, "Organized Production, Unorganized Labor: Management Strategy and Labor Activism at the Link-Belt Company," in the same collection.

28. As cited by Strom in *Beyond the Typewriter*, an account of management's assumption that women's work was hardly differentiated can be found in Harry Braverman, *Labor and Monopoly Capital: The Degradation of Work in the Twentieth Century* (New York: Monthly Review Press, 1975), as well as in works by labor economists, including David Gordon, Richard Edwards, and Michael Reich, *Segmented Work, Divided Workers: The Historical Transformation of Labor in the United States* (Cambridge: Cambridge University Press, 1982).

29. William Gill, who headed up this Interagency Committee and its task force for producing procurement guidelines (described below) referred to the technology as "automatic data processing." While this phrase also appears consistently throughout the committee's reports and correspondence, I will use the phrase "electronic data processing" (EDP) for the sake of clarity. Paperwork Task Force (Hoover Commission), "Report of Electronic Data-Processing Machines Subgroup, Business Machines Group," 15 October 1954. CBI 45, box 1, folder 25.

30. Both also had representation on the Hoover Commission's EDP subgroup. A history of the BoB and GAO can be found in Mosher, *A Tale of Two Agencies*; and Roger Trask, *GAO History, 1921–1991* (Washington, D.C.: U.S. General Accounting Office, History Program, 1991).

31. This series of studies by the Interagency Committee can be found in CBI 45, box 2, folder 19: Projects—Automatic Data Processing, Interagency Committee, 1958–72.

32. Mosher, *A Tale of Two Agencies*, 63–76, 99–107.

33. Given that the Bureau of the Budget was given general authority to establish an "organizational plan" for the entire executive branch, Gill simply argued that the action "was taken as a means of rounding out and further clarifying a total Executive Branch organizational plan for the ADP program"; William Gill, "A Report of Findings and Recommendations Resulting from the ADP Responsibilities Study," draft, January 1959. CBI 45, box 2, folder 19.

34. Ibid.

35. The responses to Gill's draft report, including Bureau of the Census, "Comments on a Report of Findings and Recommendations Resulting from the ADP Responsibilities Study," 16 February 1959, can be found in CBI 45, box 2, folder 19.

36. R. D. Huntoon to O. H. Nielson (Director, Office of Budget and Management, Department of Commerce), 16 February 1959. CBI 45, box 2, folder 19.

37. This is the copy of Gill's report that appears in CBI 45, box 2, folder 19; Huntoon to Nielson, 16 February 1959.

38. [Anonymous] Handwritten note. CBI 45, box 2, folder 19.

39. E. W. Cannon and F. L. Alt to R. D. Huntoon (deputy director, NBS), 13 February 1959; various handwritten notes; Huntoon to Nielson, 16 February 1959. All in CBI 45, box 2, folder 19.

40. Stevens's changing responsibilities can be seen in the division's annual reports, CBI 45, box 1, folder 4, and in S. M. Newman to Mary Stevens, 8 April 1958, and other documents in box 1, folder 11.

41. Executive Office of the President, Bureau of the Budget, "Guidelines for Studies to Precede the Utilization of Automatic Data Processing Equipment. Recommendations Constituting the Report of the Interdepartmental [*sic*]

Committee on Automatic Data Processing," draft, 9 October 1959. CBI 45, box 2, folder 19.

42. Stevens's "Projects" files are a substantial part of the Margaret Fox Papers, and are found in CBI 45, boxes 3–4; see also box 1, folder 10, NBS Activity Reports—Monthly, RICASIP, 1964.

43. Howard Becker et al., *Boys in White: Student Culture in Medical School* (Chicago: University of Chicago Press, 1961).

44. John Johnson, Harvey Farberman, and Gary Fine, eds., *The Cutting Edge: Advanced Interactionist Theory* (Hampton Hill, U.K.: JAI Press, 1992); Howard Becker and Michal McCall, eds., *Symbolic Interaction and Cultural Studies* (Chicago: University of Chicago Press, 1990).

45. This comment, illustrated by the next paragraph, generally follows along the lines of a semiotic analysis of social interaction as outlined by Norman Denzin in "On Semiotics and Symbolic Interaction," in Johnson et al., eds., *The Cutting Edge*, 283–302. Also important is the dramaturgical model of Victor Turner, which was brought into symbolic interactionist studies. This has received comments by Bruce Wilshire in "The Dramaturgical Model of Behavior: Its Strengths and Weaknesses," which appears in the same volume, 243–254.

46. For Peirce, see Charles Peirce, *Peirce on Signs: Writings on Semiotics*, edited by James Hoopes (Chapel Hill: University of North Carolina Press, 1991).

47. See, for example, C. J. Koomen, *The Design of Communicating Systems: A System Engineering Approach* (Boston: Kluwer, 1991); Amer Hassan et al., *The Elements of System Design* (Boston: Academic Press, 1994); and the array of perspectives that can be found in the proceedings of any systems engineering conference, such as Arthur Morrison and John Wirth, eds., *Systems Engineering for the 21st Century: Proceedings of the Second Annual International Symposium of the National Council on Systems Engineering* (Seattle: National Council on Systems Engineering, 1992).

48. David Jardini's chapter in this volume explores these issues.

8

The World in a Machine: Origins and Impacts of Early Computerized Global Systems Models

Paul N. Edwards

Introduction

The idea of "the world" is probably as old as language, with as many meanings and connotations as there are cultures. For much of human history "the world" must have seemed boundless, beyond the grasp of mortal understanding. Even long after the Scientific Revolution, when "the world" had become for many an immense but finite globe, comprehending the forces that act upon it as a whole—as a system—remained for the most part beyond reach.

Although scientists began to develop theories of world-scale processes at least as early as Copernicus, grounded empirical knowledge of geophysical features and processes remained in a rudimentary state until the Second World War. This was even more the case in the social arenas of politics, economics, and culture. The short-lived era of the League of Nations notwithstanding, it is really only since the Second World War that "the world" has *become* a system in political, economic, and cultural terms. Phrases like "global economy," "global village," "world system," and "global warming," mundane elements of modern discourse, would have been little more than rhetoric even in 1939.

Without the articulated, developed concepts of global systems that underlie these phrases, "the world" might remain in an important sense merely a symbol or icon. Yet in our society, we are regularly exhorted by bumper stickers and political leaflets to "think globally," to see ourselves as part of a larger, *knowable* whole. How did "the world" become a system? What kind of science made it possible to know the planet as a unit, to disentangle the vast array of interlocking forces that determine its characteristics as a system? As the politico-economic world became increasingly integrated, from where did the conceptual tools for grasping that integration come? This chapter begins to address these questions by looking at the first concerted efforts to simulate complex, global dynamic systems. I focus on *comprehensive mathematical simulation models* of *global dynamic processes*. Unlike maps, globes, and other simple models with one or a few dimensions, these models attempt to capture all of the major elements of time-evolving, world-scale systems.[1]

The chapter explores the origins of two types of simulation models: (1) numerical models of weather and climate, and (2) world dynamics models (offshoots of a general method known as "system dynamics"). In the 1950s, the forecasting of weather by means of mathematical models became feasible for the first time with the advent of electronic digital computers. By the 1960s, increasing computer power made possible detailed simulations of the general circulation of Earth's atmosphere. This, in turn, allowed scientists to simulate weather and climate—genuinely global systems. During approximately the same period, computer pioneer Jay Forrester created techniques for simulating the dynamic behavior of large socio-technical systems. He began in the late 1950s with factories, proceeded to cities, and published a book on the general "principles of systems" in 1968. Finally, in the early 1970s, he modeled "world dynamics." Under Forrester's aegis, the MIT System Dynamics Group used world-dynamics models as the basis for the controversial best-seller *The Limits to Growth* (1972).[2] According to the System Dynamics Group, imminent global collapse could be prevented only by sweeping, long-term, world-scale planning, based on computer modeling. In the 1980s these two kinds of models became directly linked, in part through concerns over human impacts on the global environment, such as global warming and ozone depletion.

While their historical origins and purposes are largely distinct, these two kinds of world-modeling practices overlapped in some interesting and surprising ways. Here I outline these origins and overlaps, focusing primarily on the period 1945–72. Bringing them together reveals patterns in the construction of "the world" through a systems approach to science and science-based policy.[3]

The notion of "system" played a more important role in the world dynamics case, where it served as a central, explicit organizing concept (even, some have said, an ideology), than in weather and climate modeling. But the actual models had much in common. In both cases, feedbacks among system components lay at the core of the models' design. In both cases, computers made possible the numerical solution of nonlinear dynamic equations and the vast iterative processing required to simulate the interaction of many closely linked factors over long periods of time. In both cases, the acquisition and coding of global data for model validation, calibration, and forecasting became a central, critical, and controversial issue. Finally, in both cases, computer models became the core not only of an epistemology of global systems, but also of a new policy analysis paradigm.

PART 1. 1945–56: COMPUTERS FOR SIMULATION AND CONTROL

Simulation: Weather Prediction, Weather Control, and Climate Models

Military agencies became deeply interested in the universe of possibilities opened up by digital computers after World War II. As was the case in many other areas, this interest developed through a process I have called "mutual orientation." Scientists and engineers who had worked on the military projects of the war continued, after the war, to consider military problems as part of their research agenda. The fact that most research funding in the postwar period came from military agencies amplified this effect. Scientists and engineers oriented their military sponsors toward new techniques and technologies, while the agencies oriented their grantees toward military applications. This "mutual orientation" mostly produced general directions rather than precise goals. Often there was only a very loose linkage between funding and military utility, rather than a tightly focused dollars-for-products regime.[4]

One area of mutual orientation was numerical weather prediction (NWP). Weather prediction has great military value, since weather affects virtually every aspect of battlefield operations (especially, in World War II and after, air warfare). Indeed, the data networks that made possible the first numerical weather predictions came into existence because of the need of World War II military aviators for information about conditions in the upper atmosphere. The much-increased extent and more systematic character of upper-air observations—using radiosondes (weather balloons incorporating radio transmitters), aircraft, rockets, and other techniques—gave weather forecasters the crucial ability to map activity at several atmospheric levels. Radar, another World War II product, also increased the observational abilities of meteorologists, for example, through rawinsondes (radar-tracked radiosondes).[5]

If weather could be predicted, it might also be controlled, and this too could have profound military implications. General George C. Kennedy of the Strategic Air Command claimed, in 1953, that "the nation which first learns to plot the paths of air masses accurately and learns to control the time and place of precipitation will dominate the globe."[6] Cloud seeding by dry ice and silver iodide, discovered and developed in 1946–47, seemed to some—including Nobel Prize winner Irving Langmuir, a major proponent—to offer the near-term prospect of complete control of precipitation. Respected scientists, both American and Soviet, believed in the mid-1950s that a new struggle for "meteorological mastery" had become a salient element of the Cold War arms race. Proponents of weather control frequently drew analogies between the energy released by atomic weapons and the (even larger) energy contained in weather systems, and they sought the ability to alter climate as a possible weapon of war. By January 1958, in the aftermath of Sputnik, *Newsweek* magazine warned readers of "a new race with the reds" in weather prediction and control.[7]

Numerical Weather Prediction

The key to numerical weather prediction (the only part of this story I will pursue here) was the digital computer, itself a product of World War II military needs. The chief American computer project—the University of Pennsylvania ENIAC—was designed for the Army's Aberdeen Proving Ground, where ballistics table production had fallen far behind schedule, but it was not completed until the fall of 1945. John von Neumann, a member of the ENIAC team and also a consultant to the Manhattan Project, suggested the ENIAC's first application: a mathematical simulation of a hydrogen bomb explosion.

Von Neumann also foresaw the computer's application to weather prediction. The hydrogen-bomb problem and the issue of weather prediction were conceptually linked, as both were essentially problems of fluid dynamics, an area of particular scientific interest to von Neumann.[8] But von Neumann's concern with the weather prediction issue was connected to the bomb in other ways as well. Like many other scientists in the postwar years—and as a Hungarian emigré with bitter memories of communism—von Neumann remained dedicated to the application of science to military problems as the Cold War intensified.[9]

Von Neumann became deeply interested in weather prediction after World War II–era encounters with Carl-Gustav Rossby, a leading meteorologist, and Vladimir Zworykin, an RCA electrical engineer involved in meteorological instrumentation work.[10] Early in 1946, while the ENIAC was still churning out the Los Alamos H-bomb calculation, von Neumann attended a meeting of the U.S. Weather Bureau and began to plan for work on the weather prediction problem at Princeton's Institute for Advanced Study (IAS). Under grants from the Weather Bureau and the navy and air force weather services, he assembled a group of theoretical meteorologists at the IAS.

Among the IAS Numerical Meteorology Project's initial goals was "to examine the foundation of our ideas concerning the general circulation of the atmosphere, with the intention of determining the steady state of the general circulation of the atmosphere and its response to arbitrarily applied external influences."[11] The group spent the next three years developing numerical methods for various aspects of general-circulation and weather problems. Under von Neumann's direction, in late 1949 the group prepared to perform the first computerized weather prediction using the ENIAC. In March and April 1950, members of von Neumann's team carried out two 12-hour and four 24-hour retrospective forecasts using observational data. The calculations required about 800 hours of ENIAC computer time, with each 24-hour forecast taking about 24 hours to perform once the methods had been settled and the programs debugged.[12] Results, while far from perfect, seemed to justify further work. By 1954 the civilian Weather Bureau, the Air Weather Service of the Air Force, and the navy's Naval Weather Service had established computerized weather prediction programs. Routine computerized national weather forecasting began in Sweden in 1954, and in the United States in 1955.

NWP models, then as now, work by breaking up the atmosphere into a set of "grid boxes," tens to hundreds of kilometers square and hundreds to thousands of meters deep. Within each grid box, conditions such as temperature, humidity, and pressure are assumed homogeneous. The models simulate what happens to the air mass in each grid box on a periodic "time step" (today, typically about twenty to thirty minutes; in early models, sometimes as much as three hours). For example, part of a warm air mass might move upward into an adjacent grid box, becoming cooler in the process. Thus the model as a whole simulates how the specified initial conditions will change over time. However, the chaotic nature of weather processes limits forecasts based on this method to a maximum of about two weeks.

The 1950 ENIAC forecasts used some 270 grid points approximately 700 km apart, laid out in a two-dimensional 15 × 18 grid covering North America and much of the surrounding oceans at a single high-altitude atmospheric level. The forecasts used a three-hour time step. Wind speed, wind direction, vorticity, and barometric pressure were the sole forecast elements. A subsequent experimental forecast, using a somewhat more refined model, employed a grid of 361 points spaced approximately 320 km apart, while the initial Weather Bureau production forecast models used a 600-point grid at three altitudes.[13] These resolutions meant that only gross factors affecting weather (regional barometric pressure, temperature, and high-altitude wind speed and direction, for example) could be predicted. As computer models slowly matured, human forecasters used a combination of the computer-produced regional weather maps, radar images, and especially their own experiential, "subjective" knowledge of local conditions in preparing their forecasts.

Gathering Global Data

One of the great challenges of this era turned out to be the collection of data and, especially, its entry into the computer in a form suitable for calculations.

International agreements to share weather observations date to the 1878 founding of the International Meteorological Organization (IMO), one of the oldest intergovernmental organizations in the world. Telegraph, and later telephone, radio, and teletype, allowed rapid transmission of information. Long before World War II, standard coding systems had been worked out that allowed relatively smooth, coordinated transfer of this information. In 1947, as part of the general organization of the United Nations, the IMO became the World Meteorological Organization (WMO)—from the point of view of this chapter, a significant name change that marks a new, global approach.[14] A central purpose of both organizations was to develop and promote data distribution systems and standardized observational techniques.[15]

With the advent of computerized NWP came new needs for data and for ways to handle them. As computer models grew in sophistication, they required information about the state of the atmosphere from ground level to very high altitudes and from locations as near as possible to points on their regular grids, which covered very large areas (continents, eventually hemispheres). Data collection for global forecasting suffered particularly from a lack of observation stations over the oceans, which cover some 70 percent of the Earth's surface, and from irregularly spaced stations, especially in sparsely populated areas such as Siberia, Canada, and much of the Southern Hemisphere.

As for data entry, at the outset of computerized weather prediction "gathering, plotting, analyzing and feeding the necessary information for a 24-hour forecast into a computer [took] between 10 and 12 hours," with another hour required for computation, plus additional time for transmitting the results from the central computer facility at Suitland, Maryland, to local forecasters.[16] Well into the 1960s, much weather data were hand recorded and hand processed before being entered into computers.[17] Much of this work, such as the interpolation of grid-point data from hand-drawn maps, was difficult, time-consuming, and error-prone. Despite international standards for data coding, data distributed in potentially machine-readable form, such as teletype, often arrived in a Babel of different formats, necessitating conversions.[18] A great deal of available data was never used at all, since the time required to code it for the computer would have delayed forecasts beyond their very short useful lifetimes. In 1962, a WMO report predicted that "the principal meteorological benefits of high-speed automatic computing machines during the next few years will lie as much in the processing of large assemblages of data as in numerical forecasting."[19]

Meteorologists had always engaged in the "smoothing" of data, in which errors and anomalous data points were eliminated from the data. In the 1950s, this process was based both on "subjective" readings of observations (using the meteorologist's judgment to reject probable errors) and on explicit theories of large-scale atmospheric behavior. Another standard practice was the interpolation of intermediate values (in both time and space) from known ones. Both smoothing and interpolation were now increasingly automated; as computerization continued, virtually all of it was automated. The methods themselves did not really change, but their automation forced meteorologists to develop explicit, computer-programmable theories of error, anomaly, and interpolation. The effect was simultaneously to render invisible the data "massage" necessary for forecasting.[20]

In other words, the data themselves were subjected to processing based on models of physical behavior (for interpolation), and/or on the needs of NWP models for correctly gridded and time-stepped data. Eventually, the NWP models themselves were actually used to "produce" standard data sets. Today, for example, the widely used twice-daily atmospheric analyses of the National Meteorological Center (in the United States) and the European Center for Medium-Range Weather Forecasts "incorporate observational data from both the surface and from satellites into a 4-D data assimilation system that uses a numerical weather prediction model to carry forward information from previous analyses, giving global uniformly gridded data."[21] Thus the twice-daily periods of actual observation are transformed into 24-hour data sets *by computer models*; these data sets, in turn, become inputs to other weather and climate models.[22]

Models as an Experimental Domain

In the period I have been discussing (1945–56), the utility of digital computers—huge, power-hungry, unreliable, and expensive—remained an open question for many. According to William Aspray, von Neumann "regarded [the computer's] application to meteorology as the crucial test of its scientific value, in large part because the hydrodynamics of the atmosphere is a prime example of those complex, nonlinear phenomena that were previously inaccessible to mathematical study."[23] The success of NWP proved this value, which involved changes much more fundamental than faster calculation. As one of its key members, Jule Charney, explained in an address to the National Academy of Sciences in 1955, "the radical alteration that is taking place [in meteorology] . . . is due . . . to [the computer's] ability to serve as an *inductive device*. . . . The machine, by reducing the mathematical difficulties involved in carrying a physical argument to its logical conclusion, makes possible the making and testing of physical hypotheses in a field where controlled experiment is still visionary and model experiment difficult, and so permits a wider use of inductive methods."[24] In other words, modelers had already begun to experience the appeal of computer models as an alternative experimental domain.

Having successfully fostered the emerging program of numerical weather prediction, von Neumann and his colleagues turned their attention to modeling the atmospheric general circulation (i.e., its global motion and state) in 1953. By mid-1955 Norman Phillips had completed the first, two-layer "quasi-geostrophic" model.[25] Von Neumann and Harry Wexler of the U.S. Weather Bureau immediately proposed a substantial research program on numerical methods for the general circulation problem. In response, the Weather Bureau created a General Circulation Research Section, under the direction of Joseph Smagorinsky; his group eventually became the Geophysical Fluid Dynamics Laboratory (GFDL), now located at Princeton University. Starting in 1959, the laboratory developed a nine-level hemispheric general circulation model (GCM).[26] Groups at UCLA and Lawrence

Livermore National Laboratory also began building GCMs around the same time.[27] In addition to their weather prediction applications, GCMs would later become the crucial proving grounds—the experimental domain—for theories of anthropogenic climate change.

As the grid scales of weather models and their time-steps began to shrink, and as meteorologists sought to model the entire globe, the lack of global uniformly gridded data increasingly became a problem. By the end of the period discussed here, it was beginning to dawn on atmospheric scientists that the core issue of their discipline had been turned on its head by the computer. Whereas in the very recent past, through the data networks built for World War II military aviation, they had rather suddenly acquired far more data than they could ever hope to use, now they could see that in the not-too-distant future they might not have enough—at least not in the standard formats (computer processable), from the right places (uniform grid points), at the right times (uniform time-steps). The computer, which had created the possibility of NWP in the first place, now also became a tool for refining, correcting, and shaping data to fit the models' needs.

Control: Computers for Military Systems

While von Neumann was leading the drive to build weather and climate models, another computer pioneer of similar stature, Jay Forrester, was learning to apply computers to a much different kind of problem. Like von Neumann's, Forrester's project had military sponsors. But his became much more directly focused on a Cold War military problem: the defense of North America against nuclear-armed Soviet bombers. Forrester was among the first to envision the application of computers to an integrated information and control system on a continental scale.

The Semi-Automatic Ground Environment (SAGE) system, conceived by Forrester and others in the early 1950s and completed by 1961, was capable of using radar data to plot intercept courses for fighter aircraft automatically, of remote control of the latter's autopilots to guide them to their targets, and even in principle of controlling the release of air-to-air missiles. SAGE marked the first effort to apply computers to large-scale problems of real-time *control*, as distinct from calculation and information processing.[28]

Control has since become one of the primary applications of computers. But in the 1940s, when Forrester initiated the digital computer project known as Whirlwind, this was far from an obvious use of the new technology. It was dismissed or resisted by many for reasons ranging from reliability problems and expense to the availability of alternatives (primarily analog techniques).[29] Long before Whirlwind became the core of SAGE, Forrester had sought out and studied possible military applications, in a process of "mutual orientation" like that described above vis-à-vis weather prediction and control. His source of funding (the Office of Naval Research and the Navy Special Devices Center), the political climate, and their personal experiences oriented Forrester's group toward military applications, while the group's research eventually oriented the military toward new concepts of command and control. When the first Soviet atomic test and the outbreak of the Korean War shook the nation's confidence in its defenses, Forrester was ready. In the new political context, Forrester's then-unusual agenda of digital computers for control could suddenly fill a vast, newly perceived gap.

In building the SAGE system, Forrester and his engineers were dealing with issues quite similar to the problem faced by meteorologists of this period, albeit in a much different domain. They too needed to gather data on an enormous (continental) scale, and they needed to find a way around the immensely time-consuming and error-prone activity of human reading and interpretation of data. They solved the former problem with a widely dispersed network of radar stations as far north as the Arctic Circle; eventually, just as with meteorology, much of the data for the system came from satellites. Like the meteorologists, the SAGE engineers solved the problem of automatic data conversion by creating new ways—sensors, interpretation programs, modems—to acquire and input data in numerical form. In both cases time pressures were intense; although weather prediction is not *quite* a "real-time" problem, like air defense, every hour's delay between data collection and forecast output rendered weather predictions less useful.

In the late 1950s and beyond, SAGE spawned dozens of similar computerized real-time command-control systems, including SACCS (the Strategic Air Command Control System), the many computer systems built for NORAD (the North Atlantic Air Defense Command), NADGE (the NATO Air Defense Ground Environment), and WWMCCS (the World Wide Military Command and Control System). These projects extended the SAGE concept to create a world-encompassing surveillance, communications, and control system. It is perhaps, then, not surprising that Forrester—like von Neumann before him—eventually turned to modeling the world, as we will see below.

Part 2. 1957–69: Building Global Models

By 1957, computers were exhibiting the remarkable combination of steep climbs in capability and declines in price that have characterized their development ever since.[30] As capabilities increased, so did expectations. Computer-based numerical weather prediction had become a routine element of U.S. weather forecasting, although NWP models were still continental (vs. hemispherical or global) in scale. Von Neumann's long-term goal of modeling climate was beginning to seem feasible. Meanwhile, the SAGE system was on the verge of its first operational tests, in 1958. Both NWP and SAGE proved the value of computers, in different but related ways: NWP for near-real-time simulation models of physical processes, SAGE for real-time data analysis and control of complex human-machine systems. Within a decade, computer simulation techniques were adopted by many other sciences—including the social sciences, where models of economies and cities were just two of the myriad applications. Part 2 discusses the maturation of global general circulation models, on the one hand, and the precursors of world dynamics in Forrester's turn to socio-technical modeling, on the other, during this period.

Climate Models Mature

The Geophysical Fluid Dynamics Laboratory (GFDL) was the direct descendant of von Neumann's vision for weather and climate modeling. During the 1960s, this laboratory developed the crude models of Phillips and Smagorinsky into the first global, three-dimensional atmospheric GCMs. The National Center for Atmospheric Research (NCAR) in Boulder, Colorado, founded in 1960 under NSF sponsorship, developed expertise in climate modeling a few years later.

GCMs are simply global versions of the NWP models discussed above. When used to forecast weather, they are initialized with observational data and run to simulate short periods (up to two weeks). But GCMs can also be used to study climate. In this case, they are usually *not* initialized with actual weather data. Instead, they start with physical constants such as the amount of solar energy striking the atmosphere, the speed of Earth's rotation and wobble on its axis, and the atmosphere's chemical composition. The models are run until they achieve a stable ("equilibrium") state, understood to be their model "climate."[31] This simulated climate may then be compared to observed climatological averages of the actual Earth. Physical parameters may be altered to simulate other planets, to simulate climates of ancient times ("paleoclimate"), or to project future trends. Thus climate models are simulations in a more profound sense than are numerical weather forecasts.

Even more than weather prediction, climate modeling demands enormous computational power. A typical model must compute changes for many thousands of grid boxes some 20,000 times to simulate the global climate for a single year. In 1971, GFDL's Syukuro Manabe estimated that a model with a 500-km grid and nine atmospheric layers required about 120 hours of computer time to simulate a single year of climate.[32] By the early 1980s, supercomputers had reduced the time to about twelve hours per simulated year.[33] Computer speed limitations meant that early climate simulations were usually run for only a simulated few months—at most, for a year or two. But because climate is by definition a long-term (multiyear) statistical average, because of the strong effect of the ocean "heat sink" on atmospheric behavior, and because the models take several "years" to settle into reliable patterns, runs of 20–100 years are necessary to determine their equilibrium states.[34] Even in the early 1980s, a single complete GCM run thus required 1,200 hours—fifty continuous days—of expensive supercomputer time. Despite vast increases in computer power, full runs of today's state-of-the-art GCMs still require hundreds of supercomputer hours, since modelers add complexity to the models even more rapidly than computers improve.

By the end of the 1960s, GCMs had become the central tool of climate science, despite their still-primitive state. By that point, as we will see below, the issue of carbon dioxide accumulation in the atmosphere had started to acquire some urgency among a small but influential group of scientists. This combination set the stage for the still-continuing debate over human-induced global climate change.

The International Geophysical Year

As computerized NWP models began to dominate weather forecasting in the late 1950s, with GCMs looming on the scientific horizon, the need for global data sets became increasingly apparent. The first attempts to construct genuinely global data networks for meteorological observation came with the International Geophysical Year (IGY). During the "year" between July 1957 and December 1958, scientists from some fifty nations conducted global cooperative experiments to learn about world-scale physical systems, including Earth's atmosphere, oceans, ionosphere, and geological structure. Among the major scientific participants and organizers was the WMO.

As the IGY began, the theory of carbon dioxide (CO_2)-induced global warming was becoming the focus of considerable scientific attention. Scientists—primarily oceanographers—had begun to explore the fate of carbon dioxide released into the atmosphere by fossil-fuel consumption. They knew that the oceans absorbed CO_2, but whether and how fast the oceans could absorb all of the fossil-fuel carbon remained in question. In the early 1950s Hans Suess used the radioactive carbon injected into the atmosphere by nuclear blasts to trace the circulation of carbon from fossil fuels, concluding that some but not all of this excess carbon remained in the atmosphere.[35] A widely cited article by Gilbert Plass, published in 1956, aroused renewed interest in carbon dioxide as a factor in climate change.[36] Suess and Roger Revelle, head of the Scripps Institute of Oceanography, predicted that fossil fuels might soon induce a rapid change in world climate. In a now-famous phrase, they wrote that humanity was conducting, unawares, "a great geophysical experiment" on the Earth's climate.[37]

To track the "experiment's" progress, Revelle proposed to build a monitoring station for atmospheric CO_2 at Mauna Loa, Hawaii, as part of the IGY. This station, initially opened in Antarctica, has operated on a continuous basis ever since. It is the chief source of what is probably the only undisputed fact in the global warming debate: the steady rise in the atmospheric concentration of CO_2, from about 280 ppm at the beginning of the industrial era in the mid-nineteenth century to about 370 ppm today. Revelle and others also sought to build a global network of monitoring stations for atmospheric chemistry.[38]

The IGY's meteorological component focused most of its attention on the global general-circulation problem. Three pole-to-pole chains of atmospheric observing stations were established along the meridians 10°E (Europe/Africa), 70°–80°W (the Americas), and 140°W (Japan/Australia). Dividing the globe roughly into thirds, these stations coordinated their observations to collect data simultaneously on specially designated "Regular World Days" and "World Meteorological Intervals." An atmospheric rocketry program retrieved information from very high altitudes. An extensive effort was made to gather information about the Southern Hemisphere from commercial ships, as well as (for the first time) from the Antarctic continent. Data from all aspects of the IGY were deposited at three World Data Centres: one in the United States, one in the Soviet Union, and a third divided between Western Europe and Japan.[39]

The IGY efforts thus represent the first global data networks for constant, consistent, structured observation on a scale and grid to match the emerging atmospheric models. Some, but not all, of these efforts continued after 1958; rising Cold War tensions during this period undoubtedly contributed to their incomplete success. Within a few years, Revelle's hopes for a global atmospheric-chemistry network fell by the wayside, and even the Mauna Loa station experienced severe funding difficulties.[40]

Nevertheless, the cooperative activities of the IGY began a trend toward global programs such as the World Weather Watch (WWW) and the Global Atmospheric Research Program (GARP). Conceived about 1960, WWW took another decade to enter into practice. It coordinated global data collection from satellites, weather rockets, ocean buoys, ocean-launched radiosondes, and commercial aircraft as well as conventional observing stations. This was necessary, according to one participant, because "currently conventional methods ... will never be sufficient if the state of the atmosphere over the whole globe is to be observed at reasonable cost with the time and space resolution which can be used with advantage in computer-assisted research and forecasting."[41] GARP had its roots in a 1961 proposal by John F. Kennedy to the United Nations for "further co-operative efforts between all nations in weather prediction and eventually in weather control."[42] It, too, took most of the decade to implement. With the participation of American, Soviet, and many other nations' scientists, GARP sponsored a series of regional and global observations and experiments.[43] Data gathered by these programs were especially important for global weather and climate models, because they included the first detailed observations of equatorial weather, which (in part since it is so constant and predictable) had never been carefully observed.

Mirror Worlds: Models and Data

By the end of the 1960s, global, three-dimensional climate models had emerged as the central tool of climate science. Modelers had begun to speak routinely of "experiments with the models." Just as with weather prediction, acquisition of global uniformly gridded data became a necessity.

But because of the long-term nature of climate processes, climate models posed especially severe data problems. How could they be empirically validated? The seasonal cycle, because of its extreme variability compared with climate change, provides a well-known, reasonably well-understood benchmark against which to test climate models, but this is not enough. Unlike weather models, which are easily checked against observations over very short periods, comparing climate models with reality demands data on long time scales—ideally, at least 100 years.

Records of land and sea surface temperature exist for large areas over the last hundred years. But changes over time in thermometer quality, location, number, and measurement technique create uncertainties. For example, most thermometers are located on land and clustered in and near urban regions, where "heat island" effects raise local temperatures above the regional average. Temperature records at sea come primarily from shipping lanes, ignoring the globe's less traveled areas. Records from the atmosphere above the surface exist only for the last few decades, but until quite recently these too came mostly from populated land areas in the Northern Hemisphere. Paleoclimatic data from a variety of "proxy" sources (tree rings, ice cores, fossilized pollen, etc.) are also available, though naturally the accuracy and level of detail in this data is far lower than in direct instrumental observations. Model inputs can be set to the different conditions (orbital precession, trace gas concentration, etc.) of past periods and validated by how well they simulate the paleoclimatic record.

During the 1960s, orbiting weather satellites began to provide the first truly global pictures of the atmosphere. The global coverage of satellite data makes them almost fatally attractive to climate modelers. "We don't care about a beautiful data set from just one point," one modeler recently told me. "It's not much use to us. We have one person whose almost entire job is taking satellite data sets and putting them into files that it's easy for us to compare our stuff to."[44] Yet even the satellite data are problematic. They provide only proxy measurements of phenomena at low altitudes, which may be distorted by optical effects and orbital drift.[45] In addition, their lifespans are short (two to five years), and their instruments may drift out of calibration over time.

The way in which most of these problems were resolved was by filtering the actual observations through other models, which smoothed, interpolated, and gridded the scattered, uncertain, and often absent data. Some of this was done by hand or by eye, but in many situations —and routinely in the case of satellite data—computer models provided automatic conversion of instrument readings into standard data sets. Without this much lesser known and appreciated class of intermediate models, validation and calibration of NWP models and GCMs could not proceed. In this seemingly paradoxical mirror world, data used to validate one class of models are themselves partly the product of other models. There was (and is) no real alternative.

From Industrial Dynamics to World Dynamics

Jay Forrester's "world dynamics" models were to gain an enormous influence as the basis for *The Limits to Growth*, a popular best-seller that predicted the catastrophic collapse of socio-economic systems by 2050.[46] Their origins lay in Forrester's mid-career shift from computer engineering to management science, where he brought his expertise with comprehensive computerized systems to bear on the problems of factories, cities, and eventually the world as a whole.

As the research phase of the SAGE project drew to a close in the mid-1950s, Forrester grew restless. He had enjoyed a virtually unique position as one of the most important pioneers of digital computing. Now he sought some new area in which he could once again be a pioneer.[47] The opportunity came when he was invited to join MIT's Sloan School of Management, whose mission was to develop a "scientific" approach to management. There he began to explore the problem of the causes of cyclical change in factory production and employment. He published his first study (of General Electric factories in Kentucky) in 1958. There he argued that a company should be viewed "not as a collection of separate functions but as a system in which the flows of information, materials, manpower, capital equipment, and money set up forces that determine the basic tendencies toward growth, fluctuation, and decline."[48] By 1961 he had refined this approach into a comprehensive, model-based theory of company activity he called "industrial dynamics."

Forrester's industrial models typically showed a "roller-coaster effect,"[49] later known as an "overshoot-and-collapse" mode. Production and employment, for example, tended to rise high, then rapidly fall back to very low levels, creating employee layoffs and idle factories in a "boom-bust" cycle. This pattern, according to Forrester, had less to do with cyclical change in the larger economy than with the built-in "delays" and "amplifications" in the "information-feedback system" of company management. Managers, inclined to attribute such fluctuations to external factors, typically tried to dampen these swings. But the policies they introduced often, Forrester believed, actually made the roller-coaster effect worse.

The introduction to *Industrial Dynamics* (1961) demonstrates the profound influence of the first phase of Forrester's career on his later management-science work. Over and over he points to the automation *and simulation* of military systems by means of digital computers as the root of current modeling capabilities. The industrial dynamics approach was built, he wrote, on "four foundations" which were "primarily . . . a by-product of military systems research" since 1940:

- the theory of information-feedback systems
- a knowledge of decision-making processes
- the experimental model approach to complex systems
- the digital computer as a means to simulate realistic mathematical models[50]

For Forrester, the primary sources of simulation models were "the design of air defense systems," which he noted had "received tens of thousands of man-years of effort in the last decade," and engineering studies such as the simulation of river basin development in the context of hydroelectric power plant construction. Recognizing that simulations of economic systems, electric power grids, and other complex phenomena had been attempted on analog computers starting in the 1930s, he pointed out—echoing von Neumann—that "these analog computers . . . do not readily deal with nonlinear systems" and that they were incapable of sufficient size and complexity for simulation of economic and corporate problems. For him, the development of digital computers from 1945–60 represented "a technological change greater than that effected in going from chemical to atomic explosives."[51]

Throughout the text of *Industrial Dynamics*, Forrester frequently noted that although his examples came primarily from industry, his models relied on "orderly underlying principles from which system behavior derives." Forrester argued that "systems of information-feedback control" were the essential organizing principle of all complex organized entities, from biological organisms to machines and computers.[52]

Over the next decade, Forrester refined these general principles of systems and applied them to areas of increasing complexity and scale.

He first turned his attention to modeling cities. Like *Industrial Dynamics*, Forrester's *Urban Dynamics* (1969) argued that systems as complex as cities are "counterintuitive," in the sense that policies developed to correct problematic behavior often end up making problems worse. In effect, policy makers tend to see and treat symptoms rather than causes of problematic system functions. This occurs because "a complex system is not a simple feedback loop where one system state dominates the behavior, [but] a multiplicity of interacting feedback loops ... controlled by nonlinear relationships."[53] The issue of nonlinearities and multiple feedbacks made complex systems virtually impossible, in Forrester's view, for unaided minds to grasp, since people think mainly in terms of linear relationships and simple feedbacks.

Forrester's models tended to be insensitive to changes in most parameters, even of several orders of magnitude—indicating to him that "complex systems resist most policy changes."[54] Effective policies usually followed a counterintuitive "worse before better" pattern. A system's short-term responses to change tended to be of opposite sign from long-term responses, so that policies which produced desirable effects for a couple of years would end up creating negative effects in the long run (and vice versa). In short, Forrester's slogan for policy makers was "no pain, no gain."

Forrester argued that there were only two possible solutions to the insensitivity of systems to most parameter changes. The first was to find, through modeling, those few parameters and structural changes which *could* produce desirable effects. The second was to design—again with the help of models—comprehensive policies which took into account the complex interactions among all the different elements of a system. Models thus became, for him, virtually a sine qua non of effective policy making in any complex system.

The *Urban Dynamics* models also reflected three other interesting and unusual characteristics of Forrester's approach. First was his strikingly cavalier attitude toward empirical information. Forrester wrote that "the barrier to progress in social systems is not lack of data. We have vastly more information than we use in an orderly and organized way. The barrier is deficiency in the existing theories of structure." Rather than gathering more data, Forrester thought it much more important to model as many important system relationships as could be incorporated. "It is far more serious," he wrote, "to omit a relationship that is believed to be important than to include it at a low level of accuracy that fits within the plausible range of uncertainty." He noted that this modeling approach "follows the philosophy of the manager or political leader more than that of the scientist. If one believes a relationship to be important, he acts accordingly and makes the best use he can of the information available. He is willing to let his reputation rest on his keenness of perception and interpretation."[55] These sentiments reflected Forrester's lifelong belief that tools should always be forged through actual practice, never only in academic laboratories.

Second was Forrester's insistence that models for policy should be comprehensive. This demanded interdisciplinary and cross-disciplinary work. "The barriers between disciplines must melt away. . . . Within the same system we must admit the interactions of the psychological, the economic, the technical, the cultural, and the political."[56]

Finally, in a theme that became enormously important during the *Limits to Growth* controversies, Forrester argued that growth was a developmental phase rather than a constant of urban systems. *Urban Dynamics* focused on the "life cycle" of cities over a 250-year period, in which empty land is settled and developed until fully inhabited. Then it proceeds through a series of socio-economic "realignments" until it reaches an "equilibrium" state of stagnation.[57] "Continued exponential growth," Forrester asserted unequivocally, "is impossible."[58]

These three features of Forrester's modeling techniques and modeling philosophy were indeed pioneering. Although many of his contemporaries regarded his models' lack of empirical data as an extremely serious if not a fatal problem, Forrester was among the first to insist that computer models could serve important policy purposes even in the absence of good data. Like the climate scientists, ecologists, and other systems scientists whose work also matured during the 1960s, Forrester believed that sorting out the structure and dynamics of a system using a computer model was the key to understanding. Data could come later, in part because a systems model could help reveal *which* data might be most important. His models' far-reaching, interdisciplinary character foreshadowed many later developments, such as the rise in the late 1980s of Earth systems models, which encompass atmosphere, oceans, agriculture, and ecology, and of integrated assessment models, which incorporate economics, energy, and human social systems as well. Though Forrester was far from the first to articulate the theme of inherent limits to growth—his critics invariably compared him to Malthus—his models drew attention to ways in which a many-dimensional world system might be more finite and fragile than it appeared.

PART 3. 1970–72: MODELS FOR POLICY

Between 1970 and 1972, as the environmental movement came of age in the United States, world models of both sorts rather suddenly became the focus of substantial public controversy. Climate modelers went public with claims that human activities were likely to alter the climate, and world dynamics modelers predicted imminent collapse of human societies in a world ravaged by overpopulation and over-extraction of limited resources.

I do not have space here to give the political aspects of these controversies the attention they deserve. Instead, I focus on the relationships between models and data, especially the role of modeling projects in provoking the creation and extension of global data networks. In the end, I argue, these models and data networks have played a major part in creating "the world" as we know it today.

SCEP and SMIC

Most of the series of "experiments with models" carried out in the 1960s showed a global warming somewhere in the range of 1–6 °C with a doubling of atmospheric carbon dioxide over the preindustrial era. This was in fact a very old scientific result, dating to the work of Svante Arrhenius in 1896.[59] But three things about the 1960s findings were new. First, there were the Mauna Loa CO_2 measurements, which showed exponentially increasing levels of CO_2 far from the urban pollution centers of the developed world. Second, the 1950s had seen an extremely rapid rise in the rate of fossil fuel consumption, the chief anthropogenic source of CO_2. Finally, there were the GCMs. In climate science as in many other fields, computers lent their enormous scientific and popular cachet to the GCM results. Together, these developments led to rising alarm about human tampering with the atmosphere. According to Hart and Victor, "by 1968 the notion that pollution could modify the climate was a commonplace."[60]

In 1969, the WMO called for extending the global atmospheric data network in a new direction: to monitor pollutants that might change the climate, such as CO_2 and particulate aerosols.[61] That call was soon underscored by two important scientific working groups, the Study of Critical Environmental Problems (SCEP, 1970) and the Study of Man's Impact on Climate (SMIC, 1971). The SCEP and SMIC reports are widely cited by scientists and policy makers alike as the first point at which anthropogenic climate change began to become a major public issue. They also mark the point at which the climate modeling story and the world dynamics story begin to converge.

Both studies were organized by the entrepreneurial professor Caroll Wilson of MIT's Sloan School (where Forrester had worked since 1956). Wilson had been involved in science policy for decades; he began his career as Vannevar Bush's assistant during World War II and later served as general manager of the Atomic Energy Commission. He planned the month-long summer conferences on which the reports were based as contributions to the UN Conference on the Human Environment, scheduled for Stockholm in June 1972.

SCEP focused on world-scale environmental problems, conceived as "the effects of pollution on man through changes in climate, ocean ecology, or in large terrestrial ecosystems."[62] NCAR and Scripps—important centers of atmospheric and ocean modeling respectively, and both loci of important work on the carbon dioxide theory of climate change—were both heavily represented among the invitees. Smagorinsky of GFDL attended on a part-time basis, and Revelle (now at Harvard) consulted.

The climatic effects section of the report cited GCMs as "indispensable" in the study of possible anthropogenic climate change. The report pointed to two key uses of models: as "laboratory-type experiments on the atmosphere-ocean system which are impossible to conduct on the actual system," and as a way of producing "longer-term forecasts of global atmospheric conditions."[63] While noting their many deficiencies, the report argued that models were "the only way that we now conceive of exploring the tangle of relations" involved in climate. It therefore recommended an expanded program of climate, ocean, and weather modeling research.[64]

SCEP's recommendations focused heavily on the problem of uniform global data. The report noted that "critically needed data were fragmentary, contradictory, and in some cases completely unavailable. This was true for all types of data—scientific, technical, economic, industrial, and social."[65] It recommended three initiatives: (1) "new methods" for global information gathering, which would integrate economic and environmental statistics, along with "uniform data-collection standards," (2) "international physical, chemical, and ecological measurement standards," administered through a "monitoring standards center," and (3) integration of existing monitoring programs to produce an "optimal" global monitoring system.[66]

Where the majority of SCEP participants were Americans, SMIC—less than a year before the Stockholm UN meeting—included a much more international cast of characters. SMIC provided a detailed technical discussion of GCMs and other, simpler climate models. The study reached conclusions very similar to those of SCEP with respect to human impacts on climate, including possible effects of then-controversial supersonic commercial aircraft. The body of the SMIC report reflects relative confidence in the availability of global uniform data, citing the WMO data network, the Global Atmospheric Research Program, and the increasing body of observations gathered by satellites. Nevertheless SMIC recommended a global monitoring program even more elaborate than the one proposed the year before by SCEP, including a global network of 100 monitoring stations to sample air and precipitation chemistry, measure solar radiation, and gather other meteorological data. SMIC estimated that an adequate program could be established for a yearly budget of about $17.5 million. Wilson's project had at least one direct outcome: at the UN Stockholm conference, the SCEP and SMIC calls for a global monitoring network were approved "with little discussion."[67]

These urgent calls for more and more data to feed the global models illustrate the process of data/model interaction for which I have been arguing. Climate and weather models increasingly gave a picture of "the world" as a whole, an interconnected set of systems whose interactions could be understood only through a combination of simulation and observation. The needs of each one drove the other forward. Without complete global data sets, modelers could neither validate nor parameterize their models. Without computers and models, data collection on that scale would have been not only pointless but meaningless. The models were what *made sense* of the data; they made a coherent world from collections of bits. In a certain important epistemological sense, they gave us "the world" as an ecological and physical unity.

The Limits to Growth

Around the same time as Wilson's SCEP and SMIC, a new kind of global model was being built by another Sloan School professor: Forrester. In June 1970 Forrester attended the first general meeting of the Club of Rome, a small international group of prominent businessmen, scientists, and politicians organized by Italian industrialist Aurelio

Peccei. His invitation came at the behest of Caroll Wilson, himself a new member of the Club of Rome.

The Club of Rome, founded in 1968, had at this point been considering for a year a proposal to model what it called the world *problématique*—global, systemic problems—in cybernetic terms.[68] The Volkswagen Foundation had rejected the modeling proposal for methodological vagueness. Forrester suggested that his approach might overcome this objection, and invited Club members to attend a workshop on industrial dynamics. By the time the Club of Rome's executive committee arrived at MIT three weeks later, Forrester had worked up and programmed on the computer a rough model, "World 1." The model divided world systems into five major subsystems (natural resources, population, pollution, capital, and agriculture) and incorporated some sketchy data and guesses about relationships among these variables. This very rapid work-up of the world model again reflected Forrester's fundamental belief that system structure and dynamics were more important than precise data.

Perhaps not surprisingly, World 1's characteristic modes resembled those of the industrial and urban models Forrester had already developed, especially the phenomenon of "overshoot and collapse." Eduard Pestel recalled that Club of Rome founder Peccei was tremendously impressed "by the fact that all computer runs exhibited—sooner or later at some point in time during the next century—a collapse mode regardless of any 'technological fixes' employed. Peccei obviously saw his fears confirmed. . . ."[69] The Club of Rome returned to the Volkswagen Foundation with a new proposal. This time, with Forrester's clear and well-developed methodology in hand, the application for an eighteen-month modeling project at MIT was approved.

Forrester's former student Dennis Meadows led the System Dynamics Group, with Forrester acting as consultant.[70] The team developed two successor models, known as World 2 and World 3.[71] World 3 incorporated over 120 strongly interdependent variables. Attempts were made to calibrate the models by starting model runs in the year 1900 and adjusting parameters until the model results roughly matched actual historical trends. World dynamics' essential conclusion was that many existing trends (resource consumption, pollution increases, population growth, etc.) displayed exponential growth rates that a finite planet could not possibly sustain. The world dynamics models continued to show, after refinement and even on the most optimistic assumptions, that natural resources would be rapidly exhausted, that pollution would rapidly increase to life-threatening levels, and that catastrophic collapse, including massive famine, would follow around the year 2050. *The Limits to Growth* became an international phenomenon, selling over seven million copies worldwide in some thirty languages.

The System Dynamics Group's self-described "bias" followed Forrester in favoring comprehensive model structure and dynamics over precise data. Indeed, the data used in the world models were generally poor in quality. In many cases, they were simply guessed. Although reviewers attacked them savagely for this apparent sin,[72] the situation was more complicated than it first appeared.

Like the weather and climate modelers before it, the System Dynamics Group had a great deal of difficulty acquiring high-quality information in the right form. Since the central idea of the world models was to produce very long runs (up to 150 years), in order to project long-term consequences of current trends, they could be validated only with very long time series. As Meadows put it in an interview, "it was hard in those days to find the kind of comprehensive, cross-national time series data on the issues we wanted to see, except on population. So we were looking where we could, with the United Nations and the World Bank as principal sources of information."[73] But the United Nations and the World Bank were only a quarter-century old. Where were Meadows and his colleagues to find *world-scale* information on pollution, agriculture, trade, and so forth, going back more than twenty-five years? The scant available information, such as population figures, was mostly either in a highly aggregated form that prevented analysis using the model categories, or at a level of disaggregation that would have taken the small modeling group years to organize. So Meadows's group chose the path Forrester had blazed for them: model the structure and worry about the data later.

Irving Elichirigoity's analysis of this situation bears careful attention: "The Meadows team had problems finding globally oriented information because information of that type was not normally collected within a framework of scientific practice that did not conceptualize the global as an entity on which information needed to be collected."[74] Just as with weather and climate modeling, one indirect effect of *The Limits to Growth* and its successors was to create an epistemological framework in which gathering global information became necessary and made sense.

All the models drew heavy fire from some sectors of the scientific community and, especially, economists. Within a couple of years most scientists regarded them with indifference or even contempt. Many in the policy community found the world-models approach—in an era when computer simulation was far less widely understood and accepted than today—technocratic in the extreme. This impression was only amplified by the Club of Rome's elite character and by the perceived arrogance and insensitivity of some of the modelers. Nevertheless, during the rest of the 1970s the Club of Rome commanded considerable international respect. It convened a series of meetings among senior politicians to discuss global resource concerns. The meetings, held in major world capitals, sometimes included the presidents and prime ministers of such nations as Canada, Sweden, the Netherlands, and Mexico.[75]

The Limits to Growth and the Club of Rome had few, if any, *direct* policy impacts. Nevertheless, through its models, popular books, meetings, and person-to-person canvassing of politicians, the Club succeeded in communicating, to both a broad public and a policy elite, its two basic conclusions: (1) that population, pollution, and consumption levels could not continue to grow indefinitely, and (2) that attempts to control problems piecemeal, without taking into account the interconnected nature of world socio-technical-environmental systems, would not work and might actually backfire. It is safe to say that these principles achieved the status of shared background assumptions for a large subset of the world policy community. In addition, the world dynamics modelers helped to establish computer simulation as an important technique of policy analysis. In the process they—like today's global change modelers—established a hybrid science/policy community for which the models were a key focal point.[76]

SCEP, SMIC, and *The Limits to Growth* are important because they mark the moment in the history of environmentalism when global issues first became salient not only to scientists, but also to the general public. Before this point, virtually the only issue discussed *as global* was population. The Club of Rome played a major part in building awareness of the integrated character of world systems, and especially of natural resources with human economies. From this point on, a growing minority of scientists, environmentalists, economists, and concerned citizens moved beyond the "pollution paradigm" to conceive of some environmental problems as global in scope.

CONCLUSION

During the past decade, computerized global modeling has become a widely accepted paradigm in science, in policy forecasting, and especially in science-based policy analysis. Earth systems models based in the physical sciences are including human systems (such as agriculture and forestry) in climate modeling, while integrated assessment models are combining GCM outputs with models of energy and resources to analyze climate policy options. Some of these, such as the Dutch integrated assessment model IMAGE, descend directly from Forrester's world dynamics.[77] The history of these modeling techniques thus holds some important lessons for the present.

In each case, computer models represented a fundamentally new approach to the phenomena in question. Detailed simulation models of global dynamic processes were, in general, not possible before the advent of electronic digital computers, for two reasons. One of these is obvious: the scale of calculation involved was prohibitive. The second reason—not nearly so obvious as the first—is that the global data required to construct, calibrate, and validate such simulations were largely unavailable.

From the mid-1950s on, efforts to build global atmospheric models and efforts to gather global uniformly structured data sets proceeded in tandem, each one driving the other's progress. The modeling efforts provided the rationale for the creation of global data networks. In addition, the models' requirements for time-stepped, gridded data shaped the technical structure of those networks. In turn, models were increasingly used (especially in weather and climate science) to refine rough, sparse, and poorly gridded data sets, resulting in increasingly blurred boundaries between models and data.

The profound interdependence of models and data in these cases suggests an important epistemological issue, namely the question of what is "global" about global models.[78] For both climate modeling and world dynamics, none of the available data sets remotely approach what might be construed as a minimal requirement for truly "global" data. Instead, coverage is spotty, inconsistent, poorly calibrated, and temporally brief. Rather, it is the *models* that are "global"; the data, with the exception of the (problematic) satellite measurements and population figures, are local or regional at best. Part of the work of the models, then, is to make those data *function as* "global" by providing an overarching reference frame.

As in many other areas of science, the U.S. high-technology Cold War strategy made immense financial and technical resources (especially computers and satellites) available to scientists and influenced research directions, particularly in the 1950s. In the case of weather and climate modeling, both the necessary equipment and the requisite data networks were direct products of military efforts during World War II and the early Cold War. Contests to achieve accurate weather prediction and weather control developed as part of the general Cold War arms race. In the case of the world dynamics models, Forrester's systems thinking had early precursors in his 1944–56 work on World War II and Cold War military projects, especially the SAGE continental air defense system. Forrester and his research group envisioned comprehensive solutions to world-scale military problems featuring digital computers as the focal technology. These concepts and technologies of global systems control had a profound influence on Forrester's later world modeling projects.

In both cases, international cooperative ventures were catalytic. The UN-sponsored World Meteorological Organization, the World Weather Watch, the Global Atmospheric Research Program, and other projects produced a variety of data and data networks that became crucial to weather and climate model development. The 1957 International Geophysical Year, which produced the first global uniform meteorological data sets, was a hybrid scientific/political response to Cold War tensions. The world dynamics models grew out of the Club of Rome's efforts to promote understanding of the world "*problématique*." They employed data gathered by various UN agencies, but also focused a glaring light on the inadequacy of world-scale socio-economic information sources. The importance of uniform data for these kinds of models has only grown with time.

Thus the second crucial aspect of the "global" character of global models is their role in generating and extending global data networks. The worldwide spread of scientific instrumentation for atmospheric observations, and of the knowledge and practices required to make such instrumentation function reliably, is one form of globalization rarely mentioned in the modern litany of "global" activities. This is equally true for the less developed, but also extremely extensive, data networks for collecting information about population, energy use, agriculture, trade, and other socio-economic activities. Global data networks require high levels of standardization, creating new commonalities in practice and understanding worldwide.[79]

With the emergence of global environmental politics, these data networks are among the forces creating concepts of global systems, global problems, and global common interests. From the perspective of this chapter, these are among the most important globalizations of all. For it is with models and data networks that modern concepts of "the world" have been built.

NOTES

1. An ordinary globe is a "model" of the world, and global maps were very important in colonial empire building. Scientists engaged in a variety of efforts at global mapping (e.g., of vegetation and average climatic conditions) in the nineteenth century, and possibly before. See, e.g., Jane Camerini, "The Physical Atlas of Heinrich Berghaus: Distribution Maps as Scientific Knowledge," in *Non-Verbal Communication in Science Prior to 1900*, ed. Renato G. Mazzolini (Florence: Olschki, 1993): 497–512. In the nineteenth century, Malthus speculated on the problem of the world's maximum population; Fourier and Arrhenius developed theories of global temperature as determined by the chemical composition of the atmosphere. In the twentieth century, meteorologists (among others) built a variety of analog models of global processes, such as "dishpan" experiments with rotating trays of colored liquids exposed to a heat source. "Weather Now Computed," *Science News Letter* 64 (1953): 196.

2. Donella H. Meadows et al., *The Limits to Growth: A Report for the Club of Rome's Project on the Predicament of Mankind* (New York: Universe Books, 1972).

3. This chapter is a preliminary study for a book, tentatively entitled *The World in a Machine: Computer Models, Data Networks, and Global Atmospheric Politics* (Cambridge: MIT Press, in preparation).

4. See chap. 3 of Paul N. Edwards, *The Closed World: Computers and the Politics of Discourse in Cold War America* (Cambridge: MIT Press, 1996).

5. Joseph J. Tribbia and Richard A. Anthes, "Scientific Basis of Modern Weather Prediction," *Science* 237 (1987): 493–499.

6. Kennedy, cited in James R. Fleming, " 'Cloud Wars': Weather Modification and the U.S. Military, 1947–1977" (forthcoming), ms. p. 10. Ms. available from James Fleming, Colby College, Waterville, Maine, 04901. On this subject see also James R. Fleming, ed., *Historical Essays on Meteorology, 1919–1995* (Boston: American Meteorological Society, 1996), and R. G. Fleagle and others, *Weather Modification in the Public Interest* (Seattle: University of Washington Press, 1974).

7. Fleming, "Cloud Wars," ms. p. 5.

8. In addition to his work on shock waves and aerodynamics for Los Alamos, von Neumann was "consultant to BRL (Ballistics Research Laboratory) on hydro-

dynamical and aerodynamical problems associated with projectile motion." J. G. Brainerd, ENIAC project supervisor, cited in William Aspray, *John von Neumann and the Origins of Modern Computing* (Cambridge: MIT Press, 1990), 37.

9. Aspray quotes von Neumann, during hearings on his nomination to the AEC: "I am violently anti-Communist, and I was probably a good deal more militaristic than most. . . . My opinions have been violently opposed to Marxism ever since I remember, and quite in particular since I had about a three-month taste of it in Hungary in 1919." Ibid., 247.

10. Rossby was among the early promoters of cloud seeding for military weather control. He chaired a 1947 Panel on Meteorology for the Joint Research and Development Board of the Department of Defense, which recommended an "intensive research and development effort." Fleming, "Cloud Wars," ms. p. 2. Zworykin had previously proposed building an *analog* computer for weather prediction.

11. Aspray, *John von Neumann*, 132, quoting Rossby.

12. J. G. Charney, R. Fjörtoft, and J. von Neumann, "Numerical integration of the barotropic vorticity equation," *Tellus* 2 (1950): 237–254; G. W. Platzman, "The ENIAC computations of 1950—Gateway to numerical weather prediction," *Bulletin of the American Meteorological Society* 60 (1979): 302–312.

13. "The Man Who Works the Weather Machine," *U.S. News & World Report*, 20 May 1955, 16.

14. See Clark A. Miller, "Scientific Internationalism in American Foreign Policy: The Case of Meteorology, 1947–1958," in Paul N. Edwards and Clark A. Miller, eds., *Changing the Atmosphere: Science and the Politics of Global Warming* (Cambridge: MIT Press, in preparation).

15. "Convention on World Meteorological Organization, 1947," in *The Encyclopedia of the United Nations and International Agreements*, ed. Edmund Jan Osmanczyk (Philadelphia: Taylor and Francis, 1985), 923–925.

16. "Long-Range Weather Forecasts by Computer," *Science News Letter*, 12 June 1954, 376.

17. Lt. Col. John R. Collins Jr., "Automated Data Processing at the United States Air Force Environmental Technical Applications Center," in *Data Processing for Climatological Purposes: WMO Technical Note No. 100*, ed. World Meteorological Organization (Geneva: World Meteorological Organization, 1969), 26.

18. B. Bolin et al., *Numerical Methods of Weather Analysis and Forecasting: WMO Technical Note No. 44* (Geneva: World Meteorological Organization, 1962), 3.

19. Ibid., 2.

20. On these methods see, for example, ibid.; Collins, "Data Processing"; and V. V. Filippov, "Quality Control Procedures for Meteorological Data," in *Data Processing for Climatological Purposes: WMO Technical Note No. 100*, ed. World

Meteorological Organization (Geneva: World Meteorological Organization, 1969), 35–38.

21. Jeffrey T. Kiehl, "Atmospheric General Circulation Modeling," in *Climate System Modeling*, ed. Kevin E. Trenberth (Cambridge: Cambridge University Press, 1992), 367–368.

22. See Naomi Oreskes, Kristin Shrader-Frechette, and Kenneth Belitz, "Verification, Validation, and Confirmation of Numerical Models in the Earth Sciences," *Science*, 4 February 1994, 641–646; Paul N. Edwards, "Data-Laden Models, Model-Filtered Data: Uncertainty and Politics in Global Climate Science," *Science as Culture* (forthcoming); and Stephen D. Norton and Frederick Suppe, "Epistemology of Atmospheric Modeling," in Edwards and Miller, *Changing the Atmosphere*.

23. Aspray, *John von Neumann*, 152.

24. Charney, cited in ibid., 152–153.

25. Norman A. Phillips, "The General Circulation of the Atmosphere: A Numerical Experiment," *Quarterly Journal of the Royal Meteorological Society* 82, 352 (1956): 123–164.

26. Joseph Smagorinsky, "General Circulation Experiments with the Primitive Equations," *Monthly Weather Review* 91, 3 (1963): 99–164.

27. Cecil E. Leith, "Numerical Simulation of the Earth's Atmosphere," in Berni Alder et al., eds., *Methods in Computational Physics*, vol. 4 (New York: Academic Press, 1965), 1–28; Yale Mintz, "Design of Some Numerical General Circulation Experiments," *Bulletin of the Research Council of Israel* 76 (1958): 67–114.

28. See chap. 3 of Edwards, *The Closed World*.

29. The interpretation of SAGE offered here is detailed in ibid. Also see Kent C. Redmond and Thomas M. Smith, *Project Whirlwind: The History of a Pioneer Computer* (Boston: Digital Press, 1980); and Kenneth Schaffel, *The Emerging Shield: The Air Force and the Evolution of Continental Air Defense 1945–1960* (Washington, D.C.: Office of Air Force History, United States Air Force, 1991).

30. Kenneth Flamm, *Creating the Computer: Government, Industry, and High Technology* (Washington, D.C.: Brookings Institution, 1988).

31. "Weather," for the atmospheric sciences, refers to actual meteorological events, while "climate" refers to long-term averages such as seasonal temperature change, annual rainfall, and so forth.

32. Syukuro Manabe, private communication, cited in Study of Man's Impact on Climate (SMIC), *Inadvertent Climate Modification* (Cambridge: MIT Press, 1971), 143.

33. Robert M. Chervin, "High Performance Computing and the Grand Challenge of Climate Modeling," *Computers in Physics* (May–June 1990): 235.

34. SMIC, *Inadvertent Climate Modification*, 143.

35. Hans E. Suess, "Natural Radiocarbon and the Rate of Exchange of Carbon Dioxide between the Atmosphere and the Sea," in *Nuclear Processes in Geologic Settings*, ed. National Research Council Committee on Nuclear Science (Washington, D.C.: National Academy of Sciences, 1953): 52–56.

36. G. N. Plass, "The Carbon Dioxide Theory of Climatic Change," *Tellus* 8 (1956): 140–154.

37. R. Revelle and H. E. Suess, "Carbon Dioxide Exchange between the Atmosphere and Ocean and the Question of an Increase of Atmospheric CO_2 during the Past Decades," *Tellus* 9 (1957): 19.

38. David M. Hart and David G. Victor, "Scientific Elites and the Making of US Policy for Climate Change Research," *Social Studies of Science* 23 (1993): 643–680. This article provides an insightful look into the surprisingly slow development of a public debate over the climate change issue.

39. Sir Harold Spencer Jones, "The Inception and Development of the International Geophysical Year," in *Annals of the International Geophysical Year*, ed. Sydney Chapman (New York: Pergamon Press, 1959), 393–395, 410–411 et passim; Comité Spécial de l'Année Géophysique Internationale, "The Fourth Meeting of the CSAGI," in *Annals of the International Geophysical Year*, ed. M. Nicolet (New York: Pergamon Press, 1958), 300–313, 350–353 et passim. The geopolitical structure of this division is intriguing; note especially the absence of a "World" Data Centre in the Southern Hemisphere.

40. C. D. Keeling, "Atmospheric Carbon Dioxide Variations at the Mauna Loa Observatory," *Tellus* 28 (1976): 538–551. See also Hart and Victor, "Scientific Elites," 651, and n. 36.

41. G. D. Robinson, "Some Current Projects for Global Meteorological Observation and Experiment," *Quarterly Journal of the Royal Meteorological Society* 93, 398 (1967): 410.

42. Kennedy, cited in ibid., 409.

43. See for example "Report of the Fifth Session of WMO Executive Committee Inter-Governmental Panel on the First GARP Global Experiment," GARP Special Report No. 26 (World Meteorological Organization Executive Committee, 1978).

44. Anonymous interviewee, interviewed by Paul N. Edwards at NCAR in 1994.

45. James W. Hurrell and Kevin E. Trenberth, "Spurious Trends in Satellite MSU Temperatures from Merging Different Satellite Records," *Nature* 386, 6621 (1997): 164–166.

46. Meadows et al., *Limits to Growth*.

47. Jay W. Forrester, interviewed by Marc Miller, 1977. MIT Archives, *Computers at MIT Oral History Collection,* Acc. no. 78–106, transcript pp. 44–45, 48 et passim.

48. Jay W. Forrester, "Industrial Dynamics: A Major Breakthrough for Decision Makers," *Harvard Business Review* 36, 4 (1958): 52.

49. Jay Forrester, *Industrial Dynamics* (Cambridge: MIT Press, 1961), 41.

50. Ibid., 14.

51. Ibid., 18–19.

52. Ibid., 15.

53. Jay Forrester, *Urban Dynamics* (Cambridge: MIT Press, 1969), 9.

54. Ibid., 110.

55. Ibid., 115.

56. Ibid., 114–115.

57. Ibid., 38.

58. Ibid., 40.

59. Svante Arrhenius, "On the Influence of Carbonic Acid in the Air upon the Temperature of the Ground," *Philosophical Magazine and Journal of Science* 41 (1896): 237–276.

60. Hart and Victor, "Scientific Elites," 665.

61. Ibid.

62. Study of Critical Environmental Problems (SCEP), *Man's Impact on the Global Environment* (Cambridge: MIT Press, 1970), 5.

63. Ibid., 78.

64. Ibid., 88.

65. Ibid., 6.

66. Ibid., 7.

67. Hart and Victor, "Scientific Elites," 664.

68. Irving Fernando Elichirigoity, "Towards a Genealogy of Planet Management: Computer Simulation, *Limits to Growth*, and the Emergence of Global Spaces" (Ph.D. diss., University of Illinois, Urbana, 1994), 149ff.

69. Eduard Pestel, *Beyond the Limits to Growth* (New York: Universe Books, 1989), 24, cited in ibid., 174.

70. Forrester himself moved on to other projects and was not deeply involved in the work of the System Dynamics Group. In 1971 he published a technical report on the world models as *World Dynamics* (Cambridge: MIT Press, 1971).

71. In addition to the rudimentary discussion in *Limits to Growth*, the System Dynamics Group's world models are exhaustively described in Dennis L. Meadows et al., *Dynamics of Growth in a Finite World* (Cambridge, Mass.: Wright-Allen Press, 1974).

72. The canonical critique is H. S. D. Cole et al., eds., *Models of Doom: A Critique of* The Limits to Growth (New York: Universe Books, 1973), but there were many others. A reasonably comprehensive retrospective look at the debate is Brian P. Bloomfield, *Modelling the World: The Social Constructions of Systems Analysts* (Oxford: Basil Blackwell, 1986).

73. Meadows, interviewed by Irving Elichirigoity, quoted in Elichirigoity, *Planet Management*, 175–176.

74. Ibid.

75. Alexander King, "The Club of Rome and Its Policy Impact," in *Knowledge and Power in a Global Society*, ed. William M. Evan (Beverly Hills: SAGE Publications, 1981), 205–224.

76. Paul N. Edwards, "Global Comprehensive Models in Politics and Policy-making," *Climatic Change* 32 (1996): 149–161.

77. Jan Rotmans, *IMAGE: An Integrated Model to Assess the Greenhouse Effect* (Boston: Kluwer Academic Publishers, 1990).

78. An especially insightful look at this problem appears in Peter Taylor and Fred Buttel, "How Do We Know We Have Global Environmental Problems?" *Geo-forum* 23, 3 (1992): 405–416.

79. Peter M. Haas, "Obtaining International Environmental Protection through Epistemic Consensus," *Millennium* 19, 3 (1990): 347–364; Peter M. Haas, Robert O. Keohane, and Marc A. Levy, eds., *Institutions for the Earth: Sources of Effective International Environmental Protection* (Cambridge: MIT Press, 1993).

9

The Medium Is the Message, or How Context Matters: The RAND Corporation Builds an Economics of Innovation, 1946–1962

David A. Hounshell

When the United States and the former Soviet Union were deeply engaged in their Cold War arms race, the RAND Corporation emerged as the nation's leading "think tank" devoted to national security studies. If RAND could have trademarked anything about itself, its leaders might have selected "systems analysis." Established by the air force in 1946, RAND created systems analysis as it invented itself, and soon this new methodology and RAND became synonymous. Systems analysis brought into play numerous methodologies developed earlier (including operations research from World War II, game theory that flowed from Oskar Morgenstern and John Von Neumann's 1944 book, probability and statistics, econometrics, and comparative worth assessment of military technologies, among others). Seeking to refine and complement these methods, RAND also fostered the development of linear programming and dynamic programming. RAND bound these analytical tools together under the rubric of systems analysis. In doing so, RAND created a mystique about itself, systems analysis, and the power of quantitative analysis. RAND, its brand of systems analysis, and its potent mix of mathematicians, logicians, economists, political scientists, and engineers helped to foster a major quantitative revolution in the social sciences in the United States.[1]

RAND helped the air force—and the United States more broadly—to fight the Cold War. It also helped both the men in blue and the larger Department of Defense to plan for World War III in the event that the Cold War turned hot. Both in fighting the Cold War and in planning for World War III, the development of new technology occupied a central, yet highly problematic role in RAND's calculus. In attempting to optimize military systems (e.g., offensive bombing delivery systems or air defense systems), RAND's researchers quickly found that technological change introduced vast uncertainties, often forcing them to make simplifying assumptions that seemed ridiculous to those who had actually fought a war. For example, to assume for the sake of manageability in a model that no technological change would occur did not ring true to a pilot who had fought in World War II. If technological change itself introduced too many contingencies and too much uncertainty into systems analysis to derive a meaningful optimization of war-fighting systems, this problem itself was compounded by the questions of how one could optimize technological change itself. That is, how could the air force or any service in the Department of Defense derive maximum benefit from public and private investment in research and development? Where in the continuum from basic research to development should monies be invested to obtain, so to speak, the biggest bang for the buck? And how much should be invested?

This chapter documents and analyzes the development at RAND of a body of pioneering literature in what has come to be known as the economics of technical change. Although undertaken in the 1950s and early 1960s to help RAND fight the Cold War and plan for World War III, RAND's research on technical change remains part of the foundation of knowledge in the economics of technical change. As this chapter will show, RAND's research on technical change emerged from a growing sense within RAND—especially by a group from RAND's economics department—that RAND's main stock in trade (systems analysis) could not effectively deal with technological change. The dynamics of technological change seemed to some RAND researchers central to understanding the dynamics of complex technological systems, such as modern air warfare. The optimization promised by RAND's systems analysts either would have to be limited to more stable systems with lower orders of uncertainty, or it would have to incorporate a fundamental understanding of technological change into its calculus.

Simultaneous to the growing internal pressures within RAND to understand the nature of technical change, a development external to RAND but to which RAND was both privy and in which it had strong vested interests also provided an important source of stimulation to RAND's research on the economics of technical change. That development was the United States' decision to build its first generation of intercontinental ballistic missiles (ICBMs) under a very different management structure. Promoted by the air force as an alternative to the bureaucratic methods of the principal airframe manufacturers, the new structure, ironically, came with the label of "systems engineering." It sought to circumvent the deeply ingrained development practices of the entrenched U.S. airframe industry, an industry out of which some of RAND's top managers, including its president, Frank Collbohm, had emerged. As this chapter will demonstrate, RAND's pioneering work in the economics of technical change thus arose not only in response to the problems of RAND's systems analysis but also as a critique of systems engineering and its attendant ideas, such as concurrency in research, design engineering, and production engineering, as they were promoted by air force General Bernard Schriever and by Simon Ramo and Dean Wooldridge, the systems engineers anointed to manage the ICBM project.

Context matters. This chapter seeks to demonstrate how profoundly the environment in which RAND operated, conditioned as it was by both internal dynamics and external events, shaped the production of knowledge in the economics of technical change at RAND. In more than one way, the medium was the message.

RAND AS A RESEARCH CENTER, 1946–1962

The early history of RAND is generally well known, and space limits prevent anything but the barest sketch of the institution's rise.[2] Some generalizations are in order, however. First, RAND represented an institutional innovation by General Henry H. ("Hap") Arnold, the commanding officer of the Army Air Forces, made at the end of World War II to achieve several objectives: (1) to secure for the air force a continued, instrumental role of scientific and engineering research in the service's operations, (2) to retain within the service's control the nation's top scientific, engineering, mathematical, and social-scientific talent that had come out of the university to meet the wartime emergency and that seem poised to reenter the university as "normalcy" returned in the postwar era, and (3) to develop a complete "science of warfare" that had just started to emerged from the work of scientific and technical organizations such as the Statistical Research Group of the Applied Mathematics Panel, which operated within the Office of Scientific Research and Development (OSRD), and the Office of Statistical Control, which operated within the Army Air Force's headquarters during the war.

As had become abundantly clear to any Washington insider who read Vannevar Bush's July 1945 report to the president on postwar science and technology policy—*Science, the Endless Frontier*—a fundamental tension in American science was reasserting itself between the necessity for and value of basic, unfettered research conducted in the university and more applied research carried out in governmental laboratories and industrial organizations. As Daniel J. Kevles has argued, the pure science ideal has been one of the dominant themes in the history of American science; leaders of the scientific establishment have asserted, or reasserted, an ideology of unfettered research at important junctures in the political and institutional development of American science. Leading scientists asserted the pure science ideal when they believed that social, political, and economic forces were pushing science too far toward utilitarianism. Bush's *Science, the Endless Frontier* pointed to that condition, a product of the war when scientists ceased doing basic research and devoted all their efforts to developing new tools of war.[3]

Arnold's genius in the design of RAND was to cut through this tension by creating RAND as a "university without students," an organization devoted to research whose major product was not weapons or technologies but the *research report.* Although RAND's name derived officially from the acronymic combination of *R*esearch *AN*d *D*evelopment, there was real truth to the more satirical explanation that the acronym stood for *R*esearch *A*nd *N*o *D*evelopment. With generous funding, the RAND university without students succeeded in attracting some of the nation's finest scientists, engineers, and social scientists. Brought together just as the Cold War emerged, these researchers made for a potent formula, the envy of many university departments and a pioneering organization that led research across a wide front of applied mathematics, engineering, and the social sciences. In addition to a deeply internalized belief that their work was saving the world from a totalitarian threat, RAND's researchers espoused one other important ideal: the sacredness of RAND research's independence. RAND's researchers shared an ideology of pure research that was as strong as that of any university-based scientific leader of the day. At one point in the early 1960s, RAND's leaders threatened to dissolve the organization unless RAND had its way with the air force. Having become highly dependent on RAND's research across an array of areas, the air force capitulated, thereby recognizing all the more RAND's commitment to independent research. On the other hand, throughout its early history, RAND had only one real customer, so pronouncements about independent research could go only so far. These pronouncements may have served RAND's internal needs more than its external relationship with the air force and the Pentagon more broadly.

Toward an Economics of Technical Change

Although its founders charged RAND with the goal of developing a robust general science of warfare in which winning strategies and necessary tactics could be derived to achieve global superiority over an aggressor, the organization met with extreme difficulty in realizing this objective. Two of the organization's first large-scale systems analyses, one focused on offensive forces and the other on defensive warfare, met with strong criticism not only from RAND's patron but also from within RAND itself. These studies had employed extremely large numbers of variables that were treated in a series of simultaneous equations, but as RAND's researchers realized, the way each of these variables related to the others had to be specified in the equations, thus leaving room for errors in outcome reflecting flawed assumptions. Moreover, using methods to maximize some variables and minimize others raised a more general question of precisely what one wanted to maximize or minimize. Researchers labeled this question the "specification problem." After considerable pointing of fingers and gnashing of teeth, RAND's principal proponents of systems analysis decided that the method would work on more tightly bounded problems wherein the specification problem could be kept under control. By making this early decision to limit the scope of problems on which systems analysis would be employed, RAND's leaders set the stage for the organization's successful deployment of systems analysis and its meteoric rise to importance as a national security think tank by the early 1950s.

Despite the newfound success of systems analysis and many RAND researchers' desire to extend the method and diffuse it more broadly in the military establishment, some researchers continued to express concern about systems analysis. Much of this concern came from researchers who were formally trained as economists. Among them was Armen A. Alchian.

Alchian became a consultant to RAND in 1947, shortly after he had joined the Department of Economics at the University of California at Los Angeles.[4] Soon he was spending each afternoon and almost every Saturday at RAND; by late 1949, his association with RAND changed from that of consultant to formal employee.[5] He was one of several such economists at UCLA who held "joint" positions at the two institutions. Alchian's "boss" at RAND, mathematician John Williams, gave him no specific problem or set of problems on which to work; like so many researchers who joined RAND, he was simply encouraged to work on the problems he found interesting.[6] Although Alchian engaged in some rather conventional work at RAND, such as the development of a policy tool for making decisions about equipment replacement by taking present-value costs into consideration,[7] he also undertook an important analysis of "experience curves" or "progress functions" developed from data on World War II bomber and fighter aircraft production. Issued as a classified RAND Research Memorandum in late 1949, Alchian's analysis statistically tested the experience curves of various aircraft manufacturers against one another to develop a method of judging the reliability of learning curves in predicting the costs over time of serial manufacture.[8] Alchian's work constitutes some of the earliest formal work done on learning curves—on "learning by doing"—but this work was not declassified until 1963, so he lost much credit for his research to scholars such as Harold Asher and Werner Z. Hirsch.[9]

In early January 1951 Alchian joined with five other RAND economists in offering an extensive critique of RAND's early systems analyses, the problems of which have already been discussed above. The six economists argued that "specialized knowledge and extensive computations are insufficient to indicate which systems are better or worse than others." They continued: "The best of many alternative systems cannot be determined unless a valid criterion is employed." To gain analytical power, they maintained, systems analysts must specify operational objectives and then use measurable criteria—such as cost—to judge the merits of alternative systems.[10] Alchian would continue to hammer away at the champions of systems analysis within RAND and within the air force. Soon his thinking encompassed not only the problems of systems analysis for choosing the "best" system, but also the challenge that systems analysis posed to the management of R&D.

Alchian's involvement in a specific systems study, known as the Dynamic Offensive Bombing Systems, or DOBS, appears to have led directly to his pursuit of questions about R&D. The DOBS problem, he noted, could be stated simply:

> Given (1) the pattern of Research and Development expenditure among the weapons in the current Air Force program, and (2) the implied dates at which they will probably become available for operational use, what time phasing and selection of weapons (and quantities) will yield a specifiable time pattern of kill potential at minimum cost?[11]

Analysis of this "phasing" problem suggested to Alchian that a critical aspect of the study was the air force's R&D budget allocation because R&D could affect the performance of weapons systems. Yet decisions about procurement of weapons systems for future deployment appeared to Alchian to be fundamentally different from R&D allocation decisions. As he stated the matter: "These two decisions are very different both in their effects, in their timing, and in the information required." To demonstrate this difference, Alchian assembled data for ten bomber and fighter aircraft development projects, which showed the length of time (in months) from initial design requests to initial testing of the planes by their respective service. Analysis of these data showed that R&D decisions and procurement decisions were, in fact, quite different. A definite pattern of intervals between these two decisions also emerged from the data.

"That these two decisions are separable in time," Alchian concluded, "is very fortunate. . . ." He explained his reasoning as follows:

> [E]ven though the type of information required for each [decision] is similar, the accuracy of available predictions has a different effect. The information required pertains to the feasibility of actually producing a weapon which will have specified performance characteristics. It is doubt about this that necessitates R and D work. A second kind of information required is the type of performance that is wanted. But the kind of performance that is wanted depends upon the kind of war to be fought and the enemies' capabilities. If this were known with certainty we could begin design on the optimal weapon. But since we suffer from predictive myopia we can either guess what it will be and then design what we hope will be an optimal weapon,— or we can admit we don't know and obtain insurance by designing several alternative weapons, one for each possible contingency. Thus the R and D decisions—which we would have, under ideal foresight, confined to a single weapon,—are, in fact, made with only a range of possibilities to characterize the future. We therefore recommend the development of several alternative weapons—guaranteeing that an ignorant critic will be able to show that a large majority of them were "useless"—but assuring ourselves flexibility and safety with optimal weapons systems in actual use. The essential conditions are first that the cost of this flexibility be small relative to the cost of maintaining the operational weapon system; and second that the cost, minute though it is, enable us to save enormous amounts by choosing optimal systems rather than being forced to use the only system available; and third, that we be not completely blind in our foresight so that we can select sensible R and D allocations.[12]

Procurement decisions differed from R&D decisions, argued Alchian, in that they must be based on such things as "immediate costs, enemy capabilities, and the type of war th[at] is possible." He continued:

> With these necessarily narrower or pinpointed conjectures it [i.e., the air force] selects from the menu of available weapons—weapons that have already been through research and development. The analogy in the title ["The Chef, Gourmet, and Gourmand"] applies here. R and D decisions are those of the chef, who concocts new dishes and plans a menu of available alternative dishes, from which the gourmet at a later time has the privilege of choosing in light of his tastes, companions, and income. The task of the chef and the gourmet must be kept strictly distinct. To confound the two is as disastrous in the military as in the restaurant business.

Alchian went on to argue that decisions about R&D—those of the chef—required considerable care. As he put the matter, "R and D not only advances us technically—it also is our only assurance of flexibility of choice in the future. An intelligent R and D program must satisfy *both* objectives." Although R&D decision making was critical, Alchian maintained that it "may well not be getting the attention it deserves." He suggested that perhaps the comparatively small amount of money devoted to R&D as compared to the enormous sums spent for weapons procurement might have conditioned the lack of attention to R&D. "I would only argue that from what I have seen or read of the process by which R and D funds are allotted," Alchian quipped, "I fear we shall all soon cease to be economizing gourmets with à la carte menus and become expensive table d'hôte gourmands."[13] Apparently convinced that decision making about military (especially air force) R&D was fundamentally important, Alchian undertook to learn more about it. He and his colleagues at RAND were motivated all the more by the significant changes both in air force R&D and in the larger organization and management of R&D within the Department of Defense (DOD), including the DOD reorganization of 1953 that abolished the Research and Development Boards and transferred responsibility for R&D to a newly created position of Assistant Secretary of Defense for Research and Development.[14]

In mid-1953, Alchian posed a series of questions about how the air force determined its spending on R&D and how it chose to allocate those funds across the R&D spectrum. He confessed that during the two previous years, he had "examined budget directives, research planning guides, organizational charts, research project prospectuses, progress reports, research project histories, Congressional hearings, development planning objectives, etc." Together, these sources pointed to one firm conclusion: "prescript and practice differ." Alchian had just spent a week at the Pentagon discussing air force R&D with a broad spectrum of officers ranging from those in the air force up through the Department of Defense's Research and Development Board, the agency that ultimately established the department's R&D budget. At the Pentagon he had found a bewildering array of data and reports, all of which pointed to the fact that in assembling the military's R&D budget, no one at the Research and Development Board level (or really down several levels into the various services' R&D organizations) knew what constituted a good project or a bad project. As Alchian concluded, "The missing ingredient seems to be a basis for determining the usefulness of research and development projects.... It was literally impossible to ascertain what criteria are used in selecting projects." As a consequence, the Department of Defense (and certainly the Bureau of the Budget) had simply imposed a ceiling on the total defense budget, and the services and units within the services fought over resources within these budget constraints. "The research and development effort," Alchian concluded, "will continue to be determined by a process describable as essentially haphazard in rationale despite beautiful looking organization charts."[15]

While at the Pentagon, Alchian had "suggested that perhaps a study of some past research and development work might enable us to ascertain the savings consequent to this work as contrasted to what the situation would have been in the absence of such research and development...." But, he reported, "cold stares and 'impossible' was [*sic*] the reaction" (to which he admitted, "and they could be right"). The young economist then set out a series of answers to the question "What can be done?" One approach, he suggested, would be to find a way to improve the coordination of what were really isolated units in the services' R&D organizations. A second avenue would be not to mess with coordination but simply "to modify the various criteria used by the categories of agencies and to modify the working constraints within which these agencies must work." Most promising, however, would be to pursue a series of "ad hoc suggestions in various aspects of research and development." One approach, for example, would be to explore whether the R&D budget ceiling could be moved up or down given that no rigid ceilings existed on other military functions such as "procurement, personnel, training, [and] construction." Alchian concluded his exploration by raising a final example of a problem to be worked on: "Or to suggest a completely different ad hoc problem, are systems analyses, which are supposed to be good for procurement decisions, good or bad for evaluating research and development decisions?"[16]

Alchian raised this question about whether systems analyses were the proper means of evaluating R&D decisions for at least two reasons. The first has already been stated explicitly: RAND had distinguished itself as the pioneer of systems analysis, and despite the problems the institution had encountered with the approach and doubts about the method raised by RAND's economists, many RAND researchers continued to promote systems analysis as the principal tool for evaluating decisions about how the air force should allocate its resources for weapons.

The second reason was no less important. Indeed, it motivated some economists at RAND even more strongly than it did Alchian. From its formal establishment as a separate service branch through the early 1950s, in response to both internal and external pressures, the air force went through numerous reorganizations of its scientific and technical units. By 1953, these reorganizations had resulted in a more tightly controlled and centralized organization known as the Air Research and Development Command (ARDC), which had been formally established in 1950.[17] Among the air force officers who played roles in the establishment of ARDC, one would come to symbolize many of the ideals espoused by advocates of systems analysis—Bernard A. Schriever. Some of Alchian's first musings about the importance of separating R&D decisions from procurement decisions were directed at the principles Schriever (then a colonel) advocated in his role as a planning officer for strategic bomber development.[18] Schriever became an increasingly effective and powerful advocate for an approach to weapons development that eventually bore the name "concurrent engineering" or, more simply, "concurrency." Stated in its purist form, the fundamental idea of concurrent engineering is that a final product (such as a bomber or a missile) can be so well specified in terms of its performance that all aspects of its engineering, design, and production can be pursued essentially in parallel rather than in series. The ideal of concurrency and the objectives of systems analysis (optimization of some system or identification of the best system from a welter of possibilities) fit well together, at least ideologically. With the rise of Schriever within the air force's R&D establishment, concurrency gained ascendancy. This was especially true when in late 1953 and early 1954, the air force determined to build an intercontinental ballistic missile system, eschewing traditional aircraft manufacturers and placing responsibility for delivery of the weapon in the hands of Schriever and an upstart firm known as Ramo-Wooldridge Corporation. Together, Schriever, as the head of the ARDC's new Western Development Division, and Ramo-Wooldridge, as the firm solely responsible for the ICBM's "systems integration and technical direction," championed the principles of concurrent engineering as the best means not only to develop the ICBM, but also to win the Cold War.[19]

Ironically, although RAND's advocates of systems analysis shared much in common with Schriever and Ramo-Wooldridge, RAND's president, Frank Collbohm, objected to the plan for the ICBM's development. When the air force's Atlas Scientific Advisory Committee was asked to review and approve the Schriever-Ramo-Wooldridge plan for ICBM development and procurement, Collbohm was the only member to object formally to it. The Atlas Scientific Advisory Committee was the immediate successor to the "Teapot Committee" convened in 1953 by Trevor Gardner, Special Assistant for Research and Development to the Secretary of the Air Force. Headed by one of RAND's most famous consultants, John von Neumann, the Teapot Committee had concluded that an accelerated ICBM program should be pursued and that the program should use a radically different approach to the one already in place in the air force, which had been relying upon the aircraft manufacturer Convair for the missile's early development.[20] After reaching these conclusions (guided in part by the findings of one of RAND's own researchers),[21] the Teapot Committee had been reconstituted in early 1954 as the Atlas Scientific Advisory Committee, with new members—including Collbohm.[22] Collbohm raised formal objections to the plan, writing von Neumann, "Our aim is to deliver an operational capability as soon as possible. If this is to be achieved, the Project must be pushed into production-engineering as rapidly as possible."[23] He rejected the idea that the system could be optimized a priori through advanced systems engineering and research and development. In response, Von Neumann appointed a special three-man subcommittee headed by the famous aviator Charles A. Lindbergh and consisting of George Kistiakowsky and Jerome Wiesner. When the subcommittee reported its findings—supporting the vesting of responsibility for the entire Atlas program in the hands of Schriever and Ramo-Wooldridge—Collbohm again objected, saying specifically that "too much time was being devoted by the present Atlas management complex toward optimizing the intercontinental ballistic missile rather than getting into design and production on a system which would give an early operational capability."[24] When Simon Ramo offered an informed, point-by-point critique of Collbohm's position, Collbohm reluctantly capitulated. But he argued for the initiation as soon as possible of parallel development programs based on his belief that an advanced weapons systems could not be optimized a priori and that numerous comparatively cheap experimental projects would yield better results than a single grand strategy.[25]

Voiced clearly in the meetings of the Atlas Scientific Advisory Committee, Collbohm's ideas about advanced weapons systems development must surely have been known within RAND. Indeed, Bruno W. Augenstein, a distinguished RAND researcher who fully briefed the Teapot Committee on RAND's findings about the feasibility and effectiveness of an ICBM armed with a hydrogen bomb, maintains that the air force originally asked RAND to assume the systems integration and technical direction of the Atlas project before it was given to Ramo-Wooldridge. Despite Augenstein's urging that RAND accept the challenge, Collbohm turned the opportunity down in short order. Precisely why Collbohm demurred is unclear. From Collbohm's statements to von Neumann, we know that he believed there was a misfit between the air force's new religion of systems engineering and what Collbohm thought was the best way to get an ICBM built and fully operational. Also, Collbohm believed that RAND's undertaking such a task would not be consistent with his institution's fundamental mission for the air force.[26] Unquestionably Collbohm's views matched quite closely the ideas that Armen Alchian had been developing since the late 1940s about the importance of diversity in technological development; the critical differences between research, development, and procurement; and the inherent problems in employing systems analysis to optimize the performance of an advanced weapons system that had yet to be developed. Thus, it is no coincidence to find Alchian and other RAND researchers raising fundamental questions about weapon system innovation at precisely the same time that Collbohm was confronting the policy questions posed by the air force's ICBM program.

In mid-1953, Armen Alchian provided his own tentative answers to his question about whether systems analyses were good or bad for R&D decision making in an essay entitled "Systems Analyses—Friend or Foe?"[27] Alchian penned his analysis in response to a speech that had recently been delivered to the Society of Automotive Engineers by Major General Donald L. Putt in which he laid out the air force's new commitment to "the system approach."[28] A protégé of CalTech aerodynamicist and principal air force scientific consultant Theodore von Kármán, Putt had been in the air force's R&D Directorate since 1948 and had just been promoted to Lt. General and given charge of the ARDC, the new organization responsible for all the air force's internal and external R&D programs, including those of the RAND Corporation.[29] As historian John Lonnquest has stressed, Putt had been at the forefront of the reorganization of the air force's R&D programs. Putt and his "kitchen cabinet" of young officers, many of whom would play instrumental roles in the ICBM program (including Schriever), wielded the "systems approach" as their principal weapon in the struggle for power within the air force.[30] Although Armen Alchian might not have been aware of the bureaucratic infighting, he was certainly aware of Putt's new emphasis on the "systems approach." Juxtaposed with RAND's own embrace of systems analysis, this embrace of the systems approach raised fundamental questions in Alchian's mind about the management of R&D.

Before addressing Putt's paean to systems thinking, Alchian set out some definitions. *Research*, he argued, was "that endeavor designed to increase our knowledge, or the state of the arts (knowledge[,] not applied technique)." *Development*, on the other hand, "connotes those activities designed to utilize that knowledge in producing manufacturable and operational weapons." If research and development could meaningfully be so distinguished from each other, Alchian mused, then he had grave doubts about Putt's "advocacy of the systems approach for research and development." The reason was simple. Research was "a search process—a shot in the dark—a gambling process" in which society as a whole benefits enormously but in which the payoffs from "individual studies vary from minus to plus." Thus, Alchian argued, "to think that we can estimate which particular studies will be the ones that will pay off the most and hence should be the only ones pursued is to misread the history of research." Making research decisions to yield the "optimum" weaponry seemed to Alchian to be flawed:

> We simply do not have the required degree of foresight nor the ability to determine what we shall be able to learn and know. Therefore when I get the impression that General Putt is saying we

> should choose now the weapon system we ought to research and develop I have to dissent because I submit that this is a doubly dangerous doctrine. My contention, for the sake of clarifying the issues, is that we should neither make *a* preferred choice now and that we should certainly not be thinking of systems in research and probably not in development.[31]

The methods used to optimize the overall performance of systems in which the performance of individual components in the system is known—operations research and even the more flexible methods that RAND had pioneered, such as linear and dynamic programming and systems analysis—were not applicable. Research was simply too uncertain a process. Even methods on which RAND was at work to aid in decision making under conditions of uncertainty were inappropriate given the nature of research and development. "For procurement," Alchian concluded, "the systems approach is essential." For research, it had "no usefulness at all" and could even be "disastrous."[32]

Knowing full well that he had "ungraciously used a statement by [Air Force] Research and Development's best friend to caution against inappropriate applications of a kind of analysis at which RAND is itself a strong exponent," Alchian appealed to his colleagues in the economics department to consider his essay the opening of a dialogue on how to deal with research and development as an economic problem—that is, as a problem in the allocation of scarce resources to gain a maximum return on investment. With the completion of this "D" essay—a document that would be seen only within RAND and that could not be "quoted or cited in external RAND publications or correspondence"[33]—Alchian moved RAND toward the development of foundational literature on the economics of innovation.

One of RAND's foremost champions of systems analysis, Edward Quade, furthered the discourse when he proposed that RAND create a course entitled "Appreciation of Systems Analysis," in which the method's limitations as well as its benefits would be discussed. Quade pitched his proposal on the basis of the Wright Air Development Command's (WADC—the chief aircraft development unit of the air force) having firmly embraced systems analysis. As he wrote, "Indeed, there seems to be a feeling in some parts of the Air Force that the systems approach may provide the complete answer to all questions of development, procurement, and operation as well as those of design. . . . Lately, WADC design contracts for major components have required (or strongly suggested) that the contractor make his proposal as the result of a systems study."[34] Armen Alchian and his colleague Reuben A. Kessel used Quade's proposal to challenge in a more thorough manner the suitability of systems analysis and other optimization methods for handling the allocation of resources for R&D. As they baldly stated, "Our motive in writing this is to emphasize . . . [that] systems analyses, unless used in an environment different from that which apparently now prevails, will do more harm than good in the development decision."[35]

Systems analysis sought to minimize cost or to maximize damage, performance, or some other criterion. The critical question was this: under what conditions would such maximization/minimization calculations be reasonable? Returning to the themes Alchian had articulated earlier, he and Kessel maintained that the method worked fine "if the predictability of technological capabilities were known"—that is under conditions of certainty. But scientific and technological change altered weapons' performance characteristics—characteristics upon which development and procurement decisions were based. The fact that scientific and technological change were highly uncertain made systems analysis not only dubious but outright dangerous. Research and development, which presented a unique "class of problems," were best pursued by a policy "in which diversity is the optimal principle of choice." Moreover, Alchian and Kessel argued, selection of weapons systems based upon preordained routes of research and development (in other words, under conditions in which "the amount of diversity will probably be too small") would surely not only result in suboptimal outcomes but also yield outcomes that reflected "organizational and institutional biases."

Therefore, concluded the economists, because of the nature of scientific and technological change, which yields conditions of uncertainty, "the criterion of decisions is not simple maximizations; the essence of the decision process is to affect the scope of random factors so as to give a 'good' probability distribution of outcomes. . . . [This] requires diversity of investment—not variety of possible environments or flexibility of particular weapons. . . . Optimal diversity in concrete situations cannot be ascertained. But institutional arrangements, wherein biases are created against diversity and toward identification of analysis with decision, are *prima facie* evidence of a system that yields suboptimal diversity."[36]

Such a major challenge to RAND orthodoxy, coming as it did from within RAND itself, probably did not endear Alchian and his colleagues in the economics department to RAND's most enthusiastic practitioners of systems analysis. But these fundamental questions resonated with the issues surrounding the best way to develop the ICBM, issues that obviously occupied RAND's president, Frank Collbohm. By July 1954, these questions had coalesced into a formal research program within RAND's economics department (but supported enthusiastically by Collbohm and pursued jointly with the cost analysis department and the missiles department) to examine research and development decision making and budgeting.[37] Although the research findings that would emerge from this program would come too late to inform debate on the ICBM program,[38] they would nevertheless come into play in a later—and much more public—debate about national security and the best way to meet the Soviet challenge. The economists at RAND prevailed, perhaps in part because their views received strong support from Collbohm, who in opposing the manner in which the air force had elected to develop the ICBM had espoused ideas that resonated with those of the economists. Out of this project would emerge work that today still constitutes foundation stones in the economics of technical change.

RAND's Study of Research and Development

Burton H. Klein, an economist who had been at RAND since April 1952, was selected to head RAND's study of research and development.[39] Klein had devoted his dissertation research at Harvard University to the development of the German economy under national socialism, especially its design for economic and territorial expansion. Regarded by many as a brilliant, eccentric individual, Klein published his dissertation six years after its completion in 1953 in the Harvard Economic Studies series under the title, *Germany's Economic Preparations for War*.[40] Klein's early work at RAND remains obscure in the surviving records, but at some point early in his RAND career he developed what would become a career-long interest in the study of innovation. He continued to pursue this interest as head of RAND's economic department from 1963 to 1967 and then at CalTech, where he was professor of economics from 1967 until his retirement.[41]

In mid 1954 Klein and Reuben Kessel outlined the "ambitious research program" that the economics department would conduct on the air force's missile research and development budget in particular and its R&D in general. They aimed at establishing a "performance budget" for R&D and expected the project to gather and analyze empirical data on how military research and development budgets were constructed and how changes in those allocations affected the production of knowledge. By understanding the allocation of resources in R&D, the project, they hoped, would yield decision rules to guide policy makers. For the remainder of the decade and well into the next, Klein directed work on the problem of allocation of resources for R&D and how understanding the economics of R&D—or the economics of technical change—could contribute not only to the air force's mission, but also to the nation's welfare.

As with most long-term projects at RAND, the cast of researchers working on the research and development project changed over time, as did the precise focus of the project itself. (See the appendix for a tabular summary of the RAND R&D project research staff.) Changes in the composition of the research team and changes in the political contexts in which the project was being carried out altered the nature of the research and the research findings, sometimes subtly, sometimes radically. Between mid-1954 and 1958, most of the research for this project was supported directly by Project RAND (that is, under the research agreement between the air force and the RAND Corporation). As interest in the questions about R&D grew within Klein's group, however, and as researchers realized that understanding the larger processes of R&D not just in the military but also in the civilian economy was critical to long-term progress in the new research field they were creating, support also came from RAND's general research funds—that is, from funds that the corporation acquired for "pure research," research initiated by RAND's staff to explore new areas of interest and to open up new avenues of research that the corporation did not have to defend as being of immediate utility to the air force.[42]

A careful reading of Klein and Kessel's first semiformal discussion of the research and development project demonstrates that they lacked clarity on precisely how they would tackle the enormous problem of gathering the empirical data necessary to develop a model for the optimal allocation of an R&D budget. Moreover, at least initially, Klein lacked a clear vision of what analytical tools he would apply to the analysis.[43] Obviously they had to figure out how the air force in particular and the Department of Defense in general made budgeting decisions about R&D resources.

Toward this goal, the R&D team evaluated how the air force's research budget was allocated in the previous four fiscal years[44] and how the Department of Defense organized its R&D programs.[45] In addition, several of the R&D team members conducted interviews with air force R&D officers, in Washington, D.C.,[46] at the Air University at Maxwell Air Force Base in Montgomery, Alabama,[47] and at Wright Air Force Base in Dayton, Ohio.[48] Precisely what RAND's R&D team learned from these interviews is, as yet, unclear. The air force's R&D organization underwent seemingly constant reorganization in the 1950s as its officers tried to cope with the limitations on air force support for basic research at universities imposed by the Bureau of the Budget. (BoB was trying to enforce the National Science Foundation's legislative mandate as the nation's principal agency for the support of basic scientific research.) From its reorganization in 1949 until the early 1960s, the air force R&D organization also tried to find the optimal mix of civilian and military R&D scientists but without success. RAND's R&D team must have, therefore, heard a bewildering array of opinions from officers as to how best the air force should conduct its R&D. Shifting political tides, brought about by such events as the Russian launch of Sputnik, the world's first artificial satellite, also ripped through the RAND R&D project, as will be discussed below.

After almost a year into the R&D project, Klein and his small group still had little concrete work to show for their efforts. Other than arguing that systems analysis was inappropriate for the early stages of R&D and other than gaining a sense of how the air force's R&D organizations were structured, the group had produced little tangible evidence that its members had gained a firm theoretical or programmatic grip on R&D as an economic problem. This situation began to change in the summer of 1955. In early July, Alchian, Kessel, and Klein issued a jointly authored paper, "The Efficiency of the Research and Development Program: A Discussion of the Issues," which, as its title clearly suggests, sought to identify the fundamental issues in the air force's R&D program and move the program toward increased efficiency.[49]

Alchian, Kessel, and Klein first distinguished three elements in R&D that, they argued, "are essentially different in . . . character and purpose": (1) basic research, (2) "experimental development," and (3) "development of weapons systems." These three elements consumed, respectively, 5, 30, and 65 percent of the air force's R&D budget. The economists argued that keeping these three elements of R&D distinct was important for their analysis and for policy makers. Two fundamental policy questions could be raised about R&D: (1) What sort of direction should be given to each of these three classes of activities? and (2) What ground rules (financial incentives, degree of competition, rules of company disclosure of technical know-how, etc.) with respect to each of these activities is most likely to maximize the effectiveness of the R and D budget as a whole? Alchian, Kessel, and Klein maintained that no single policy would work uniformly well across all three of the distinct elements of R&D. Policies had to be specific to each of the elements.

The authors then turned to an extensive discussion of policies vis-à-vis each element. They argued strongly that in both basic research and experimental development, "the Air Force attempts to be a technical specialist" too much and that it did not "take advantage of the expertise available for making" wise decisions in each of these two domains. With respect to weapons systems development, Alchian, Kessel, and Klein posited that "the Air Force appears to go too far in tying down the developer. Much more is specified than desirable operational characteristics; in effect the limits within which the designers may use their own ingenuity are very narrowly prescribed." After discussing the various institutions within the air force system of R&D—ranging from the air force's own R&D organizations, to universities, private research institutions, and firms—and considering restructuring incentive systems to maximize the performance of each of these institutions, the economists formulated their basic hypothesis:

> It is our hypothesis, therefore, that both the direction and the implementation of the research and development program involve major biases against research and experimental development. If this hypothesis is correct, the technological progress being achieved with the current Research and Development budget is much less than it might be.

The economists called for more work on the problem and a thorough testing of their hypothesis. Toward this end, they suggested as an example that they would "like to make some case studies of what we call experimental development to determine better the actual constraints placed on the developer." They were especially interested in "explor[ing] the practical possibilities of making experimental development more profitable, and at the same time shifting some of the risk [from government] to business."[50] This paper makes clear that the authors had already begun to consider several case studies of weapons development programs and technologies, ranging from aircraft engines to radar systems. Also, Alchian, Kessel, and Klein's paper clearly indicates that they considered the improvement of incentive structures—preferably market incentives—to be fundamental to the advancement of the air force's R&D program. But beyond advocating case studies and the refinement of incentives, the economists made little headway toward the development of a usable economic theory of R&D. They nevertheless sought reactions to their "discussion of the issues" from both their economist colleagues at RAND and the air force's R&D managers.[51]

Alchian, Kessel, and Klein's paper clearly evoked an extended response from their colleague and Stanford University economist Kenneth J. Arrow, a long-time RAND consultant who often spent part or all of his summers at RAND, including that of 1955. Although Arrow no longer recalls the specific circumstances under which he wrote it,[52] a few weeks after the Alchian, Kessel, and Klein paper appeared, Arrow produced a richly provocative essay, "Economic Aspects of Military Research and Development," which both took up the issues Klein and his colleagues were struggling with and contained kernels of some of the fundamental ideas that he later published in his foundational article, "Economic Welfare and the Allocation of Resources for Invention" (1962).[53]

RAND played no small role in the intellectual development of Arrow, who would later receive the 1972 Nobel Prize for Economic Sciences for his work in general-equilibrium theory and welfare economics. Awarded a master's degree in mathematics from Columbia University in 1941, Arrow spent the war years as a captain in the Army Air Corps' Weather Division. Following the war, he returned to Columbia with the goal of completing a Ph.D. in mathematics, but he soon switched to economics. Not long after finishing his course work, Arrow moved to the University of Chicago to join the Cowles Commission, which has been described by one writer as "the epicenter of mathematical economics and econometrics in America."[54] The direction of Arrow's work changed permanently, however, after he spent the summer of 1948 at RAND contemplating the problems of applying game theory to Soviet-U.S. relations.[55] According to Arrow, this intellectual problem led him to question whether a nation could have a utility function. Stimulated directly by several RAND researchers and regular consultants, Arrow produced during the summer of 1949 his answer, RAND RM-291, bearing the title "Social Choice and Individual Values." This report not only fulfilled the requirement for his Columbia University dissertation in economics (the degree was awarded in 1951) but also appeared in 1951 as a book by the same title in the Cowles Commission's monograph series with John Wiley & Sons.[56] Arrow's biographer characterizes Arrow's book—and his work—as a "breakthrough in the theory of social choice."[57]

With RAND's R&D group facing in 1955 the problems of analyzing military R&D and determining the optimal allocation of resources, Arrow demonstrated what his biographer has termed a "knack for surveying an existing body of literature and then reformulating it to furnish fresh insight."[58] The existing body of literature was limited at the time, and much of what he surveyed was obviously Alchian's, Kessel's, and Klein's respective and collective musings on R&D. By thinking about the problem of military R&D within the framework of his own social-choice theory, Arrow generated propositions that would eventually emerge as elementary to the economics of R&D. Through social welfare economics Arrow brought a disciplined framework to the problem of the economics of R&D. Certainly it served to encourage much of the subsequent work of the RAND group—especially the case studies that Klein, Kessel, and Alchian had advocated. It also appears to have inspired the work of individual members of the group in fundamental ways.

Following out Alchian's initial thinking, Arrow discussed the problem of development under conditions of uncertainty. At the beginning of the process, uncertainty is greatest; as a project moves forward, however, knowledge or "information" improves, resulting in "the reduction of uncertainty." As Arrow noted, "Development is after all a special case of learning. . . ."[59] He continued:

> In technical language we may say that in development the *a priori* probability distribution of the true state of nature (the unknown performance characteristics of possible models) is relatively flat to begin with. On the other hand the successive *a posteriori* distributions after more and more studies have been conducted are more and more sharply peaked or concentrated in a more limited range, and therefore we have better and better information for deciding what the next step shall be. This implies that at the beginning the preferences among alternative possible lines of investigation are much less sharply defined than they are apt to be later on. This suggests [a la Alchian], at least, the importance of having a wide variety of studies to begin with, gradually eliminating the less promising as more and more information is accumulated. At each stage we have information which will suggest expansion of certain lines of development and the curtailment or elimination of others.[60]

"It would be very interesting," Arrow followed, "to examine some of the major innovations of the past, both in the civilian and the military fields, to see how this principle has operated." In at least two other places in his eighteen-page essay, Arrow advocated undertaking case studies of innovations to address questions that required empirical data to answer. Burton Klein and the members of his team had already advocated such an approach and were perhaps already on their way to developing several such case studies. Arrow also urged that a survey of the literature on invention be done, including the literature on the psychology of the inventor. As will be discussed below, this is precisely what Klein later commissioned a young, Yale-trained economist named Richard R. Nelson to do.

Arrow also pointed toward one of the major precepts of his classic 1962 paper when he argued that most of the literature he had seen on invention suggested that inventors or technological innovators were very rare. Thus a "scarcity of first-rate talent" existed, creating "a kind of indivisibility." Arrow did not pursue this idea of "indivisibility" further; he would not do so until he drafted his now-classic paper for the 1960 NBER conference on "The Rate and Direction of Inventive Activity."[61] In the latter paper, however, ideas or knowledge, which he equated with invention, would be indivisible rather than inventors per se.

In looking at the supply side for invention, Arrow also called attention to what he called in a subtitle the "Inappropriability of the Product":

> The basic incentive problem in the field of development, which arises both in the civilian and the military spheres, is that the product is not very well adapted to appropriation as property. Patent laws provide only a very meagre amount of protection to the developer of an idea. It has frequently been remarked by commercial firms that they regard their advantage derived from research as being rather that of being ahead of their competitors than the advantages obtainable by taking patents on the results of their research. This probably should be investigated somewhat more carefully.[62]

In his sweeping essay, Arrow developed other propositions that neither he nor Klein's research team followed up on to any great extent—at least in the period before 1962. These propositions addressed such questions as whether the technological capabilities of firms resided in the organization (i.e., management) or were merely the result of management's skill or luck in picking creative inventors; what the implications were of the government's being the sole customer for military R and how information flowed within and across firms.[63]

Arrow joined his discussion of firms' incentives for research with his admittedly tentative conclusion that inventors were more important to firms' success than managers by positing the ideal institution in which to carry out scientific and technical advances:

> If we agree that it is the skilled workers and not the managerial talent that is primarily important in [invention and] development, the idea of setting up research institutes seems to be a very attractive one. This, indeed, has been very important in Germany[,] and the establishment of private nonprofit corporations by the Air Force, such as RAND itself, shows that this method of organization is regarded as having some potentialities even here.... An organization which has as its primary object the generation of new ideas is apt to be a much more receptive place than an organization whose primary concern is production.[64]

Finally, Arrow returned to Armen Alchian's theme that had helped to spur the creation of the RAND R&D study: how should the air force (or, more generally, the government) behave as a R&D decision maker? Without mentioning systems analysis, Arrow nonetheless concluded:

> It seems very possible that a considerable decentralization of decisions on development would be in order—avoidance of duplication is not necessarily a virtue. Differing judgements as to the desirability of particular developments should be allowed full chance to operate without excessive coordination or control, which is better advised as more information has accumulated rather than in the preliminary stages. The free entry of ideas should not be unnecessarily interfered with by the government any more than by private firms.

Arrow's paper offered a rich, powerful analysis of the economics of military (and, less explicitly, civilian) R&D because it placed the subject within the framework of welfare economics. Precisely how Klein and the economics-of-R&D group reacted to Arrow's paper remains unclear, but the subsequent actions of the group suggests that the paper had a substantial impact on how the RAND economists began to tackle the seemingly intractable problems they were addressing.[65] One indication of the impact of Arrow's paper on the group is the specific nature of the work undertaken by Richard R. Nelson after he joined RAND in mid-1957.

Klein recruited Nelson specifically to work on the economics-of-R&D project. The son of an engineer-turned-government-economist,[66] Nelson was a comparatively unorthodox economist, at least in his interests if not in his training. After finishing his B.A. at Oberlin in 1952, he entered Yale University's Ph.D. program in economics, where he swiftly finished his formal training and dissertation, "A Theory of the Low Level Equilibrium Trap in Underdeveloped Economies."[67] Yale awarded Nelson the Ph.D. in 1956. Rather than immediately taking a teaching post, Nelson received a postdoctoral fellowship that allowed him to study engineering for a year at Massachusetts Institute of Technology, the university that had been the largest contractor for military R&D during World War II and that was unquestionably the dominant Cold War research university.[68] Following his year at MIT, Nelson took a visiting professorship at Oberlin for the academic year 1956–57. During the spring term, Nelson interviewed with Burton Klein when the latter was on one of his periodic visits to the air force's R&D facilities in Ohio, and Nelson soon joined Klein's R&D team at RAND as a member of the economics department.

Between July and December of 1957, Nelson produced a survey of the literature on the economics of invention, research, and development, a survey that Arrow had urged in his paper, "Economic Aspects of Military Research and Development."[69] When it appeared in print in April 1959, Nelson's work constituted the most comprehensive review of the literature in this domain that had been published to date.[70] Nelson's survey incorporated the main finding of RAND's economics of R&D research group into his review—that the "more successful development efforts have been marked by the early exploration and test of several alternatives."[71]

After completing his literature review, Nelson began to model the economics of parallel R&D programs. Parallel R&D programs, it will be recalled, had been championed in principle by RAND's president, Frank Collbohm, when he opposed placing complete responsibility for the air force's ICBM program in the hands of the Ramo-Wooldridge Corporation, and they were simply one aspect of Alchian's argument for greater diversity in the early stages of R&D programs. Parallel R&D programs were also related to, but not identical with, Alchian, Reuben Kessel, and Burton Klein's emphasis on devoting more resources to experimental development rather than putting all R&D resources into a single, highly specified R&D and procurement program. If Nelson and the group could find a way to model the allocation patterns of running parallel R&D programs, they might be able to prove on theoretical grounds what they had come to conclude on the basis of both their instincts and empirical findings from their growing number of case studies. The title of Nelson's January 1958 paper, "Parallel Development Projects—Good or Bad Strategy?" reflects the environment in which it was produced.[72] Over time Nelson developed his ideas and model, and these appeared in print as a now-classic article in *The Review of Economics and Statistics* in 1961.[73]

Nelson's article gave the work of RAND's economics-of-R&D group considerable play, for the bulk of the empirical data he presented drew from the numerous case studies that had been and were continuing to be pursued by the group. As Nelson noted, "This paper is one product of a continuing RAND study of Research and Development management. The idea of a parallel development strategy has a long history at RAND. Burton Klein, William Meckling, Emmanuel Mesthene, Leland Johnson, Thomas Marschak, Armen Alchian, William Capron, and others have all contributed to its evolution."[74] These case studies gave Nelson's paper a distinctly military and Cold War flavor, but as he acknowledged, the Cold War had created "heightened interest in inventive activity" owing to "the growing awareness that our national security may depend on the output of our military research and development effort."[75] Even then, however, Nelson sought to generalize the problem beyond military R&D. Following a paragraph in which he asked the reader to consider how the air force should make decisions about the development of "an advanced fighter aircraft," he wrote, "Substitute the phrase 'industrial research laboratory' for 'Air Force,' the phrase 'new product' for 'fighter aircraft,' and the problem is a very common one facing R and D management."[76]

The final programmatic aspect of Klein et al.'s large-scale study of R&D derived from the case studies of R&D projects that Alchian, Kessel, and Klein and Arrow had advocated in their respective discussion papers written in the summer of 1955. Soon, case studies began to flow from various members of the research group (see table 9.1). Klein wrote in 1960 that he and his research team had "examined in considerable detail some fifty development programs, attempting to find out as much as we could about the nature of the decisions that were made, and the kinds of evidence on which they were made."[77] These case

Table 9.1
RAND'S case studies of research and development projects, 1956–1962 (listings in chronological order)

Author(s)	Name of study and date	Publication (if any)
Capron, William M.	"The Klystron Story: A Case Study in R&D," D-3432, 23 January 1956	Not known
Klein, Burton H., and Edward Sharkey	"A Case Study in Exploratory Development—Side-Looking Radar," D-3672-PR, 6 June 1956	Not known
Klein, Burton H.	"A Common Sense Approach to Airborne Radar Development," D-4141-PR, 22 February 1957	Not known
Klein, Burton H., and Edward Sharkey	"A Suggested Approach to Airborn Radar Development," RM-1935, 11 July 1957	Not known
Meckling, William	"Are We Overplanning Aircraft Development?" D-4082-PR, 24 January 1957 (on B-52 development)	Not known
Johnson, Leland L.	"Development of the F-102A and F-106A Interceptor Aircraft," D-4634-PR, 9 October 1957	Shows up in Johnson's RM-2549, 20 May 1960
Alchian, Armen A.	"Organizing for Development Duplication," D-4688, 1 November 1957 (on Manhattan Project)	Not known
Meckling, William	"Development Costs of the B-58," D-4320, 1957 (*Note:* Meckling and Thomas K. Glennan, Jr., issued a summary study, "Development of the B-58," 15 August 1963, D11615-PR.)	Not known

Table 9.1 (continued)

Author(s)	Name of study and date	Publication (if any)
Marschak, Thomas A.	"An Initial Investigation of Aircraft Engine Development," D-5114-PR, 16 April 1958	Appears as RM-2283, 5 November 1958
Johnson, Leland L.	"Development of the North American F-108 Long-Range Interceptor," D-5177, 13 May 1958	Not known
Johnson, Leland L.	"Development of the McDonald F-101 Voodoo Series Fighters," D-5200, 20 May 1958	Shows up in Johnson's RM-2549, 20 May 1960
Johnson, Leland L.	"Development of the North American F-100 and F-107A Aircraft," D-5227, 28 May 1958	Shows up in Johnson's RM-2549, 20 May 1960
Johnson, Leland L.	"Development of the Republic F-105A/B," 1958	Shows up in Johnson's RM-2549, 20 May 1960
Marschak, Thomas A.	"An Initial Investigation of Aircraft Engine Development," RM-2283, 5 November 1958	Partly shows up in Marschak's "The Role of Project Histories in the Study of R&D," in Marschak, Thomas K. Glennan, Jr., and Robert Summers, *Strategy for R&D: Studies in the Microeconomics of Development* (New York: Springer-Verlag, 1967).
Marschak, Thomas A.	"Flexibility in Missile Development: The Case of the Sidewinder," D-6098-PR, 6 March 1959	Partly shows up in Marschak's "The Role of Project Histories in the Study of R&D," in Marschak, Thomas K. Glennan, Jr., and Robert Summers, *Strategy for R&D: Studies in the Microeconomics of Development* (New York: Springer-Verlag, 1967).

Table 9.1 (continued)

Author(s)	Name of study and date	Publication (if any)
Meckling, William H.	"On Alternative B-70 Development Programs," D-6701-PR, 19 August 1959	Not known
Nelson, Richard R.	"The Link between Science and Invention: The Case of the Transistor," D-6752-RC, 1 September 1959 (on Bell Labs and the discovery of the transistor)	Published under same title in Richard R. Nelson, ed., *The Rate and Direction of Inventive Activity* (NBER, 1962).
Marschak, Thomas A.	"Strategy and Organization in a Systems Development Project," P-1901-RC, 3 February 1960* (On Bell Labs and its development of a transcontinental high-frequency microwave transmission system)	Published under same title in Richard R. Nelson, ed., *The Rate and Direction of Inventive Activity* (NBER, 1962).
Johnson, Leland L.	"The Century Series Fighters: A Study in Research and Development," RM-2549, 20 May 1960	Partly shows up in Marschak's "The Role of Project Histories in the Study of R&D," in Marschak, Thomas K. Glennan, Jr., and Robert Summers, *Strategy for R&D: Studies in the Microeconomics of Development* (New York: Springer-Verlag, 1967).
Mesthene, Emmanuel G.	"The Titanium Decade," RM-2915-PR, January 1962	Not known
Mesthene, Emmaneul G.	"The Development of Titanium Metal: A Case-Study of Government Policies for Advancing Basic Technology," D-9970-PR	Not known

---

Note: William Meckling also wrote a report on strategic bomber development that I have been unable to find in the RAND system. Allusion to it shows up as part of Thomas Marschak's "The Role of Project Histories in the Study of R&D," in Marschak, Thomas K. Glennan, Jr., and Robert Summers, *Strategy for R&D: Studies in the Microeconomics of Development* (New York: Springer-Verlag, 1967). Meckling also wrote a document, D-6579, "The Savior," on an R&D project, but its contents are classified. Possibly these are the same study.
* The initial draft of this report probably first appeared as a "D" in RAND's report series, but I have not yet located it in the RAND system.

studies—analytically structured narrative histories—of various technological developments deeply informed the conclusions about the management of R&D reached by Klein and his colleagues.[78] The case studies covered both military and civilian projects, and they included some in which RAND had been technically involved. When analyzed together, they confirmed the hypotheses that Armen Alchian had first made in 1953 about the dangers of applying systems analysis to weapons acquisition programs in which research and development figured heavily. Moreover, these case studies supported the Klein group's hypothesis that funding numerous small, comparatively inexpensive experimental programs was superior to funding a single, highly specified development program run on the basis of concurrent engineering principles.

The R&D group's collective findings first manifest themselves in a highly visible medium (an article in *Fortune* in May 1958)[79] and in classified form (two air force briefings, the first delivered initially in February 1958 and the second presented for the first time in June 1958).[80] Under the provocative title "A Radical Proposal for R. and D.," Klein told the readers of *Fortune*, "We need *more* competition, duplication, and 'confusion' in our military research and development programs," not less.[81] His theme harkened back to Alchian's first musings:

> Better planning, stricter control from the top, elimination of the "wasteful duplication" of interservice competition—this sums up the general belief on what we must do about military research and development if we are not to be fatally out-distanced by the Russians.
>
> The truth is precisely the reverse. The fact is that military research and development in this country is now suffering from too much direction and control. There are too many decision makers, and too many obstacles are placed in the way of getting new ideas into development. R and D is being crippled by the official refusal to

> recognize that technological progress is highly unpredictable, by the delusion that we can advance rapidly and economically by planning the future in detail.[82]

Klein marshaled evidence from the numerous case studies his group at RAND had carried out or were in the process of completing. These included Armen Alchian's exegesis of H. D. Smyth's history of the Manhattan Project, in which Alchian highlighted the multiple parallel R&D paths the United States followed in its development of the atomic bomb.[83] Alchian had prepared this analysis in the white heat that followed the Soviet launch of Sputnik, 4 October 1957, and the charges being leveled in Washington that interservice rivalry had led to the United States' falling behind in the space race. Moreover, Alchian was particularly worried about the Eisenhower administration's organizing a single, tightly planned and controlled project to counter the Russian threat.[84] Klein's article also spoke to the United States' R&D policy in the wake of Sputnik. Faced with enormous scientific and technological uncertainties, the only sound policy for R&D in space exploration and advanced military weapons systems was to pursue multiple paths simultaneously. In addition, Klein illustrated the highly vertical nature of the Department of Defense's R&D organizations and argued that the hierarchy extracted severe penalties in terms of efficiencies. He closed his analysis by warning *Fortune*'s readers, "Unless research and development is substantially decontrolled, we are likely to be sadly disappointed in where we stand five to ten years from now."[85]

William Meckling and Emmanuel Mesthene delivered the R&D team's first formal (secret) briefing to the air force in February 1958.[86] It provided the men in blue with a clear indication of what they would soon read in Klein's *Fortune* critique of the air force's and Department of Defense's R&D policies and hear more thoroughly in Klein, Meckling, and Mesthene's classified briefing of RAND's Military Advisory Group, "Research and Development Management," presented 12 June 1958, several weeks after the *Fortune* article appeared.[87] Out of this briefing process would emerge the trio's report, "Military R&D Policics," in December 1958, which offered not only a critique of current policies but also suggested five principles which would yield more effective development policy.[88]

At almost exactly the time that *Fortune* rushed Klein's ideas into print, Richard Nelson put the finishing touches on what became his classic essay "The Simple Economics of Basic Scientific Research."[89]

Although Klein unquestionably moved swiftly to publish his research group's main findings about military R&D projects as a way of interjecting those findings into the raging post-Sputnik national debate about how the United States should organize large-scale R&D projects such as the space program, the findings themselves were not significantly shaped by the Sputnik-induced hysteria in the United States. Klein's group consistently pushed and developed the basic ideas first broached as early as 1953 in Armen Alchian's critique of systems analysis. If the Sputnik crisis gave voice to Klein et al.'s long-held ideas about R&D, the crisis itself seems to have given rise to a formulation in an economics of R&D that had not been a central concern of the RAND group at least as exemplified by the group's case studies: the problem of appropriating the returns to scientific research. As noted above, Kenneth Arrow's once-classified 1955 essay, "Economic Aspects of Military Research and Development," had identified appropriability as a possible problem. With considerably less precision and to a much lesser extent, Alchian, Kessel, and Klein had alluded to the same issue when they earlier laid out a discussion of the issues in research and development. But Nelson's "Simple Economics ..." essay was profoundly shaped by the post-Sputnik hysteria that gripped RAND, the air force, the Pentagon, the White House, and the entire nation. Context matters.[90]

The Impact of the Sputnik Crisis on RAND's Economics of R&D

"Recently," Nelson began his essay, "orbiting evidence of un-American technological competition has focused attention on the role played by scientific research in our political economy." He continued:

> Since Sputnik it has become almost trite to argue that we are not spending as much on basic scientific research as we should. But, though dollar figures have been suggested, they have not been based on economic analysis of what is meant by "as much as we should." And, once that question is raised, another immediately comes to mind. Economists often argue that opportunities for private profit draw resources where society most desires them. Why, therefore, does not basic research draw more resources through private-profit opportunity, if, in fact, we are not spending as much on basic scientific research as is "socially desirable"?[91]

In the wake of the Soviet Union's launch of Sputnik I on 4 October and Sputnik II on 3 November and the United States's public humiliation when an American satellite launcher (sponsored by the Office of Naval Research) failed miserably at Cape Canaveral on 6 December 1957, the scientific community in the United States laid the blame for what was termed variously as the "missile gap," the "science gap," and the "technology gap" on the Eisenhower administration's increasingly restrictive policies on basic research. The expression "the missile gap" stuck. It sent panic throughout the United States; if the Soviets could launch an earth-orbiting satellite with a missile, why could they not also use missiles to send atomic and hydrogen bombs to hit targets in the United States? Secretary of Defense Charles E.—"Engine Charlie"[92]—Wilson, the former president of General Motors, who had once said "what's good for General Motors is good for the United States," became the object of scorn by the scientific community for his derogatory views on basic research: "Basic research is when you don't know what you are doing."[93] Bowing to pressure from the Bureau of the Budget, which sought both to add funds and to strengthen the very young National Science Foundation, and from Eisenhower to trim the federal budget, Wilson had reduced the Department of Defense's R&D budget by 10 percent, which when inflation was taken into account meant a real reduction in the department's R&D expenditures of some 25 percent between 1952 and 1957. And Wilson had promised additional cuts.[94]

The physicist Edward Teller, "father of the hydrogen bomb," sent chills down the spines of Americans when he said, following Sputnik's launch, "Ten years ago there was no question where the best scientists in the world could be found—here in the U.S. Ten years from now the best scientists in the world will be found in Russia."[95] Another old hand from World War II R&D projects, I. I. Rabi, persuaded President Eisenhower to elevate the stature of science within his administration by announcing on radio and television on 7 November 1957 the creation of a Special Assistant to the President for Science and Technology and the appointment of MIT President James R. Killian to that post.[96] A week before this announcement, the Office of Naval Research, which had been the largest funder of academic basic research in the United States between 1946 and 1952 (until overtaken by the new NSF and brought under control by Wilson), moved immediately to restore its budget for academic science. According to Harvey Sapolsky, at its 29 October 1957 meeting, the Naval Research Advisory Committee, which consisted mostly of academic scientists, "enthusiastically endorsed an ONR proposal for a 40 percent increase in the agency allocations to academic science."[97] He goes on to note that "[d]espite one member's admonition that the number should not be pulled 'out of the air,' it is obvious from the discussion that none of the members present had a clear notion of the basis for the percentage increase requested. It [the 40 percent figure] was thought a number, as another member suggested, that people [would] swallow easily."[98] Between 1957 and 1960, ONR's Contract Program (the organization's academic research contract vehicle) moved from a budget of $20 million to one of $82 million.[99]

The air force was not far behind the navy. During the summer of 1957 the Air Force Office of Scientific Research had been in what has been described as a panic mode as it struggled to determine how to meet a newly imposed, immediate 5 percent reduction in its research budget. By September 1957, the office had determined that it would have to cancel 600 research contracts with university scientists. As a reporter for *Aviation Week* put it, "The entire structure of [the] Air Force advanced research program is in danger of collapse."[100] Pleas up the ranks of the air force and into the Office of the Secretary of Defense for exempting basic research from the newly imposed budget cuts went unheard until the air force brought in heavy hitters like I. I. Rabi to work the White House directly. Sputnik transformed the atmosphere at the Air Force Office of Scientific Research (AFOSR) from one of panic to one of political entrepreneurship. As Komons writes, "From the $11-million expenditure rate imposed in early September, AFOSR's expenditure rate for the first six months of the fiscal year climbed to $19 million. Its fiscal year 1958 budget, pegged at $16.3 million in June [1957] and reduced by five percent in July [1957], ended up at a level of $22.5 million—a thirty-eight percent increase over fiscal year 1957."[101] In January 1958, Brigadier General H. F. Gregory, Commander of the AFOSR, wrote his directors, "The Sputnik influence along with the announced national policy of more emphasis in research has been extremely stimulating to the scientific community."[102]

RAND itself had been deeply affected by this sea change in research funding policy. During the summer of 1957—the same period in which Richard Nelson was getting his grounding in the economics of R&D in Santa Monica—RAND had become aware that the Wilson-imposed budget cuts would affect its research programs. Indeed, one of the 600 telegrams prepared by the Air Force Office of Scientific Research in September announcing the AFOSR's cuts to and cancellations of research contracts was addressed to RAND. Sputnik thus saved RAND from an immediate, sharp cutback in its funding.

The Department of Defense as a whole also responded to what *Fortune* magazine called the "research crisis."[103] The watch at the department had changed; Wilson was out, and Neil H. McElroy had replaced him as secretary of defense. As Kevles points out, "McElroy rescinded his predecessor's order to reduce basic defense research funds by 10 percent and issued a directive endorsing the support of basic science as a key element in Defense Department policy."[104] Roger Geiger succinctly captures how good things became for academic scientists in the wake of Sputnik:

> Scientists attained representation in the nation's corridors of power and there propounded the ideology of basic research. The NSF was the principal beneficiary of this new environment (aside from NIH, which continued to mushroom); but the defense establishment also played a role. Support of academic research thus grew to flood stage with increases from every tributary. NSF added $100 million to the university research economy in the initial post-Sputnik era—more than a sevenfold rise. The defense establishment, however, contributed $283 million in additional support.[105]

Although the Department of Defense increased basic science funding, it also established a new directorate, headed by a Director of Defense Research and Engineering, and created the Advanced Research Projects Agency (ARPA) within the Office of the Secretary of Defense. Both these moves suggested strongly centralizing tendencies within the higher reaches of the DOD, which clearly threatened the research autonomy of the ONR, the Air Force Office of Scientific Research, and, by extension, RAND.[106]

These moves were intended to address in part the criticism that quickly gathered in the wake of Sputnik that the Pentagon had badly managed its R&D program. Addressing the National Science Foundation in May 1958, RAND's Charles Hitch captured the sentiment quite nicely:

> Judging from the press there is general agreement that all is not well with military R and D, and rather surprising unanimity in diagnosing what is wrong. Apart from a small but distinguished group of dis-

> senters (mainly from science and industry) practically everyone charges that military R and D is uncoordinated and inadequately planned, and plagued by duplication, competition, secrecy, and waste. The remedies are alleged to be obvious: there must be strong central direction and coordination; more and better central planning; tough-minded decisions to eliminate duplication; suppression of inter-service and other competition; probably some "Czars" to knock heads together.[107]

As head of RAND's economics department, Hitch had followed RAND's project in the economics of R&D, so it is not surprising that he went on to espouse the Klein group's main argument about the importance of diversity in research and early development owing to the extreme uncertainties inherent in R&D. Unquestionably Hitch and the entire RAND economics-of-R&D team saw as their mission heading off any substantial effort to centralize R&D processes in the United States in general and in the Pentagon in particular. What had begun as a cleverly offered critique of system analysis by Armen Alchian and of concurrent engineering by Klein and his colleagues had now become a high-stakes effort in national R&D policy formulation. Klein's *Fortune* article, the corpus of his group's published scholarship, and his team's briefings for the air force made their position on diversity in approaches, low-cost experimental development, and parallel development paths quite clear.[108] But on what empirical basis did the RAND group deal with basic research, especially the question of whether the United States was spending enough on basic research?

In the entire corpus of idea-pieces, essays, documents, research memoranda, and scholarly publications that the RAND economics-of-R&D group produced before 4 October 1957, there is virtually no attention given to the question of what incentives private firms have for investing in basic research. As has been noted already, Alchian, Kessel, and Klein's July 1955 essay, "The Efficiency of the Research and Development Program: A Discussion of the Issues," distinguished "research for basic principles" from "experimental development" and "development of weapons systems" in their articulation of the importance of parsing the elements of the overall R&D program of the military.[109] But in discussing basic research, the economists did not suggest that firms lack incentives for undertaking basic research, although they clearly recognized that incentives structures should vary across all three aspects of R&D if overall R&D efficiency were to be maximized.

Certainly the economists did not suggest that the nation was systematically underinvesting in basic research and that the welfare of the nation was therefore fundamentally threatened. Richard Nelson's sweeping survey of the literature on invention, research, and development[110] does not address the issue at all, and certainly it does not suggest that firms systematically underinvest in basic research owing to their inability to capture adequate returns on their investment in basic research (the "appropriability" problem, as it later became known). None of the case studies of military R&D projects discusses the appropriability problem. Of course, all the explorations on the benefits of parallel research and development and what Alchian, Kessel, and Klein called "experimental development" (which, as demonstrated above, emerged from the economists' critique of system analysis and concurrent engineering practices) assumed that military contractors (i.e., weapons suppliers) had low incentives to engage in high-risk R&D absent the guarantee of securing a procurement contract from the air force or some other branch of the military.

None of these formulations discusses "high-risk R&D" in the same manner as, say, Kodak's Kenneth Mees analyzed fundamental scientific research in industry in his textbook on the management of industrial research.[111] The RAND group's fundamental objective was to correct what they perceived as a major problem with systems-analysis-driven procurement practices, concurrent engineering regimes, and the lack of diversity (or an inadequate amount of diversity) in the early stages of military R&D. Their work advocated the development of policies to ensure that the Pentagon would engage in more early "experimental development," more parallel development projects, and less concurrent engineering. Deep study and analysis of the emergence of new technologies and systems—both military and civilian—led the RAND group to these policy recommendations.

CONCLUSIONS

Ironically, although the Soviet launch of the Sputnik artificial satellites led the RAND group to go public with these policy recommendations based on the group's extensive case studies, Sputnik also provided both the catalyst and context for the group's most long-lasting contribution to the economics of technical change. The Sputnik-inspired panic led Richard Nelson to write in 1958 and publish in 1959 his foundational article, "The Simple Economics of Basic Scientific Research."[112] This article and the continued concern in the United States about a missile gap subsequently inspired Kenneth Arrow in 1959 to return to his 1955 secret document and to transform some of its tentative ideas into his classic paper "Economic Welfare and the Allocation of Resources for Invention."[113]

Welfare economics—the type of economics that Arrow brought to RAND—provided the framework for both Nelson's and Arrow's analyses. The central question in welfare economics concerns the conditions under which an optimal allocation of resources throughout an economy is achieved. The Sputnik-induced fear (capitalized on and perhaps exploited by the American scientific community) that Russian missiles could as easily deliver hydrogen bombs to American cities as they could satellites into space, while the United States lagged in such capabilities because of capitalism's "market failure" of not pursuing enough basic research, became the starting point for Nelson's (and Arrow's) analysis. To those panicked by Sputnik, the United States obviously suffered from a suboptimal allocation of resources. Given this condition, "simple" welfare economics could pinpoint the failure by working backward to determine why resources throughout the economy were not being allocated for basic research. Assuming that firms could not adequately appropriate the full economic benefits of basic research (most of which accrued to society as a whole, not to the firm footing the bill), it was clear that firms had inadequate incentives to invest in basic research; hence, the larger social welfare was suffering.

These, then, were the "simple economics" of basic research. As Nelson admitted frankly, however,

> The assumptions on which the preceding argument is based rest but shakily on fact. Basic research is certainly not a homogeneous commodity.[114] ... Thus one cannot make an airtight statement, based on welfare economics, that we are not spending as much on basic scientific research as we should. But I believe that the evidence certainly points in that direction.[115]

Welfare economics provided clarity on why the United States was behind the Russians, why more public support for basic research was necessary, and why universities and nonprofit research organizations in particular should be the recipients of government largesse. As an abstractor at CalTech somewhat excessively characterized Nelson's piece a few years later, "The author's basic suggestion is to collectivize research."[116]

Importantly, the core work of the RAND economics-of-technical-change group did not, however, draw on welfare economics. The numerous case studies undertaken by the group, however, provided a rich empirical basis on which to draw conclusions about the importance of maintaining a diversity of approaches to technological development, especially in the early stages when scientific and technical uncertainty is greatest. This work, which emerged both from an internal critique of systems analysis at RAND and in response to the way the nation decided to build its first ICBM, points to work that Richard Nelson and another RAND researcher, Sidney M. Winter, undertook later in their pioneering studies in evolutionary economics.[117] More immediately, however, it provided the careful reader of the day with an empirically grounded rationale for understanding why certain elements of the U.S. response to Sputnik—such as the Eisenhower administration's decision (made in spite of the president's personal preferences) to allow each of the military services to continue simultaneous development of its own intermediate range ballistic missiles and to push multiple follow-on programs to the Atlas ICBM—made sense, given the objective of national security. In this respect, the objectives of the economics-of-technical-change group—especially vis-à-vis countering the Soviet threat with the development of ballistic missiles—were fully realized by the early 1960s. Ironically, this core work—deeply grounded in case studies—has not survived in the way that Nelson's "Simple Economics of Basic Scientific Research" and Arrow's "Economic Welfare and the Allocation of Resources for Invention" remain, rather unfortunately in my view, as fundamental to the (neoclassical) literature in the economics of technical change.[118]

Appendix

RAND's Economics of Innovation Project associates, 1955–1962 (listed alphabetically)

Name	Dates at RAND and on project	Comments and output
Alchian, Armen A.	Employee: 12/16/49–2/16/60 Consultant: 7/11/47–12/15/49; 2/16/60–9/30/64 Project: 1955(?)–1960(?)	• Ph.D., Economics, Stanford, 1944 • Carried out early analysis of experience curves • Initial critique of systems analysis in early R&D appears to have stimulated RAND R&D project • Unclear if he was ever formally part of R&D project • Left RAND's employ in 1960
Arrow, Kenneth J.	Consultant: 6/15/48– Project: 1955(?)–1962(?)	• Ph.D., Economics, Columbia, 1951 • Formal relationship with RAND R&D project tenuous since he was a consultant rather than a staff member
Capron, William M.	Employee: 7/16/51–9/26/56 Consultant: 9/27/61–6/8/62 Project: 1955–1956(?)	• M.P.A., M.A., Harvard, 1947, 1948 • Left RAND for Stanford Economics Department
Hitch, Charles	Employee: 7/19/48–1/15/61 Project: 1955(?)–1961(?)	• Master's, Economics, Oxford, 1935 • Head of RAND Economics Department, which was the home of the R&D project • Left RAND in 1961 to implement RAND program budgeting methods in McNamara Pentagon

Name	Dates at RAND and on project	Comments and output
Johnson, Leland L.	Employee: 6/10/57–12/1/67 Project: 1957–1962	• Ph.D., Economics, Yale University, 1957
Kessel, Reuben A.	Employee: 9/29/52–9/3/56 Consultant: 9/4/56–9/30/57 Project: 1955–1956(?)	• Ph.D., Economics, Chicago, 1954 • Returned to Chicago, 1957
Klein, Burton H.	Employee: 4/21/52–8/31/67 Consultant: 1/1/49–4/20/52 Project: 1955–1967(?)	• Ph.D., Economics, Harvard, 1953 • Many of Klein's ideas from the R&D project are embodied in *Dynamic Economics* (Cambridge: Harvard University Press, 1977)
Lieberman, Vivian	Employee: 3/1/54–3/30/56 Consultant: 8/1/57–9/30/58 Project: 1955–1956(?)	• Credentials unknown to author
Marschak, Thomas A.	Employee: 5/3/54–1/22/60 Consultant: 1/23/60–9/30/73 Project: 1955(?)–1960(?)	• Ph.D., Stanford University, 1957 • Left RAND for U.C., Berkeley, 1960 • Employed many of RAND's R&D case studies in his article, "The Role of Project Histories in the Study of R&D," in Marschak, Thomas K. Glennan, Jr., and Robert Summers, *Strategy for R&D: Studies in the Microeconomics of Development* (New York: Springer-Verlag, 1967)

Name	Dates at RAND and on project	Comments and output
Meckling, William H.	Employee: 11/7/56–3/29/57 Consultant: 6/1/62–4/30/63 Project: 1956–1957	• M.B.A., Denver, 1947; Economics study, Chicago, 1949–1952 • Left RAND to direct Naval Warfare Analysis Group, Center for Naval Analyses
Mesthene, Emmanuel G.	Employee: 3/2/53–7/29/64 Consultant: 12/20/52–2/28/53 Project: 1955–1964	• M.A., Philosophy, Columbia, 1949; Ph.D., Columbia, 1964 • Later directed Harvard University Program on Technology and Society
Nelson, Richard R.	Employee: 7/16/57–6/1/61; 1/29/63–9/2/68 Consultant: 9/3/68–9/30/87 Project: 1957–1961	• Ph.D., Economics, Yale, 1956 • Left RAND in 1961 for position with Council of Economic Advisors but later returned
Winter, Sidney G., Jr.	Employee: 10/1/59–9/21/61; 8/8/66–8/16/68 Consultant: 7/9/62–8/7/66; 8/19/68–9/30/72 Project: 1959–1961(?)	• Ph.D., Economics, Yale, 1964

Acknowledgments

I am deeply grateful to David R. Jardini for his enormous help with the research for this chapter. Without his initial work in Santa Monica on my behalf, this paper could not have been written. I also thank Rick Bancroft, librarian in RAND's Classified Library, for his help in securing several documents critical for this chapter's findings. Wesley Cohen, Steven Klepper, Richard R. Nelson, F. M. Scherer, Ashish Arora, Gustave Shubert, Louis Galambos, William Lazonick, Mary O'Sullivan, Glenn Bugos, Burton Klein, and Brad Bateman provided extensive commentary on early drafts of this chapter, for which I am most grateful. I thank also colleagues who offered comments and criticisms when I presented this research at the Dibner Conference and in seminars at Carnegie Mellon University, U.C.L.A., and the University of

Gothenburg. The interpretations contained in the chapter are those of the author and do not represent those of the RAND Corporation or present and former members of its research staff. The research for this chapter was supported by the Science and Technology Studies Program of the National Science Foundation (SBR-9421048).

NOTES

1. Indeed, RAND was one of the principal sites for the quantitative revolution, another one being the Cowles Commission at the University of Chicago, with which RAND had important ties. As one Nobel Prize–winning economist later wrote, "For centrality to the postwar quantitative social sciences, the Cowles Commission and the RAND Corporation were definitely the places to see and to be seen." Herbert A. Simon, *Models of My Life* (New York, 1991), 116.

2. For a recent review article about RAND and its early Cold War history, see David A. Hounshell, "The Cold War, RAND, and the Generation of Knowledge, 1946–1962," *Historical Studies in the Physical and Biological Sciences* 27, 2 (1997): 237–268.

3. Daniel J. Kevles, *The Physicists: The History of a Scientific Community in Modern America* (New York: Knopf, 1978).

4. Alchian had earned his Ph.D. at Stanford University in 1944, finishing his dissertations, entitled "Some Observations on the Contemporary Analyses of the Effects of a General Change in Money Wage Rates," after he had begun his teaching career at the University of Oregon. *Dissertation Abstracts International*, CD-ROM version. During and immediately after the war, Alchian served for three years in the army's statistical office for three years. Based in Ft. Worth, Texas, he wrote intelligence tests for screening and placement of recruits. Interview with Armen A. Alchian, Los Angeles, California, 9 January 1997.

5. RAND Corporation Employee Number File (by Alpha), s.v., Alchian, Armen.

6. Interview with Armen A. Alchian, 9 January 1997.

7. Armen A. Alchian, "Economic Replacement Policy," R-224, 12 April 1952, RAND Corporation.

8. Armen A. Alchian, "Reliability of Progress Curves in Airframe Production," RM-260-1, 7 October 1949 (revised 3 February 1950), RAND Corporation. Alchian delivered a related paper at the American Economic Association in New York, 27–30 December 1949, entitled "An Airframe Production Function," P-108, 10 October 1949, RAND Corporation. This paper was abstracted in *Econometrica* in June 1950. RM-260-1 was initially classified because it was based on classified data.

9. Armen A. Alchian, "Reliability of Progress Curves in Airframe Production," *Econometrica* 31 (1963): 679–693; Harold Asher, "Cost-Quality Relationships in

the Airframe Industry," R-291, 1956, RAND Corporation (also issued as Ph.D. diss., Ohio State University, 1956); Werner Z. Hirsch, "Manufacturing Progress Functions," *Review of Economics and Statistics* 34 (1952): 143–155. Prior to Asher and Hirsch, Kenneth Arrow and Selma Schweitzer Arrow carried out analyses of progress curves using World War II aircraft production data. See their two RAND reports, "Methodological Problems in Airframe Cost-Performance Studies," RM-456, 20 September 1950, RAND Corporation" and (with H. Bradley) "Cost-Quality Relations in Bomber Airframes," RM-536, 6 February 1951, RAND Corporation. The foundation paper in progress functions is, of course, T. P. Wright, "Factors Affecting the Costs of Airplanes," *Journal of Aeronautical Science* 3 (1936): 122–128.

10. A. A. Alchian, G. D. Bodenhorn, S. Enke, C. J. Hitch, J. Hirshleifer, and A. W. Marshall, "What Is the Best System?" D-860, 4 January 1951, RAND Corporation.

11. Armen A. Alchian, "Phasing Problems, II: The Chef, Gourmet, and Gourmand," D-1136, 8 January 1952, RAND Corporation. This paper and two other papers by Alchian opened an important dialogue regarding the validity of a particular systems analysis that RAND had carried out in a report, "Economic Comparison of Intercontinental Airplane Systems for Strategic Bombing" (R-229, RAND Corporation), which Alchian later noted was being heavily used by the Strategic Air Command Development Planning Objective to chart strategy on the development of its next intercontinental bomber. For evidence on the dialogue, see Diran Bodenhorn, "Research and Development Decisions," D-1651, 21 April 1953, RAND Corporation. On the use of R-229, see Armen A. Alchian, "Some Impressions about the Organization and Objectives of Research and Development," D-1766, 20 July 1953, RAND Corporation.

12. Ibid.

13. Ibid. Alchian went on to develop his idea about chefs, gourmets, and gourmands and R&D and weapons procurement decision making into a regular RAND Research Memorandum: Armen A. Alchian, "The Chef, Gourmet, and Gourmand," RM-798-PR, 24 March 1952, RAND Corporation. These same ideas show up in a comprehensive—and still classified—report he wrote in 1953, "Elementary Principles for Developing and Procuring a Superior Air Force," D-1629, 3 April 1953, RAND Corporation. The title of this report is unclassified, but the report itself remains classified because it contains restricted data.

14. David Milobsky, "Leadership and Competition: Technological Innovation and Organizational Change at the U.S. Department of Defense—1955–1968" (Ph.D. diss., Johns Hopkins University, 1996), 17, 38–39. As Milobsky goes on to demonstrate, these changes in 1953 foreshadowed ever greater centralization of R&D decision making that would be made in 1958 in the wake of Sputnik. The air force also restructured its R&D organizations, beginning the process in 1949 with the reports of the Ridenour Committee (Scientific Advisory Board, September) and the Anderson Committee (Air University, November 1949), which led in 1951 to the Air Research and Development Command (ARDC). On these

changes, see John Clayton Lonnquest, "The Face of Atlas: General Bernard Schriever and the Development of the Atlas Intercontinental Ballistic Missile, 1953–1960" (Ph.D. diss., Duke University, 1996), 172.

15. Alchian, "Some Impressions About the Organization and Objectives of Research and Development."

16. Ibid.

17. The reorganization of the air force's diverse research and development programs is treated in Nick A. Komons, *Science and the Air Force: A History of the Air Force Office of Scientific Research* (Arlington, Virginia: Office of Aerospace Research, 1966); Robert Sigethy, "The Air Force Organization for Basic Research, 1945–1970: A Study in Change" (Ph.D. diss., American University, 1980); and Michael H. Gorn, *Harnessing the Genie: Science and Technology Forecasting for the Air Force, 1944–1986* (Washington, D.C.: Office of Air Force History, 1988).

18. Alchian's "Phasing Problems, II: The Chef, Gourmet, and Gourmand," singles out Schriever's principles as ones that are wrong because they do not distinguish between R&D decisions and procurement decisions. As Lonnquest notes, "By early 1953, the weapon system management concept [advocated by Schriever and approved by his commanding officer Donald Putt] had emerged as the centerpiece of the Air Force's efforts to restructure its approach to R&D." Lonnquest, "Face of Atlas," 172.

19. The most recent historical analysis of the Atlas development program is Thomas P. Hughes, *Rescuing Prometheus* (New York: Pantheon, 1998), 69–139. Hughes views the Atlas program as a pivotal step in the development of techniques for managing the development of large-scale technological systems. A recent, critical examination of Schriever, the ICBM, and the ideal of concurrent engineering is Lonnquest, "Face of Atlas." Lonnquest suggests that concurrency was largely an ideology rather than a reality of the Atlas program, but it obviously provided Schriever and Ramo-Wooldridge with important political power.

20. The original Teapot Committee included Hendrik W. Bode (Bell Labs), Louis G. Dunn (CalTech), Lawrence A. Hyland (Bendix Aviation), George B. Kistiakowsky (Harvard), Charles C. Lauitsen (CalTech), Clark B. Millikan (CalTech), John von Neumann (Princeton University), Allen E. Puckett (Hughes Electronics), Simon Ramo (Ramo-Wooldridge Corporation), Jerome B. Wiesner (MIT), and Dean Wooldridge (Ramo-Wooldridge Corporation). The history of the Teapot Committee and the air force's ballistic missile program is reviewed in Hughes, *Rescuing Prometheus*, 84–93; Lonnquest, "Face of Atlas," 85–102; Jacob Neufeld, *The Development of Ballistic Missiles in the United States Air Force, 1945–1960* (Washington, D.C.: Office of Air Force History, 1990); and Edmund Beard, *Developing the ICBM: A Study in Bureaucratic Politics* (New York: Columbia University Press, 1976).

21. Bruno Augenstein, *A Revised Development Program for Ballistic Missiles of Intercontinental Range* (Santa Monica: The Rand Corporation, 8 February 1954).

Augenstein's report made clear that hydrogen bombs could be made compact enough to be carried on a missile and would be powerful enough to make missiles highly effective weapons even if such missiles could not deliver their payloads precisely to their intended targets. His report is usually credited as one of two findings that convinced the air force to move ahead with all rapidity to the development of the ICBM. Augenstein's role in the ICBM story is told in Fred Kaplan, *Wizards of Armageddon*, 112–116; Donald MacKenzie, *Inventing Accuracy: A Historical Sociology of Nuclear Missile Guidance* (Cambridge: MIT Press, 1990), 105–110; and Lonnquest, "Face of Atlas," 98–99.

22. Some members of the original Teapot Committee did not stay on with the Atlas Scientific Advisory Committee. New members included Norris E. Bradbury (Director, Los Alamos National Laboratory), Frank Collbohm (RAND), James W. McCrae (Sandia National Laboratory), Robert R. McMath (University of Michigan), Charles A. Lindbergh, Herbert York (Director, Livermore National Laboratory), Jerrold R. Zacharias (Los Alamos National Laboratory), and Carroll L. Zimmerman (Strategic Air Command).

23. F. R. Collbohm to J. von Neumann, 24 August 1954, as quoted in Lonnquest, "Face of Atlas," 127.

24. Collbohm is quoted by Davis Dyer, "Necessity as the Mother of Convention: Developing the ICBM, 1954–1958," paper delivered at the Business History Conference Annual Meeting, 19 March 1993, citing Report of Atlas Scientific Advisor[y] Committee Panel Meeting of 3 December 1954. See also *idem*, *TRW: Pioneering Technology and Innovation since 1900* (Boston: Harvard Business School Press, 1998), note 41, 442–443. Edmund Beard goes much further than Dyer and suggests that Collbohm's opposition might have stemmed from his close connections with the aircraft industry and "a fear that industry would not like the idea and refuse to participate in the ICBM program." Beard, *Developing the ICBM*, 177. One of Collbohm's communications with von Neumann implies, to me at least, that Collbohm believed that the aircraft industry possessed considerable capabilities and that "the forces of industry competition" were good for technological development. See Collbohm to von Neumann, 14 October 1954, quoted in Hughes, *Rescuing Prometheus*, 132.

25. Dyer, "Necessity as the Mother of Convention." As Dyer points out, to some extent, Collbohm's advocacy of parallel development programs for the ICBM proved effective. Early in 1955 the air force initiated development of a second ballistic missile, the Titan, and by the end of the year it had started a program for an intermediate range ballistic missile, the Thor, and had undertaken a highly secretive program to develop both the launcher and the satellite for space reconnaissance.

26. Bruno W. Augenstein, interview with David A. Hounshell and David R. Jardini, 7 December 1994, Santa Monica, California. See also Bruno W. Augenstein, "Roles and Impacts of RAND in the Pre-Apollo Space Program of the United States," RAND Reprint-630, 1997, 4. My unrecorded telephone interview with

Augenstein on 14 April 1999 led to further information about the air force's offer to RAND, but Augenstein was unable to recall who in the air force specifically asked RAND to assume responsibility for the Atlas project. Augenstein pointed out that with respect to RAND's basic mission, the organization soon undertook a major development project on the SAGE early warning system, which led to the creation of the System Development Corporation's being spun out of RAND and which did not undermine RAND's basic mission.

27. Armen Alchian, "Systems Analyses—Friend or Foe?" D-1778, 24 July 1953, RAND Corporation.

28. Putt's speech, "The Systems Approach to Air Weapons Development," was delivered 15 January 1953. An excerpt of it appeared in the *SAE Journal*, February 1953, 60.

29. On Putt's career, see Komons, *Science and the Air Force*; Sigethy, "The Air Force Organization for Basic Research"; and Gorn, *Harnessing the Genie*.

30. Lonnquest, "Face of Atlas," 30–40.

31. Alchian, "Systems Analyses—Friend or Foe?"

32. Ibid. Essentially Alchian is saying that optimization methods and other analytic tools that RAND had developed for decision making were not applicable to R&D.

33. From the official language used by RAND to describe the security controls on its Document—"D"—series.

34. Edward S. Quade, "The Proposed RAND Course in Systems Analysis," D-1991, 15 December 1953, RAND Corporation.

35. Armen A. Alchian and Reuben A. Kessel, "A Proper Role of Systems Analysis," D-2057, 27 January 1954, RAND Corporation.

36. Ibid.

37. Minutes of the Management Committee, RAND Corporation, 14 July 1954.

38. The Minutes of the Management Committee of 1 and 22 December 1954 suggest that RAND hoped the new project would have some influence on decisions concerning the ICBM project. But in fact RAND's work on the economics of technical change had little or no influence on the ICBM program.

39. Klein had been a consultant to RAND since the beginning of 1949, so he was no doubt intimately aware of the problems RAND had faced with systems analysis, which were conveniently lumped under the rubric of "specification problems."

40. Burton H. Klein, *Germany's Economic Preparations for War* (Cambridge: Harvard University Press, 1959). Data on Klein's employment at RAND derives from RAND's employee and consultant list.

41. Biographical material on Klein has been derived from *Who's Who in America*, 43rd ed., 1984–1985, s.v., "Klein, Burton Harold." Klein is perhaps best known for his work, *Dynamic Economics* (Cambridge: Harvard University Press, 1977).

42. In an interview with this author, Richard Nelson noted that he had grown increasingly concerned about working on strictly military R&D topics. He therefore welcomed (and perhaps even encouraged) the broadening of the R&D project to include civilian R&D. Richard R. Nelson, interview with David A. Hounshell, 9 April 1996, Carnegie Mellon University.

43. I make this inference based on my reading of Kessel and Klein's early work and also on the Minutes of the Management Committee for 22 December 1954, which suggest that the Management Committee believed that Klein should involve George Dantzig in "seeing if he can contribute either in developing new methodology for handling the problems or finding problems to which linear programming can be applied."

44. Reuben A. Kessel and Vivian W. Lieberman, "Some Evidence on How the Research Budget Is Allocated," D-2671-PR, 11 January 1955, RAND Corporation.

45. Vivian W. Lieberman, "Organization and Administration of the Military Research and Development Programs: Excerpts of Testimony before the Subcommittee on Government Operations," D-3225, 7 October 1955, RAND Corporation, and D. F. Baer, "Formulating Research and Development Policy," D-4116, 11 February 1957, RAND Corporation.

46. See, for example, Reuben A. Kessel and Burton H. Klein, "Trip Report of a Visit to ARDC, and Headquarters WADC," D-3087-PR, 1 August 1955, RAND Corporation.

47. Emmanuel Mesthene, "R&D Management Education at the Air University," D-4315, 24 May 1957, RAND Corporation.

48. Kessel and Klein, "Trip Report" (see note 46).

49. Armen A. Alchian, Reuben A. Kessel, and Burton H. Klein, "The Efficiency of the Research and Development Program: A Discussion of the Issues," D-3017-PR, 6 July 1955, RAND Corporation.

50. Ibid.

51. Kessel and Klein's "Trip Report of a Visit to ARDC, and Headquarters WADC" (note 46) states clearly that this trip was to vet their and Alchian's ideas laid out in D-3017-PR.

52. Interview with Kenneth J. Arrow, Stanford University, 27 February 1997.

53. Kenneth J. Arrow, "Economic Aspects of Military Research and Development," D-3142, 30 August 1955, RAND Corporation; Kenneth J. Arrow,

"Economic Welfare and the Allocation of Resources for Invention," in Richard R. Nelson, ed., *The Rate and Direction of Inventive Activity: Economic and Social Factors* (Princeton: Princeton University Press, 1962), 609–625.

54. C. Daniel Vencill, "Kenneth J. Arrow, 1972," in Bernard S. Katz, ed., *Nobel Laureates in Economic Sciences: A Biographical Dictionary* (New York: Garland Publishing, Inc., 1989), 11.

55. RAND's personnel records note that his tenure as a RAND consultant began 15 June 1948. Among Arrow's first "published" documents at RAND were "The Allocation of Resources between Bombs and Bombers: Preliminary Report," D-615, 12 August 1949, RAND Corporation, and "Allocation of Resources for Air Defense," D-634, 29 August 1949, RAND Corporation. Like his RAND colleague Armen Alchian, Arrow also devoted attention to progress curves for airframe production. He coauthored two papers on this topic with his wife, Selma Schweitzer Arrow: "Methodological Problems in Airframe Cost-Performance Studies," RM-456, 20 September 1950, RAND Corporation, and (with H. Bradley) "Cost-Quality Relations in Bomber Airframes," RM-536, 6 February 1951, RAND Corporation.

56. Vencill, "Kenneth J. Arrow," 12; Kenneth J. Arrow, *Social Choice and Individual Values* (New York: John Wiley & Sons, 1951), Cowles Commission Monograph no. 12.

57. Vencill, "Kenneth J. Arrow," 12. This characterization perhaps overstates the originality of Arrow's important work.

58. Ibid.

59. Arrow, "Economic Aspects of Military Research and Development."

60. Ibid. Compare this formulation with that of Arrow's "Economic Welfare and the Allocation of Resources for Invention," 618.

61. Arrow's draft of this NBER paper first appeared in the RAND "P" series: Kenneth J. Arrow, "Economic Welfare and the Allocation of Resources for Invention," P-1856-RC, 15 December 1959, RAND Corporation.

62. Arrow, "Economic Aspects of Military Research and Development."

63. Arrow would take up these questions in later work, especially in his book of lectures, *The Limits of Organization* (New York: Norton, 1974).

64. Ibid. This particular discussion in Arrow's essay suggests that he was addressing Alchian, Kessel, and Klein's "discussion-of-the-issues" paper, because, as noted in the text above, they had specifically examined the question of incentive structures for nonprofit research institutes.

65. Armen A. Alchian told me that soon after the economics of R&D group got started at RAND, he essentially moved to other topics because he found the

problems of the economics of R&D too difficult to make satisfactory progress. Interview with Armen A. Alchian, UCLA, 9 January 1997.

66. Saul Nelson had worked as a railroad bridge designer and engineer until the Great Depression, whereupon he transformed himself into an economist who worked in various government agencies for the remainder of his career. Richard R. Nelson, interview with David A. Hounshell, 9 April 1996, Carnegie Mellon University.

67. *Dissertation Abstracts International*, CD-ROM version.

68. On MIT's involvement in military R&D, see Stuart W. Leslie, *The Cold War in American Science: The Military-Industrial-Academic Complex at M.I.T. and Stanford* (New York: Columbia University Press, 1993). Data on MIT's World War II research funding can be found in Appendix C of James Phinney Baxter, III, *Scientists against Time* (Boston: Little, Brown and Company, 1946), 456. Under the presidency of physicist Karl Compton, MIT had been at the forefront of engineering education in the United States during the 1930s, transforming the engineering curriculum from its orientation toward skill, know-how, and practice toward grounding students in scientifically rigorous theory ("engineering science," as it was called). Compton's successor, Julius Stratton, championed the university as the rightful home of basic or fundamental research, and he worried that American universities, including his own, might be permanently warped by the vast amount of applied research they now conducted for the military. Whether as a postdoctoral "undergraduate" engineering student the economist Nelson appreciated the extent to which MIT ran on Cold War military research monies, certainly he shared Stratton's celebration of the university as the rightful home for basic research and Stratton's concerns about the domination of research by the military. Nelson's embrace of the university as the rightful home for basic research is evident in his second published work in the economics of R&D, which will be discussed below. His concerns about military domination of research emerge in Richard R. Nelson, interview with David A. Hounshell, 9 April 1996, Carnegie Mellon University, in which he notes several times his discomfort while working at RAND about carrying out strictly air force–related research. He publicly suggested his reservations about the influence of the defense budget on the American economy in Richard R. Nelson, "The Impact of Arms Reduction on Research and Development," *American Economic Review* 53 (1963): 435–446.

69. Richard R. Nelson, "Invention, Research, and Development: A Survey of the Literature," D-4799, 27 December 1957, RAND Corporation. This report later appeared in revised form as "The Economics of Invention: A Survey of the Literature," RM-2146-1 (15 April 1958 and revised 15 January 1959), as P-1604, 29 January 1959, and in print under the same title in *The Journal of Business* 32 (1959): 101–127.

70. In 1958, John Jewkes, David Sawers, and Richard Stillerman published their book *The Sources of Invention* (London: Macmillan, 1958), the first chapter of which is an idiosyncratic review, "Modern Views on Invention," which was not

really a literature survey. Later chapters in *The Sources of Invention* cite literature in the domain, but they too are by no means systematic.

71. Nelson, "The Economics of Invention," 114.

72. D-4904, 28 January 1958, RAND Corporation. This paper specifically mentions the ongoing parallel development efforts with ballistic missiles in the United States. Nelson had made one earlier attempt to model alternative R&D decision making in "Choice from amongst Alternatives in Situations in Which Commitment Can Be Delayed and Additional Information Obtained: Some Applications to R&D," D-4668, 23 October 1957, RAND Corporation. In this early paper he used a game theoretic approach, and he framed his questions based clearly on concerns that motivated Klein and the RAND economics of R&D group, such as the wastes inherent in concurrent engineering in R&D-intensive projects.

73. Richard R. Nelson, "Uncertainty, Learning, and the Economics of Parallel Research and Development Efforts," *The Review of Economics and Statistics* 43 (November 1961): 351–364. This article represented the culmination of a series of Nelson's RAND papers, including "A Theory of Parallel R&D Efforts," D-6374, 1 June 1959, RAND Corporation; "The Economics of Parallel R and D Efforts," P-1774, 24 August 1959, RAND Corporation; and "The Economics of Parallel R and D Efforts: A Sequential-Decision Analysis," RM-2482, 12 November 1959, RAND Corporation.

74. "Uncertainty, Learning, and the Economics of Parallel Research and Development Efforts," 351. Whether the set, "and others," embraced Kenneth Arrow and Arrow's 1955 essay, cited above, is not fully clear at this point. In an interview with the author, Nelson could not recall having seen or read Arrow's D-3142. Richard R. Nelson, interview with David A. Hounshell, 9 April 1996, Carnegie Mellon University. My personal feeling is that Nelson must have seen Arrow's essay. But in a self-critique of RAND's research, Albert Wohlstetter, principal author of RAND's famous basing study, noted that the RAND culture had developed in the 1950s an unhealthy prejudice against reading any report older than six months. Thus it is not out of the question that Nelson never saw Arrow's essay. Wohlstetter's critique is offered in a sixty-nine-page paper, "RAND's Continuing Program of Broad Policy Study: Problems and Incentives," 22 February 1960, attached to Albert Wohlstetter to Management Committee, M-953, 22 February 1960, Papers of Robert Specht, RAND Corporation Archives, Santa Monica, California. Wohlstetter comments in his cover memorandum that his paper was "an outgrowth of discussions" with several RAND staff members, including William Meckling, Burton Klein, Charles Hitch, and Armen Alchian from the R&D project.

75. "Uncertainty, Learning, and the Economics of Parallel Research and Development Efforts," 351.

76. Ibid., 352.

77. Burton H. Klein, "The Decision-Making Problem in Development," P-1916, 19 February 1960, RAND Corporation. A nearly identical version of this paper was published in Richard R. Nelson, ed., *The Rate and Direction of Inventive Activity: Economic and Social Factors* (Princeton, N.J.: Princeton University Press, 1962), 477–497.

78. The discrepancy between the number of case studies enumerated in table 9.1 and the "fifty" mentioned by Klein might be explained by the team's use of existing case studies of invention and development, many of which Nelson reviewed in his literature review discussed in the text above. In his annotated bibliography that appeared originally in D-4799, Nelson cited in particular S. C. Gilfillan's *Inventing the Ship* (1935), R. Schlaifer and S. D. Heron's *The Development of Aircraft Engines and Fuels* (1950), and the case studies in J. P. Baxter's Pulitzer Prize–winning history of World War II R&D, *Scientists against Time* (1946), as exemplary studies of technological developments. Additionally, Klein could well have been alluding to the relevant case studies that appear in Jewkes, Sawers, and Stillerman's *Sources of Invention* and the numerous case studies of weapons systems development projects that Robert L. Perry (who later joined RAND, in October 1964) had begun to produce for the air force. Klein and his colleagues may have had access to these. Prior to his joining RAND, Perry had been the Chief Historian of the Wright Air Development Center and then Historian of the USAF Space Systems Division. See, e.g., Robert L. Perry, "The Atlas, Thor, and Titan," *Technology and Culture* 4 (1963): 466–477. Together, these studies number well above fifty, but there is no doubt that the RAND group itself had not written fifty case studies. Indeed, F. M. Scherer has commented to me that while working on the Harvard Weapons Acquisition Project, he travelled to Santa Monica in 1959 to read what he was told were all of RAND's case studies, and although he found them useful, he was for the most part "underwhelmed" by their number and their impact. F. M. Scherer to David A. Hounshell, 1 July 1996. Eventually, Thomas Marschak published a large-scale review article, "The Role of Project Histories in the Study of R&D," which drew from virtually every case study the RAND group produced in the 1950s and early 1960s. The article appeared in Thomas Marschak, Thomas K. Glennan, Jr., and Robert Summers, eds., *Strategy for R&D: Studies in the Microeconomics of Development* (New York: Springer-Verlag, 1967), 49–139.

79. Burton Klein, "A Radical Proposal for R. and D.," *Fortune* 57 (May 1958): 112–113, 218, 222, 224, 228.

80. The group's ideas also were presented in two other forums in the first half of 1958: Burton Klein and William Meckling, "Application of Operations Research to Development Decisions," P-1054, 3 March 1958, RAND Corporation, which appeared in print under the same title in *Operations Research* (May–June 1958), and Charles Hitch, "The Character of Research and Development in a Competitive Economy," P-1297, 13 May 1958, RAND Corporation. Hitch presented his paper at the National Science Foundation in Washington, D.C., 20 May 1958, and although Klein's group is not mentioned explicitly in the text, certainly Hitch's ideas incorporated much of the group's thinking and findings about R&D and the

value of parallel development strategies. Indeed, Hitch's paper makes clear that he considered the competitive economy the very embodiment of parallel path development and that he hoped that military R&D policy makers could develop strategies that would mimic the development patterns of the open market.

81. Klein's *Fortune* article was based largely on his paper "What's Wrong with Military R and D?" P-1267, 28 January 1958, RAND Corporation, which was written in the wake of Sputnik, as will be discussed below.

82. Klein, "A Radical Proposal for R. and D."

83. Armen A. Alchian, "Organizing for Development Duplication," D-4688, 1 November 1957, RAND Corporation. Of course, Alchian was by no means the only RAND economist to think or write about the Manhattan Project as an R&D management problem. Most of the economists—including Klein—had read Henry D. Smyth's history of the Manhattan Project, which was issued soon after the war with Japan ended in August 1945, and most of them alluded to the Manhattan Project in their various pieces of work.

84. Alchian's paper on the Manhattan Project as a supreme example of parallel development strategies would be the last piece of work he would conduct related to the economics of R&D. He, Kenneth Arrow, and William Capron did, however, carry out a study in 1958 on the supply of scientists in the United States; their study brought sensible economic analysis to a hotly debated issue in the wake of the Sputnik challenge. See "An Economic Analysis of the Market for Scientists and Engineers," RM-2190-RC, 6 June 1958, RAND Corporation. In response to Sputnik, the United States did create an organization that was supposed to centralize control over the services' competing missile R&D projects: the Advanced Projects Research Agency (ARPA). But ARPA's mission changed very rapidly following Eisenhower's capitulation to the services on ballistic missile development and his decision to create a civilian space program—the National Aeronautics and Space Administration (NASA)—in 1958. The classic study of interservice rivalry in the development of intermediate range ballistic missiles is Michael Armacost, *The Politics of Weapons Innovation: The Thor-Jupiter Controversy* (New York: Columbia University Press, 1969).

85. Klein, "A Radical Proposal for R. and D.," 226.

86. William Meckling and Emmanuel Mesthene, "Military Research and Development," B-46-PR, 18 February 1958, RAND Corporation. The language of this report clearly suggests the sense of urgency following Sputnik.

87. Burton Klein, William Meckling, and Emmanuel Mesthene, "Research and Development Management," B-63-PR, 12 June 1958, RAND Corporation.

88. Burton Klein, William Meckling, and Emmanuel Mesthene, "Military R&D Policies," R-333, 4 December 1958, RAND Corporation. Four of these principles were elaborated in the briefing cited above. This report bears a close resemblance to the above-cited briefing, which later appeared in the "D" series as Burton Klein,

William Meckling, and Emmanuel Mesthene, "Research and Development Management," D-9465, 22 November 1961, RAND Corporation.

89. Richard R. Nelson, "The Simple Economics of Basic Scientific Research," P-1288, 28 April 1958, RAND Corporation. This piece appeared under the same title in *Journal of Political Economy* 67 (June 1959): 297–306.

90. For some of the retrospective literature on Sputnik and its impact, see Barbara B. Clowse, *Brainpower for the Cold War: The Sputnik Crisis and the National Defense Education Act of 1958* (Westport, Connecticut: Greenwood Press, 1981); Robert A. Devine, *The Sputnik Challenge: Eisenhower's Response to the Soviet Satellite* (New York: Oxford University Press, 1993); Roger Geiger, *Research and Relevant Knowledge: American Research Universities since World War II* (New York: Oxford University Press, 1993); and Harvey Sapolsky, "American Science and the Military: The Years since the Second World War," in Nathan Reingold, ed., *The Sciences in the American Context: New Perspectives* (Washington, D.C.: Smithsonian Institution Press, 1979).

91. Nelson, "Simple Economics of Basic Scientific Research," 297.

92. The nickname "Engine Charlie" derived from the fact that one Charles E. Wilson served in the late 1940s and early 1950s as the CEO of General Motors, while another Charles E. Wilson headed General Electric. The latter Wilson was known as "Electric Charlie."

93. Quoted in Daniel J. Kevles, *The Physicists: The History of a Scientific Community in Modern America* (New York: Knopf, 1978), 383.

94. Wilson's and Eisenhower's budget cutting stemmed in part from their strategy of "managed competition" in the development of weapons development. See Milobsky, "Leadership and Competition."

95. Quoted in Kevles, *The Physicists*, 384.

96. Kevles, *The Physicists*, 385. See also James R. Killian, *Sputnik, Scientists, and Eisenhower: A Memoir of the First Special Assistant to the President for Science and Technology* (Cambridge: MIT Press, 1977).

97. Harvey M. Sapolsky, *Science and the Navy: The History of the Office of Naval Research* (Princeton, N.J.: Princeton University Press, 1990), 70.

98. Ibid.

99. Ibid., 71.

100. *Aviation Week* 66 (17 September 1957), as quoted in Komons, *Science and the Air Force*, 106.

101. Ibid., 108.

102. Gregory to Major General John W. Sessums, 13 January 1958, as quoted in Komons, *Science and the Air Force*, 109.

103. George A. W. Boehm, "The Pentagon and the Research Crisis," *Fortune* 57 (February 1958): 154.

104. Kevles, *The Physicists*, 386. According to Katie Hafner and Matthew Lyon, McElroy had been an executive at the Cincinnati-based consumer products company, Procter & Gamble, where he had been involved in the postwar creation of a fundamental ("blue-sky") research laboratory supported by P&G's top management and in which both he and the company took great pride. *Where Wizards Stay up Late: The Origins of the Internet* (New York: Simon & Shuster, 1996), 14–18.

105. Roger L. Geiger, "Science, Universities, and National Defense, 1945–1970," *Osiris* 7 second series (1992): 27.

106. For a recent analysis of the increasing centralization of R&D decision making in the Office of the Secretary of Defense, see David Milobsky, "Leadership and Competition: Technological Innovation and Organizational Change at the U.S. Department of Defense, 1955–1968" (Ph.D. diss., Johns Hopkins University, 1996), especially the introduction and chaps. 1 and 2.

107. Hitch, "The Character of Research and Development in a Competitive Economy," P-1297, 13 May 1958, RAND Corporation.

108. Of course, it is important to point out the profound irony that Charles Hitch would be one of the principal agents in transferring RAND's systems analysis and its system of programming, planning, and budgeting (PPB) to the Department of Defense under Secretary Robert S. McNamara—the so-called McNamara Revolution—which resulted in an extreme degree of centralization in the Office of the Secretary of Defense (OSD). On this process of transfer and its effects, see Jardini, "Out of the Blue Yonder." See also Milobsky, "Leadership and Competition." David C. Mowery pointed out the divergence between the views of RAND's group in the economics of innovation and the policies implemented in the McNamara DOD, which became known under the name Total Package Procurement; but Mowery was apparently unaware of the historical circumstances for either the development of the RAND group's conclusions or the development of policy within the McNamara Pentagon. See David C. Mowery, "Economic Theory and Government Technology Policy," *Policy Sciences* 16 (1983): 27. J. Robert Fox, *Arming America* (Boston: Harvard Business School, 1974), provides a limited history of Total Package Procurement as does his later book, *The Defense Management Challenge: Weapons Acquisition* (Cambridge: Harvard Business School Press, 1988). For a richer historical analysis, see Milobsky, "Leadership and Competition."

109. D-3017-PR, 6 July 1955.

110. D-4799, 27 December 1957. RAND Corporation.

111. Nelson had relied heavily on Mees in his survey of the literature, and both he and Burton Klein were fond of quoting Mees's famous statement that the "best person to decide what research work shall be done is the man who is doing the

research work. . . . Finally there is a committee of company vice presidents, which is wrong all the time." C. E. Kenneth Mees and John A. Leermakers, *The Organization of Industrial Scientific Research* (New York: McGraw Hill, 1950), 135. Burton Klein quoted Mees in his "A Radical Proposal for R. and D.," 226. Richard Nelson quoted Mees in his article, "The Economics of Invention: A Survey of the Literature," 125.

112. P-1288, 28 April 1958; *Journal of Political Economy* 67 (April 1959): 297–306.

113. Kenneth J. Arrow, "Economic Welfare and the Allocation of Resources for Invention," P-1856-RC, 15 December 1959, RAND Corporation. The paper first appeared in print in Richard R. Nelson, ed., *The Rate and Direction of Inventive Activity: Economic and Social Factors* (Princeton, N.J.: Princeton University Press, 1962), 609–625. Incidentally, this volume was more or less a RAND project. Charles Hitch had organized the conference, and roughly one-third of the papers published in the volume were by members of RAND's economics-of-technical-change group. Richard Nelson served as the editor of the volume and did his work while a member of RAND's economics-of-technical-change group. In his introduction to the volume, Nelson noted the Cold War as a factor in the recent interest in the economics of technical change. See Richard R. Nelson, "Introduction," in *idem*, ed., *The Rate and Direction of Inventive Activity: Economic and Social Factors* (Princeton, N.J.: Princeton University Press, 1962), 4.

114. It is interesting that in his 1955 essay on military research, Arrow had opened his analysis by stating that he would be treating research as a commodity.

115. Nelson, "Simple Economics," 160. (Page number refers to Rosenberg reprint pagination.) Interestingly, in his article Nelson entertained the theoretical possibility that "perhaps non-profit laboratories are spending too much on basic research—are operating beyond the point at which marginal cost equals marginal social benefit—and therefore it is desirable to reduce research expenditure in this sector" (Nelson, "Simple Economics," 160). But he ruled this out because of his assumptions (and because it was contrary to the popular, political consensus of the day).

116. K. C. Border, Abstract E 1, in *Government Policies and Technological Innovation*, V. IV-C, *Abstracts* (Pasadena, California: Division of the Humanities and Social Science, California Institute of Technology, n.d.).

117. The economists within RAND's R&D group in the 1950s who championed diversity in R&D were, in some respects, the intellectual forbears of the more highly developed, more fully elaborated evolutionary economics that manifests itself in Richard R. Nelson and Sidney G. Winter's 1982 landmark book, *An Evolutionary Theory of Economic Change*. That this would be so can hardly be surprising. Obviously, Nelson was an important member of the group after he joined RAND in 1957. Sidney Winter came to RAND in late 1959—relatively late in the group's development. He brought to RAND a much greater interest in developing a deliberate, rigorous economic corollary to Charles Darwin's *Origin of the Species* than in helping the air force to improve its R&D programs. Winter

wrote an early paper in this vein while he was at RAND—"Economic Natural Selection and the Theory of the Firm" (P-2167, 19 December 1960, RAND Corporation)—but it would only be later that he and Nelson would join forces to produce *An Evolutionary Theory of Economic Change.* See Richard R. Nelson and Sidney Winter, *An Evolutionary Theory of Economic Change* (Cambridge: Harvard University Press, 1982). On Winter at RAND, see Personnel Records, RAND Corporation. I have also benefited from a conversation with Sidney Winter, Wharton School, University of Pennsylvania, April 1995. Winter's paper would develop into a complete dissertation bearing a close resemblance to his RAND paper's title: "Economic 'Natural Selection' and the Theory of the Firm," which was first published in *Yale Economic Essays,* 4 (Spring 1964): 225–272.

118. Over his subsequent, illustrious career, Nelson has moved away from strong adherence to neoclassical economic theory and has recognized incentives that firms have to engage in fundamental research. Some of this movement stemmed from critiques of Nelson's and Arrow's pioneering work. See, for example, Wesley Cohen, "Empirical Studies of Innovative Activity," in Paul Stoneman, ed., *Handbook of the Economics of Innovation and Technological Change* (Oxford: Blackwell, 1995), 182–264; Alfonso Gambardella, *Science and Innovation: The US Pharmaceutical Industry during the 1980s* (New York: Cambridge University Press, 1995), 1–16; Partha Dasgupta and Paul A. David, "Information Disclosure and the Economics of Science and Technology," in George R. Feiwel, ed., *Arrow and the Ascent of Modern Economic Theory* (New York: New York University Press, 1987), 519–542; and Partha Dasgupta and Paul A. David, "Toward a New Economics of Science," Center for Economic Policy Research, Stanford University, Publication no. 320. See also chap. 1 of David C. Mowery and Nathan Rosenberg, *Technology and the Pursuit of Economic Growth* (New York: Cambridge University Press, 1989), 3–17, and Nathan Rosenberg, "Why Do Firms Do Basic Research (with Their Own Money)?" *Research Policy* 19 (1990): 165–174.

10

Out of the Blue Yonder: The Transfer of Systems Thinking from the Pentagon to the Great Society, 1961–1965

David R. Jardini

Shortly after the Soviet Union detonated its first atomic weapon in 1949, the assistant director of the RAND Corporation, L. J. Henderson, wrote to Frank Collbohm, the corporation's director,

> Actions which our government may be forced to take in view of the world situation . . . may involve the necessity of some deception by us of our own population. This is of course a very touchy subject, but intuitively it seems a very important one and the inventive aspects of how to go about this are rather fascinating.[1]

RAND was then only three years old, but the World War II alliance between the United States and the Soviet Union had already dissolved, plunging the world into the grim, four-decade-long Cold War. During this long "twilight war" the U.S. defense establishment grew enormously and created new institutions for the production and manipulation of technology. At the same time, national security concerns pervaded the country's policy agenda and fundamentally altered the course and nature of American democracy. Citing the demands of national security, policy makers were able to employ "scientific" decision-making methodologies largely to insulate critical public issues from open democratic political processes.

During the 1960s, alumni of the RAND Corporation and methodologies they created fundamentally altered the nature of public policy making in the United States. Championed as objective, scientific means of policy making, in contrast to the sloppy, politics-driven methods of earlier years, systems analysis and program budgeting spread first from RAND to the Pentagon, and then across the federal bureaucratic structure. These analytical and management methods had distinct consequences for policy making where they were adopted: they created an environment of centralized, top-down decision making; they replaced political bargaining with technocratic expertise as the primary means of policy formulation; and they gave rise to a vast market for policy-oriented social science research.

The present chapter explores RAND's development of these analytical management techniques for military purposes and the diffusion of these methodologies from the military context into broader social welfare policy-making applications. Specifically, it concentrates on three issues. First, the chapter provides a brief discussion of RAND's history, focusing on the processes of intellectual production at RAND and the ways in which the creation of techniques at RAND was shaped by the organization's military context. Second, the chapter traces the dissemination of RAND's systems methodologies from the corporation's quasi-academic setting to the highest echelons of the U.S. national security structure. Here, the chapter concentrates on the role played by RAND alumni and RAND methodologies in the "McNamara revolution" of the early 1960s. In particular, this section focuses on the utility of RAND's management techniques for the centralization of control and the insulation of policy making from traditional political processes. Finally, the chapter examines the paths by which RAND's methodological innovations diffused beyond the military establishment into the programs of the "Great Society." It argues that the construction and implementation of the Great Society social welfare programs were fundamentally shaped by Cold War techniques as RAND alumni and their analytical methods can be found at the core of Great Society policy making. In general, then, this chapter traces the subtle yet momentous consequences of Cold War technical development for American democracy. For better or worse, public policy research organizations, of which RAND is the archetype, are now embedded in American public policy making, and the tools forged at RAND for policy analysis and decision making are pervasive. This chapter argues that, by reinforcing centralized, elitist policy making, their widespread adoption in the federal government may have contributed to the alienation many Americans feel concerning the national government.

Background: The RAND Corporation

At the conclusion of World War II, many of America's civilian and military leaders recognized that, in a broad sense, research and development had been the decisive weapons of the war and promised to remain so in the future.[2] The model of research that had achieved prominence during the war, however, was fundamentally unlike prewar norms. Instead, it was characterized by massive expenditures, unprecedented coordination among military, industrial, and academic interests, and conscious and aggressive government intervention.[3] This new model had led to remarkable achievements in areas such as radar and proximity fuses[4] and had gained universal notoriety with the success of the Manhattan Project—America's vast atomic weapons program.[5] At the same time, though, it was equally clear that scientists' reluctance to work directly under military control meant that the wartime organization of research could not be extended unchanged into the postwar environment.[6]

The RAND Corporation represented one solution to this dilemma and proved a model for Cold War research. RAND was formed at the end of World War II by air force and industrial leaders who wanted to harness the nation's foremost intellectual talent to military research and planning. A critical actor in RAND's genesis was General Henry H. "Hap" Arnold, the commanding officer of the Army Air Forces during World War II. Arnold recognized not only the rapid pace of technological and scientific development that had characterized the war, but also the implications this rate of change had for military planning efforts. As the war drew to a close, he wrote to the secretary of war,

> During this war the Army, Army Air Forces, and the Navy have made unprecedented use of scientific and industrial resources. The conclusion is inescapable that we have not yet established the balance necessary to insure the continuance of teamwork among the military, other government agencies, industry, and the universities. Scientific planning must be [done] years in advance of the actual research and development work.[7]

Working with MIT researcher Edward Bowles, a consultant to the Army Air Forces who had been deeply involved with operations research during the war, and Donald Douglas, president of Douglas Aircraft Company, Arnold and his staff created a broadly defined and, nominally at least, independent research organization—designated "Project RAND"—and situated it within the Douglas Aircraft Company. Although Project RAND's original statement of purpose was poorly defined, the organization very rapidly achieved considerable clarity in its objectives. The handful of individuals who comprised the project's initial staff each had extensive experience in wartime operations research and the application of mathematical techniques to military planning. For example, Project RAND's initial general supervisor, Arthur E. Raymond, and his assistant, Frank R. Collbohm, had been deeply involved in the application of operations research techniques to air force planning during the war. Raymond and Collbohm had been chosen in 1944 by Bowles to direct the first experiment in civilian scientific participation in American wartime planning—the B-29 Special Bombardment Project.[8] Using operations research methods, Raymond, Collbohm, and their team analyzed the interactions among range, speed, armament, and bomb-carrying capacity of various B-29 systems configurations and recommended unorthodox but effective changes in the bomber. With the creation of Project RAND, the air force and Douglas Aircraft placed Raymond, who was then Douglas's vice president in charge of engineering, in general supervision of the project and vested direct operational control in Frank Collbohm.

Project RAND enjoyed virtual autonomy within the Douglas Aircraft Company's corporate structure, but by 1948 Douglas Aircraft management had concluded that Project RAND created problems for their company's efforts to secure aircraft procurement contracts due to the apparent conflict of interest between Project RAND's advisory activities and Douglas Aircraft's procurement objectives. At this point, RAND was spun off from Douglas and established as a free-standing, private, nonprofit corporation—the RAND Corporation—backed by a million dollar interest-free loan from the Ford Foundation and that foundation's pledge of credit security to RAND's commercial lenders.

In broad terms, Project RAND's objective was to elaborate upon wartime experiences by connecting military research efforts with strategic planning.[9] As noted above, one of the outstanding lessons of the Second World War for American military leaders was that the accelerating pace of scientific and technological advance had fundamental implications for the strategy and tactics of warfare, and that any country failing to incorporate scientific advance into planning risked disaster. The Army Air Force's Deputy Chief of Staff for Research and Development, General Curtis E. LeMay, whose position had been created in part to oversee Project RAND, argued, "Warfare is no longer a military problem."[10] Correspondingly, in March 1947 L. J. Henderson, one of Project RAND's early leaders, commented,

> We cannot do long range planning without taking fully into account intelligent estimates of strategic advances in science and technology; conversely if we expect research and technology in this country to give the greatest support to national security, we must bring leadership in these fields into our confidence, familiarizing a select group with our estimates of future military problems.[11]

More specifically, the intent of Project RAND, in terms of its administration and operation, was to bridge the divide between the air forces and the aviation industry so as to effect an optimal coordination of long-range research and development with the military air program. To accomplish this, Project RAND's staff was committed explicitly to the creation of systematic and scientific methods for the analysis of warfare. According to Arthur Raymond, Project RAND was concerned

> with systems and ways of doing things, rather than particular instrumentalities, particular weapons, and we are concerned not merely with the physical aspects of these systems but with the human behavior side as well. Questions of psychology, of economics, of the various social sciences, so-called, are not omitted because we all feel that they are extremely important in the conduct of warfare.[12]

Within the broad field of intercontinental warfare, RAND's leaders set as their goal the development of a scientific system of national defense based on rigorous social scientific methodologies and cutting-edge research in both the physical and social sciences. Such a system would allow policy makers to base their decisions on scientific analyses that would present clearly the consequences arising from the various alternative strategies. Three months after RAND's official inception, John D. Williams, the leader of what would become RAND's mathematics department, wrote,

> The thesis that the art of war is, at least in part, amenable to scientific handling derives support from the success which attended certain applications of scientific method during World War II—to tactical, strategic, and development problems. These successes with small, isolated components of the theory of warfare suggest the possibility of similarly treating the entire subject, and justify RAND in approaching it.[13]

Indeed, many of RAND's early goals were extensions of the Applied Mathematics Panel (AMP), an organization created during World War II by the Office of Scientific Research and Development.[14] AMP was directed by Warren Weaver, an applied mathematician and the director of the Rockefeller Foundation's science program. Weaver was an early and critical participant in Project RAND who had worked closely with Frank Collbohm during the war on Army Air Forces operations research problems. During the closing months of the war, the AMP had worked unsuccessfully to develop a theory of military worth and a mathematically based science of warfare. As Project RAND got under way, former AMP members Weaver, John Williams, Olaf Helmer, and Edwin W. Paxson were early recruits and strongly influenced the project's research program.

To pursue a science of war, Project RAND was organized according to a plan authored by John Williams that proposed a two-part structure. First, the heart of Project RAND was its interdisciplinary working groups, which concentrated on developing a scientific methodology for analyzing warfare. These groups attacked the problem at three levels: immediate problems near the tactical level, intermediate-range problems that would envisage modest modifications of existing weapons systems, and long-range problems involving analyses of anticipated scientific and technological changes and the entire war-making potential of the nation. The work of these groups almost immediately came to be known as "systems analysis." The second part of RAND's organization—its "sections"—provided the personnel for the systems analysis groups. Analogous to academic departments—indeed they would come to be called departments after 1948—Project RAND's sections corresponded to the fields of study that most strongly influenced modern warfare, such as rocketry, nuclear energy, electronics, and applied mathematics. These sections supported the prosecution of the systems analyses both by supplying manpower to the interdisciplinary groups and by performing area-specific research in support of the analyses.

While RAND's extensive efforts to create a science of warfare were largely unsuccessful, RAND was an unqualified success as an institutional innovation, and, throughout the 1950s and 1960s, its research staff produced key innovations in such diverse fields as materials science, electronics, economic theory, artificial intelligence, space systems, and the social sciences.[15] Of particular significance for the present chapter were RAND's achievements in social science theories of complex systems and tools for decision making under conditions of uncertainty. The basis for these achievements lies in RAND's pioneering work in applied mathematics, which was intended to overcome the limitations of operations research—a body of techniques developed largely during World War II to estimate the optimal performance or configuration of existing systems.

As discussed above, many of RAND's early staff members had gained experience in the use of operations research and applied mathematics during World War II. This experience made them keenly aware of the limitations of these techniques, particularly in conducting nonstatic analyses. As such, RAND's mathematicians and mathematical economists concentrated their research efforts in such fields as game theory, linear and dynamic programming, mathematical modeling and simulation, network theory, cost analysis, and Monte Carlo methods, as they sought to address the dynamic nature of military optimization problems and provide mathematically rigorous solutions to those problems.[16] George B. Dantzig, for example, elaborated at RAND the simplex method of analysis, which allowed RAND researchers to employ electronic computers to solve previously unmanageable series of simultaneous linear equations.[17] Also during the 1950s, Richard Bellman and fellow RAND staff members codified a system of optimization techniques under the rubric of dynamic programming.[18] Finally, in game theory, RAND research teams frequently included consultants John von Neumann and Oskar Morgenstern—the two men credited with founding this field during the early 1940s—as these teams made crucial theoretical contributions while working on problems such as nuclear war scenarios, radar search and prediction, and antisubmarine warfare.[19]

The centerpiece of RAND's methodological innovations—to which all of the corporation's applied mathematical developments contributed—was systems analysis. RAND staff members envisioned systems analysis as a "rational," mathematically rigorous means of choosing among alternative future systems characterized by complex environments, large degrees of freedom, and considerable uncertainty.[20] Originally created to evaluate possible nuclear weapons deployment scenarios, RAND's systems analysis techniques are quintessential modern social science, incorporating both quantitative methods, especially mathematical modeling, and qualitative analysis involving a diversity of disciplines. The objective of RAND's system analyses was to provide information to military decision makers that would sharpen their judgment and provide the basis for more informed choices. The objective of RAND's staff members, meanwhile, was to have these techniques adopted at the top echelons of the American defense structure.

THE SACRED TEXT

In March 1960, after three years of preparation, Charles J. Hitch and Roland N. McKean published the magnum opus of the RAND

Corporation's "scientific" defense analysts, *The Economics of Defense in the Nuclear Age*.[21] Both Hitch and McKean were RAND economists (Hitch was then the head of the corporation's economics division) and were deeply involved in the application of mathematical and economic theory to defense policy issues within the framework of systems analysis. *The Economics of Defense* detailed a rational approach to defense policy making that sought to replace the political basis of decision making with rigorous systematic analysis.[22] Hitch and McKean argued that "the problem of combining limited quantities of missiles, crews, bases, and maintenance facilities to 'produce' a strategic air force that will maximize deterrence of enemy attack is just as much a problem in economics (although in some respects a harder one) as the problem of combining limited quantities of coke, iron ore, scrap, blast furnaces, and mill facilities to produce steel in such a way as to maximize profits."[23] However, in private sector industries such as iron and steel, the authors argued, decision making was subject to the rigors of the market, thus ensuring the efficient allocation of resources among alternative uses. This was not the case in the "production" of national security. In the public sector and—most importantly for the authors' purposes—in defense spending, the efficient use of resources was not compelled by built-in mechanisms like those found in the private sector of capitalist economies. Hitch and McKean argued, "There is within government neither [a] price mechanism which points the way to greater efficiency, nor competitive forces which induce government units to carry out each function at minimum cost."[24]

In *The Economics of Defense*, RAND's analysts argued that national policy makers must recognize the essentially economic nature of defense decisions in order to correct the deficient defense structure and maximize U.S. military power and national security.[25] Unfortunately, this could not be achieved under the existing regime of decentralized, politically determined defense resource allocation. Thus, Hitch and McKean advised the replacement of that system with a rationally designed structure that compensated for the absence of market forces by "building-in" rigorous systematic analysis of alternative allocation strategies.[26] Operationally, the authors argued, such a strategy involved two components. First, the institutional arrangements within the government must be reorganized so as to centralize both budget and military policy making and place these under the direction of the secretary of defense. Specifically, Hitch and McKean recommended a "program-based" budget structure that recast budgets and accounts based on end-product missions or programs (e.g., strategic nuclear forces, limited war forces), rather than the costs of objects. According to the RAND analysts,

> We cannot appraise the adequacy of the defense budget, either subjectively or with the aid of quantitative analysis, by thinking about the gains from such categories as paper clips, petroleum, or personnel. Nor can we go to the other extreme and think in terms of a single national security program. Such an aggregation is too broad; we have no conception of units of "national security" that could be purchased. But there are possibilities between these extremes—aggregations of activities that produce species of end-products such as capabilities for nuclear retaliation or for limited war.[27]

Such a system would serve not only to reveal more effectively the true costs and capabilities of the military establishment, but it would also concentrate decision-making authority in the hands of top policy makers who alone have the necessary detachment from parochial interests to consider objectively overall strategic issues.

Further, the RAND analysts proposed, as a second component of their program budgeting strategy, the implementation of quantitative systems analysis techniques to aid policy makers in choosing the most efficient alternative allocations and methods. Systems analysis is an approach to problem solving developed largely at RAND during the late 1940s and 1950s that sought to make scientific—and thus depoliticized—decisions in situations of great uncertainty. For the purposes of rationalizing defense policy making, RAND's strategists argued that the detached, quantitative techniques they espoused provided top-level decision makers with the means to allocate defense resources most efficiently. This meant that defense policy making need no longer be delegated to military personnel, who were experienced in military matters but whose pursuit of parochial service interests prohibited their comprehension of the total national security "system." Thus, the combination of program budgeting and systems analysis proposed in *The Economics of Defense* both centralized defense policy making and provided executive leaders with "scientific" decision tools that replaced inherently biased *experience* with rigorous, quantitative *analysis*. As they realized, the program budgeting system advocated by the RAND analysts required, rather unrealistically, the wholesale restructuring of the defense establishment and the suppression of powerful vested interests in the military branches. However, where might a force strong enough to prosecute such revolutionary changes be found?

THE MCNAMARA REVOLUTION

In the opinion of this researcher, Robert S. McNamara is one of the more outstanding counterexamples to the argument that individuals matter little to historical development. For better or worse, McNamara participated in or led the fundamental restructuring of three of the most important organizations in the modern global economy: the Ford Motor Company during the 1950s, the U.S. Department of Defense during the 1960s, and the World Bank during the 1970s.[28] His designation by President-elect John F. Kennedy as secretary of defense in December 1960 brought to the Pentagon a man of apparently unlimited energy who was committed to rational, quantitative management and the centralization of authority. Appointed at the youthful age of forty-four years, McNamara seemed to personify the vigor and power of John Kennedy's "New Frontier" leadership.

Robert McNamara was a native Californian who graduated from the University of California at Berkeley Phi Beta Kappa and, after a brief stint as a sailor and sailors' union organizer, earned a master's degree at Harvard's Graduate School of Business Administration. After a year in the accounting firm of Price, Waterhouse, McNamara returned to academia as an assistant professor of accounting at the Harvard Business School. With American entry into World War II, his special interest in statistical control techniques brought him into contact with the Army Air Forces, which had contracted with Harvard for the training of service personnel in these methods. By 1943, McNamara was in an A.A.F. uniform, working in the "Statistical Control" unit, the task of which was to form a network of statistical officers around the world that would "collect, organize and interpret facts and figures on personnel and equipment" so as to optimize Army Air Force operations.[29]

McNamara's work in the A.A.F. was decisive in redirecting his career. In the service, McNamara came into contact with the enterprising head of the A.A.F. statistical control office, Charles "Tex" Thornton, who, at the war's conclusion, negotiated an arrangement with Henry Ford II, the president of the Ford Motor Company, to hire ten Army Air Force statistical officers as a package deal in January 1946. The Ford Motor Company was, at that time, in chaos following decades of neglect to its administrative operations, and McNamara and the nine other "Whiz Kids" moved to Ford to form an "executive agency" that would apply statistical control and management methods to the stricken firm. At Ford, McNamara's rise was rapid and unorthodox. Contrary to the traditional career path that took promising young men with engineering backgrounds through Ford's plant operations and into the corporate hierarchy, McNamara and the other statistical officers lacked any technical experience in automobile production and had started near the top of the corporate structure. As such, McNamara's ascendance to the presidency of Ford was not by way of engineering or sales, but through finance and corporate control. Similarly, his success as a manager was not based on practical or technical experience, but the detached application of quantitative analytical techniques.

As he took the reins of the Department of Defense in January 1961, McNamara sought to bring to the national security establishment the rational management methods he had championed in industry.[30] In particular, he pushed to accelerate a long-term trend toward centralization and "civilianization" of authority in the DOD. This process had begun in 1947 with the passage of the first National Security Act, which established the basic postwar structure of the armed forces and placed them under the nominal authority of a civilian secretary of defense.[31] Despite its apparent vesting of authority in civilian hands, however, this act provided the secretary of defense with scant tools to assert control, and the military branches and their respective secretariats were left with near complete autonomy in policy making. While subsequent revisions of the 1947 act gradually proffered greater powers on the secretary of defense, a tradition of defense management held sway throughout much of the 1950s in which the president and secretary of defense determined total defense expenditure levels and allotted shares of this total pie to the services. The services were then largely free to spend these moneys as their leaders saw fit, with significant civilian input only in "civilian" areas such as research and finance. In this highly decentralized system, the secretary of defense assumed a rather passive role, often acting as little more than a referee between the contentious service branches.

The intense interservice rivalries and relative impotence of the central defense authority that marked defense management in the 1950s came under sharp attack in 1957 when the Soviet Union launched Sputnik I. Critics of the existing defense administration, especially Democrats such as Senator Henry Jackson of Washington, argued that while the U.S. military branches had wasted time and vast sums in competition with one another, the Soviets had seized the technological initiative. Without corrective action, the communists would soon gain a decisive military superiority over the United States and be in a position to press forward their perceived ambitions for global hegemony. An extensive and influential report on American defense posture sponsored by the Rockefeller Fund and published in 1958 was severely critical of the diffuse nature of national defense policy making.[32] The report advised that the United States move toward a defense budgetary system that "corresponds more closely to a coherent strategic doctrine" and that would overcome interservice rivalries.[33] Finally, the colossal destructive power of nuclear weapons and, by 1958, their imminent proliferation among the U.S. service branches further strengthened arguments for effective and centralized civilian control of the armed forces. As Lt. General John W. O'Neill recalled, "The top decision makers of the country recognized that we had to have a system which guaranteed absolute control of those weapons."[34]

As a result of this pressure, Congress finally gave legal teeth to the 1947 act by passing the Department of Defense Reorganization Act of 1958, which eliminated the military departments from the chain of operational command and created a direct line of authority from the president and the secretary of defense, through the joint chiefs of staff to the field commanders.[35] Executive operational authority was thus vested in the office of the secretary of defense while the mission of the three military departments and their secretaries was reduced, essentially, to that of organizing, training, equipping and supporting the combat forces assigned to the unified and specified commanders.[36] Military doctrine remained the responsibility of the separate service branches, as did budgets, programs, and, to a limited extent, research and development. However, the 1958 act greatly diminished the operational authority of the military chiefs of staff by eroding their powers of unilateral command.

Despite the intent of Congress, however, the 1958 act did not immediately alter the decentralized nature of the defense establishment. In large part, this was because the secretaries of defense who served the Eisenhower administration during 1958–1961 were neither able nor willing to assert control over the military branches. National defense remained at the center of national attention during the 1960 presidential campaign as candidate John Kennedy charged the Eisenhower administration and, by implication, his opponent Vice President Richard Nixon with surrendering American military superiority to the Soviet Union. To dramatize his commitment to reforming U.S. national defense, Kennedy formed, under the chairmanship of Missouri Senator Stuart Symington, a "Committee on the Defense Establishment" (the Symington Committee). This committee was assigned to investigate the management of the Department of Defense and its related agencies, concentrating specifically on the relationships between civilian and military decision making in the defense establishment. Kennedy further instructed the Symington group to recommend by 31 December 1960 such changes "as may be necessary to remedy present basic weaknesses in the administration and management of our national defense establishment."[37]

As the Symington Committee members exchanged memoranda concerning possible strategies for restructuring the Department of Defense throughout September and October 1960, consensus quickly emerged in support of dramatically augmented centralization of the Department of Defense. Some members, including Senator Symington, even flirted with the long-controversial notion of formal unification of the armed forces. In a memorandum dated 10 October, for example, Symington wrote, "No longer can an arbitrary division of forces into 'land,' 'sea,' and 'air' meet the realities of the present or the requirements of the future."[38] While the final report of the committee, presented to now President-Elect Kennedy on 26 November 1960, did not recommend definitive unification of the armed services, it did advocate a fundamental restructuring of authority within the Department of Defense.[39]

The Symington Committee report instantly became a political hot potato as the military branches charged that it advocated unification of the armed forces. *Army Navy Air Force Review*, for example, exclaimed that the document was

> a highly controversial blueprint that calls for major overhaul of the Pentagon and the Armed Forces.... The committee's recommendations—most of which have been repeatedly rejected in the past by the Congress—will receive a harsh reception on Capitol Hill. If the new Chief Executive accepts the proposals without extensive modification, he must expect to run headlong into one of the toughest fights his Administration will ever encounter.[40]

This front-page diatribe went on to list the powerful group of congressmen, most of whom were Democrats, who could be expected to contest the implementation of the Symington plan. While the strident response of the military and its published champions probably overstated the rancor engendered among congressmen by the report, the president-elect certainly would not have relished the thought of beginning his tenure with a legislative struggle against members of his own party. Kennedy quickly distanced himself from the report and, despite his apparent sympathies, made no formal endorsements of the committee's findings.

Upon assuming office in January 1961, McNamara's agenda was clear. Drawing from his experience at Ford, he conceptualized the national defense establishment as a massive and complex productive system, characterized by inputs and outputs that could be rationally organized and analyzed so as to achieve optimal efficiency. As was the case at Ford, he saw centralized control in the hands of a very few expert managers to be the most effective way to pursue that efficiency. His deputy secretary, Roswell Gilpatric, remarked,

> [McNamara] made it plain [in mid-December 1960] that he intended to operate with a very small, top group. He didn't want a large span of control.... [H]e always expressed a strong belief that a very few people at the top of a great pyramid like the Pentagon could make the critical decisions, based on adequate information, and then delegate down to the lowest levels possible in the pyramid the authority and responsibility for carrying out various phases of the decisions.[41]

Thus, McNamara recognized the heightened authority granted his office by the defense reorganization act of 1958 and moved to leverage that authority by adopting a new mode of management rather than a new organizational structure. That new mode would emphasize centralized control, civilian management, and the employment of quantitative analytical techniques.

Within days of his nomination by President-Elect Kennedy, McNamara had read and absorbed the ideas of the RAND scientific strategists, recently published in *The Economics of Defense in the Nuclear Age*. Here was a rich source of ideas, albeit untested and not fully defined, that meshed easily with his own predisposition for systems thinking and quantitative, rational management.[42] Essentially, the ideas of Hitch and his colleagues at RAND had offered the achievement of centralized control without conducting a politically tortuous assault on the military establishment. The combination of program budgeting and systems analysis offered a *methodological* solution to what had previously been perceived as a strictly *structural* problem. In his first meeting with Gilpatric, in mid-December 1960, McNamara expressed his desire to bring Charles Hitch onboard his administration and asked his deputy secretary to contact Hitch and offer him the job of assistant secretary of defense—comptroller.[43] As Alain Enthoven later recalled, when McNamara and Hitch met in early January 1961 "it was love at first sight."[44]

Planning-Programming-Budgeting in the Department of Defense

If there were any doubts as to Robert McNamara's will or capacity to reshape Pentagon management, these quickly evaporated by the spring of 1961. Rather than easing into the job and enjoying the customary "looking around" period, McNamara and his staff immediately embarked on an in-depth analysis of the major issues facing the Department of Defense. In early March, the secretary assigned to the joint chiefs of staff, the military departments, and various elements of his own staff ninety-six critical requirement problems, including such formerly sacrosanct issues as, "Should the United States deploy long-range manned bombers?"[45] Furthermore, these studies differed from the rather innocuous inquiries that had historically emanated from the Office of the Secretary of Defense (OSD) in that they demanded systems analyses employing both intensive economic and military examination in which respondents explicitly compared the cost effectiveness of alternative ways of accomplishing national security objectives.

The new secretary's aggressive style came as a rude shock to military leaders more accustomed to passive civilian leadership. Roswell Gilpatric, for example, recalled McNamara's unqualified rejection of a Pentagon tradition—the military briefing:

> The military ... was always somewhat on guard and skeptical, perhaps apprehensive, because of the way in which McNamara operated, and the fact that he wouldn't listen to briefings. None of the elaborate presentations which had been racked up for the secretary and the deputy secretary and the other new people at the beginning were ever listened to by McNamara. He didn't like flip charts, didn't like men in uniform with pointers reading off things. He wanted to ask his own questions, and he wanted unstereotyped answers, and that threw them off. And also, he was not very much on tact and diplomacy in the way he handled them.[46]

McNamara moved with equal sureness to reconstruct management procedures along the program basis laid out in *The Economics of*

Defense. As one of his first acts, the secretary formed three task groups within his staff to perform detailed review and overhaul of the defense budget.[47] These three groups were organized around what McNamara and his staff considered the three primary "products" of the Department of Defense: strategic offensive and defensive forces (headed by Charles Hitch), limited war forces (Paul H. Nitze), and research and development (Herbert F. York). As assistant secretary of defense—comptroller, Hitch was responsible for pulling the three projects together and using the massive volumes of data they produced to design and implement a new, centralized budgeting system that soon came to be called the Planning-Programming-Budgeting (PPB) system.[48]

Hitch and his colleagues derived the main intellectual components of the new defense management system from *The Economics of Defense* and the research that lay behind it.[49] The new system would treat the defense establishment as a whole rather than as a proliferation of autonomous units, each of which pursued its own largely self-defined mission. According to the new plan, the "production" of national security involved inputs (such as personnel and titanium), that were converted into forces (such as the strategic retaliatory force), which subsequently produced outputs (such as deterrence of Soviet nuclear attack). The complex but finite set of relations within this system could be quantitatively modeled and analyzed such that the outputs both matched national objectives and optimized the employment of available resources. The centralization of policy-making authority was crucial here because it placed decision power in the hands of those top civilian leaders who could view the national security system holistically and act according to national rather than parochial interests. This implied a more authoritarian management system that permitted top leaders to implement their policy decisions and thus eradicate the irrational, political bargaining that had previously driven resource allocation. Centralized decision making also required that the defense management system provide a range of policy alternatives to top leaders in such a way that these alternatives could be compared with precision and in relation to policy objectives. Thus, quantitative systems analysis based on cost-effectiveness rationale would be a centerpiece of the management system, and a multiyear horizon would be employed so that such analyses would harmonize budgeting, policy planning, and weapons development time frames.

By 1965, McNamara and his OSD staff had implemented an elaborate management system that, while producing considerable controversy, had decisively centralized authority within the Department of Defense. This implementation of PPB and systems analysis had been greatly facilitated by the inherent nature of the defense organization; no executive department within the federal structure of the U.S. government was as well suited to the adoption of PPB as was the military environment of the DOD. First, despite their relative autonomy, the military branches are profoundly hierarchical in both structure and culture. Even before McNamara's tenure began, the organizational character of the Department of Defense more closely resembled that of an American industrial corporation than any other federal department, a circumstance that greatly simplified McNamara's task. Also, the considerable technical content of defense problems, especially concerning weapons analyses, makes them especially amenable to precise measurement and quantitative comparison. Cost-benefit analyses can thus be performed with considerable confidence. Finally, the application of economic theory to defense management problems is made manageable by the relatively simple output of the defense establishment: security. While "security" is certainly not an easily quantifiable commodity, it can be equated, theoretically, with the economic notion of "utility" such that general economic theory can be translated rather directly into theoretical defense economics. As will be seen below, this is far trickier when PPB and systems analysis are applied in social welfare arenas such as education, where the desired "outputs" are politically contested.

Origins of the War on Poverty

When President Lyndon B. Johnson declared war on poverty in his first State of the Union address in January 1964, work on the proposed program had been under way in the White House for less than three months. At the time, scholarship concerning poverty was rudimentary, and government policy makers, including those at the highest levels, had only the sketchiest appreciation for its extensiveness or causes. According to Daniel Moynihan, then assistant secretary of labor for policy planning and research, the first official document to set forth the need for a "coordinated attack" on poverty was a Council of Economic Advisors (CEA) staff memorandum dated 29 October 1963.[50] The next day, Walter W. Heller, chairman of the CEA, wrote members of the cabinet asking for proposals that "might be woven into a basic attack on poverty and waste of human resources, as part of the 1964 legislative program."[51] In the meantime, Heller established a special task group, constituted mainly of CEA members, officials from the Bureau of the Budget, and White House staffers, to weigh alternative strategies and construct a plan of action for the 1964 legislative campaign.

In constructing a strategy for the war on poverty, White House officials could draw on experience in combating poverty gained primarily through the Ford Foundation's "Grey Areas" programs, a New York City nonprofit community action organization known as Mobilization for Youth, and the work of the Kennedy administration's President's Committee on Juvenile Delinquency and Youth Crime. The Ford Foundation's "Grey Areas" programs began in the late 1950s as the head of the foundation's Public Affairs Program, Paul N. Ylvisaker, became disenchanted with the urban renewal projects that were ongoing at the time.[52] Ylvisaker argued that "dealing with cities as though they were bricks without people" was counterproductive, and he began looking to use his fund to support systematic approaches to the social as well as the physical problems of the urban "grey areas." With the assistance of Edward Logue, the director of redevelopment in New Haven, Connecticut, Ylvisaker designed the Grey Areas grants not just to assist in the provision of social services, but rather to transform the political and social life of the community through new community organizations that would coordinate programs in areas such as youth employment and education. The new breed of "community action agencies" spawned by the Ford Foundation Grey Areas program represented an innovation in social welfare policy that would have profound implications for the later war on poverty. According to Moynihan,

> The Ford Foundation ... purposed nothing less than institutional change in the operation and control of American cities. To this object it came forth with a social invention of enormous power: the independent community agency. In effect, the Public Affairs Program of the Ford Foundation invented a new level of American government, the inner-city community action agency.[53]

By 1963, Ford grants had been awarded to Oakland, New Haven, Boston, and Philadelphia, and, in that year, a statewide program was initiated in North Carolina. Altogether, these five projects had, by then, been awarded a total of $12.1 million.

Mobilization for Youth, Inc. (MFY), was similar in many respects to the Ford Foundation Grey Areas programs except that it drew sponsorship not just from the foundation but from the City of New

York and the federal government. The planning for MFY began in June 1957, when the board of directors of the Henry Street Settlement, on the Lower East Side of Manhattan, determined to take action in the face of rising youth crime in the area. MFY was founded as a nonprofit community action corporation composed of representatives from various agencies and institutions on the Lower East Side of New York City and persons recommended by the New York School of Social Work at Columbia University.[54] The organization had as its focus the constellation of problems associated with juvenile delinquency, and its active membership quickly adopted as a theoretical guide the "opportunity theory" of delinquent behavior developed by Richard A. Cloward and Lloyd Ohlin in their book *Delinquency and Opportunity*, which focused on the alienation of youth from the broader community.[55]

The President's Committee on Juvenile Delinquency and Youth Crime (PCJD) was a crucial institutional translator of ideas from the Ford Foundation Grey Areas programs and the MFY into the executive offices of the federal government.[56] Within weeks of the 1960 election, Attorney General Robert F. Kennedy had appointed his boyhood friend David Hackett, who had served as an important campaign aide, as special assistant to the attorney general with the task of studying the problems of juvenile delinquency. On 11 May 1961, President Kennedy sent a message to Congress that reported increasing juvenile delinquency and announced the formation of the President's Committee on Juvenile Delinquency and Youth Crime to find solutions. David Hackett became the executive director of the PCJD, and its membership included Lloyd Ohlin and Richard Boone, a member of the Ford Foundation's Public Affairs Department. Accordingly, the PCJD internalized a "dysfunctional social structures" perspective of poverty, and its work focused on the role of social institutions and structures in perpetuating delinquency among lower class youth. Soon, the committee had awarded grants amounting to $30 million to Mobilization for Youth and to organizations in a dozen other communities for developing comprehensive plans of community organization to attack the causes of juvenile delinquency.[57]

By mid-1961, the PCJD and the Ford Foundation were pursuing parallel paths in their social welfare efforts and providing a basis of experience for any future expansion of federal programming efforts. At the same time, the energetic work of the PCJD attracted considerable attention, and its members became well known in government circles as the "urban guerrillas."[58] Further, the Ford Grey Areas programs produced the notion of "community action," which was to find its way to the core of the Great Society.[59] The philosophical predilections of these two agencies toward the "dysfunctional social structures" model of social problems led them to focus their attention on the transformation of local social institutions and the encouragement of involvement by the poor in those institutions. According to Sanford Kravitz, program coordinator for the PCJD, the perceived need for community action arose from the identification of a specific set of problems:

1. Many voluntary "welfare" programs were not reaching the poor.
2. If they were reaching the poor, the services offered were often inappropriate.
3. Services aimed at meeting the needs of disadvantaged people were typically fragmented and unrelated.
4. Realistic understanding by professionals and community leaders of the problems faced by the poor was limited.
5. Each specialty field was typically working in encapsulated fashion on a particular kind of problem, without awareness of the other fields or of efforts toward interlock.
6. There was little political leadership involved in the decision making of voluntary social welfare.
7. There was little or no serious participation of program beneficiaries in programs being planned and implemented by professionals and elite community leadership.[60]

Richard Boone recalled that during this period local community representatives brought to Washington to meet with the PCJD frequently repeated the same comment, "For God's sake, quit planning for us. If you really want us involved, then don't play games with us."[61] As such, the program efforts of the PCJD and Ford explicitly pursued the objective of local involvement in program design and execution and community participation in institutional change.

Immediately following Kennedy's assassination, President Johnson informed Walter Heller, chairman of the CEA, that he intended to follow through with the former leader's planned federal assault on poverty. The Bureau of the Budget (BoB) set aside $500 million to finance the antipoverty program and a joint task group of CEA and BoB representatives began to devise the contours of a legislative agenda. Surveying the situation, the CEA/BoB group recognized that in addition to new programs that might be created, the federal government already had in place a considerable number of programs that might contribute to a more systematically organized war on poverty. However, these were scattered across the federal bureaucratic structure and enjoyed no semblance of coordination or central direction. Moreover, the CEA/BoB group was dismayed by the lack of scholarly agreement as to what an appropriate strategy against poverty might be. William Capron, a member of the CEA/BoB team, recalled,

> We were groping for some way to make the existing money in the budget focus.... [W]e were bewildered by the complete disarray of the nominal professionals in the field of poverty. All kinds of social scientists, practicing social workers and the like did not seem to agree on the diagnosis; they certainly didn't agree on the cures. It was quite clear that poverty had ... many faces, that to talk of some cutoff below which everyone was a glob had no programmatic meaning at all because you were talking about widely disparate groups, and that a single magic answer was not to be found.[62]

As Moynihan comments, "Order and efficiency are the passions of lawyers and economists," and the CEA/BoB team began looking for a means to coordinate the antipoverty programs.[63]

In the course of this search, the BoB solicited the advice of PCJD Executive Director David Hackett, who wrote two memoranda to the CEA/BoB group. These and subsequent meetings with Hackett and Richard Boone convinced William B. Cannon, a key BoB staff member, that locally organized nonprofit community corporations could be used as a coordinating mechanism. Cannon enlisted the support of Assistant Budget Director Charles L. Schultze and other staff members and outlined his ideas in a memorandum dated 12 December 1963. In it, he proposed

> the establishment of a series of State and Local Development Corporations, which would develop a plan, to be approved by the President, to conduct action demonstration and testing programs simultaneously in 10 poverty areas.... The Development Corporations would involve local communities in the design of systematic, local planning for its own self-help, and, out of this experience, developing recommendations for a longer-term program for an attack on poverty which would form the basis for legislative and administrative actions that need to be taken by the various levels of Government."[64]

According to Cannon's scheme, the nonprofit "development corporations" would choose from the menu of available federal and state programs to construct a locally relevant program strategy and expend their allotted funds according to that plan. This meant that, despite their inherently decentralized nature, the corporations would provide a manageable coordinating mechanism for the welter of social programs. Furthermore, this strategy would create something of a "market" for federal programs in which local groups would "shop" for the most attractive and effective. This approach offered to soothe the ingrained concern among the BoB economists that public resource allocation suffered from its insulation from beneficial market forces.

Whereas in early December 1963 the CEA/BoB group proposed to devote to the Community Action Programs (CAPs) $100 million of the $500 million allotted to the antipoverty program, by January, when they presented their legislative strategy to President Johnson, CAPs formed the entirety of the BoB antipoverty plan.[65] Even at this early date, however, the PCJD officials who developed the CAP ideas and the BoB staffers who adopted them diverged in their understandings concerning the CAPs' purpose. On one hand, the BoB officials saw CAPs as a mechanism for coordinating antipoverty efforts. In their understanding, CAPs created a discrete set of focal points at which federal program officers could work directly with local political and community leaders and oversee program design and implementation. At the same time, the CAP structure permitted circumvention of the existing tangle of city, state, and federal bureaucracies such that meaningful, effective programs could be enacted. According to James Sundquist, who was then deputy undersecretary of agriculture, "It was certainly never thought that the poor themselves would play a substantial role in developing and administering the [community action] program."[66] Indeed, President Johnson believed that CAPs would be controlled by local or state governments rather than private nonprofit corporations—a misapprehension that lasted until after his antipoverty legislation passed Congress in August 1964.

However, the members of the PCJD, especially David Hackett, Richard Boone, and Frederick Hayes, who remained active in the formulation of the antipoverty program and would draft much of the eventual CAP legislation, had a very different perspective on community action. They saw the CAPs as a means to reintegrate the poor into the broader social structure and thereby break down the walls of alienation that trapped them in the poverty subculture. While it is unlikely that they foresaw an active role for the poor in policy formulation, they advocated "maximum feasible participation" in the community programs as a means to catalyze the correction of dysfunctional social structures. Crucially, neither the BoB nor the PCJD officials produced detailed plans by which their rather abstract notions of the CAPs would be operationalized. As such, neither of the groups was aware that its conceptualization of the CAPs was at variance with that of its counterpart.

On 1 February 1964, President Johnson appointed Sargent Shriver to head a Task Force on Antipoverty Programs, charged with designing a detailed legislative package that would be sent to Congress and laying out plans for its implementation. Shriver immediately tapped as his deputy Adam Yarmolinsky, Secretary of Defense McNamara's special assistant, and met on 2 February with CEA and BoB representatives who explained the idea of community action. To their dismay, Shriver reacted unfavorably to the idea. Shriver had little faith in the planning capabilities of local organizations and, according to Yarmolinsky, argued, "Where you need the money worst, you'll have the worst plans."[67] Shriver held a second meeting the next day, at which the BoB proponents again presented the CAP proposals. Daniel Moynihan, an attendee of this meeting recalled,

> Charles Schultze explained that projects would be initiated at the local level, with a measure of federal prodding, and approved at the federal level. William Capron touched on the problem of local leadership in the South, especially, and noted that CAP's could be used to bypass the local "power structure" with the use of Federal funds. Richard Boone insisted that the community action programs could be "manned" by the poor themselves.[68]

Shriver, however, remained unconvinced for several reasons. First, he was in close contact with Robert McNamara throughout this process and leaned heavily on advice the secretary of defense offered. Adam Yarmolinsky recalled, "Naturally I was discussing the [social welfare] program with McNamara in the early days and going back to Shriver and discussing what McNamara and I had come out with and what McNamara was saying, and McNamara's views naturally were persuasive with Shriver."[69] Given Secretary McNamara's experience at the Ford Motor Company and his embroilment in an effort to centralize the defense administration, his advocacy of a decentralized structure as embodied in CAPs seems highly unlikely. Also, Shriver continued to look with disfavor upon the direct involvement of local residents. On 13 February, William Cannon wrote to Kermit Gordon and Walter Heller, "Shriver still appears to be highly distrustful of the willingness or ability of local people to do a proper job."[70] Finally, the best time frame that could be offered Shriver for positive results from CAPs was one year—a horizon that prohibited any impact on the upcoming presidential election or even, as Moynihan notes, the selection of a presidential running mate.[71] As a compromise solution, in a meeting on 23 February Shriver and Yarmolinsky decided to retain the Community Action Program as the largest of five "titles" to be incorporated in the antipoverty legislation. As such, CAPs would be implemented in the Economic Opportunity Act that was introduced to Congress on 16 March 1964 and signed into law on 20 August, but they were no longer the be-all and end-all of the administration's strategy. Ironically, no one would be more thankful for this than the BoB staff members who had championed the plan.

The BoB staff's subsequent regret of the CAPs was due, in large part, to language included in the final drafting of the Economic Opportunity Act's Community Action Program section (Title II), which called for "maximum feasible participation" of the poor in the programs that affected them. In part, this language was intended to provided a mechanism through which federal assistance could be extended to poor African Americans in the South despite the resistance of white-dominated institutions. Indeed, this was the manner in which the language was interpreted by many of Shriver's task force members.[72] However, the authorship of Title II indicates that it also was intended to shape the implementation of CAPs in northern urban areas. Title II was written in late February 1964 by Harold Horowitz, associate general counsel at the Department of Health, Education and Welfare (HEW), with the assistance of Richard Boone, Frederick Hayes, and David Hackett—key members of the PCJD urban guerrillas and advocates of the "dysfunctional social structures" perspective on poverty. These individuals likely saw this language as assurance that the poor would participate directly in poverty programs and thus overcome the alienation from social institutions that had historically trapped them in a cycle of poverty. Certainly this was not the intent of Shriver and Yarmolinsky, but as Yarmolinsky recalls,

> It never occurred to us to cut it [i.e., "the maximum feasible participation" language] out.... [W]e accepted it without too much

> question. There was a whole line of development from the Ford Foundation to the President's Committee [on Juvenile Delinquency], and we figured they knew what they were talking about.[73]

Even this incongruity, however, might not have led to an incident had the southern Democratic congressional delegation failed to play its invariable role as a political wildcard. As the Economic Opportunity Act neared passage, Sargent Shriver was certain to assume the directorship of the new Office of Economic Opportunity (OEO), and most observers assumed that Adam Yarmolinsky would be appointed deputy director. Yarmolinsky had served as Shriver's second-in-command during both the 1960 presidential campaign and the 1964 formulation of the antipoverty program.[74] Yet, in a meeting in the office of the Speaker of the House on 8 August, the members of the North Carolina delegation demanded as the price of their support a pledge that Yarmolinsky would have nothing to do with the administration of the poverty program.[75] As McNamara's special assistant, they complained, Yarmolinsky had defended the civil rights of black servicemen in southern states, and he was generally held in suspicion due to his "liberal" views. Furthermore, Yarmolinsky was known to drive a Volkswagen and, periodically, to sport a beard—further indications of his subversive tendencies. Faced with such overwhelming evidence and anxious not to lose the seven North Carolina votes, Shriver and Johnson immediately sacked Yarmolinsky and sent him back to the Pentagon.

Yarmolinsky's departure from the poverty program altered the trajectory of the Community Action Programs. The language of the CAP legislation permitted a wide range of interpretations, some of which contrasted with the intentions of the program's leadership. The wording of the act neither defined "maximum feasible participation" nor specified how it was to be achieved, and it failed to ascertain the extent to which poor people would be involved in real decision making. Yarmolinsky, like Shriver and most other high-level administration figures, was opposed to the adoption of a decentralized decision-making structure in which the poor participated directly in policy formulation. He considered the poor to be a constituency, and "[a] constituency, after all, need not be involved in the kind of direct democracy practiced at faculty meetings."[76] Yarmolinsky's interpretation of the CAP language was such that "in each of the communities where the [Community Action P]rogram operated, the mayor and the poor people would put together a program.... And we would use the availability of some of our money as a carrot to encourage the planning, but we would have the last word."[77] Thus, despite the literal formulation of the CAPs, Yarmolinsky's administration would have retained centralized control over policy making.

However, during the summer of 1964, the character of the Urban Areas Task Force—the group appointed by Shriver to implement the CAPs—evolved such that its ideological composition began to diverge substantially from that of Shriver, Yarmolinsky, and the remainder of the White House administration. During the weeks following the submission to Congress of the antipoverty bill, the senior officials who had prepared the rudimentary design of the antipoverty legislation returned to their respective agencies and a lower-ranking contingent, whose loyalties lay exclusively with the new organization, were seated. This group incorporated significant influence from a subset of poverty warriors, including Richard Cloward and others with experience in local agencies, who adopted the perspective that for CAPs to work the poor not only must participate in decision making but should be encouraged to express criticism and dissent.[78]

The demise of Yarmolinsky as apparent head of the CAPs in August 1964 unintentionally made way for the advocates of resident mobilization to enforce their vision of community action on the fledgling federal poverty agency. By seizing on the potentially revolutionary letter of the Economic Opportunity Act rather than its rather innocuous intent, the poverty warriors threatened to introduce real democracy to federal social welfare policy making—an eventuality that was anathema to the Cold Warriors of the Johnson administration.

The Civilianization of PPB and the Centralization of Social Policy Making

When the Economic Opportunity Act formally instituted President Johnson's war on poverty in August 1964, the man placed at the head of the Office of Economic Opportunity's Community Action Program was Jack Conway, then on leave as the United Auto Workers' principal representative in the headquarters of the AFL-CIO. A labor organizer from Detroit who made his reputation during the early days of the auto workers' union, Conway sought "to structure Community Action Programs so that they would have an immediate and irreversible impact on the communities."[79] As such, he had considerable sympathy for the poverty warriors who adopted the more radical interpretation of "maximum feasible participation." Conway championed the notion of a "three-legged stool"—CAPs designed so as to involve in policy planning the three relevant segments of low income communities: private charities, public agencies, and the poor. Indeed, the CAP guidelines published early in 1965 explicitly endorsed this activist approach and stated, "A vital feature of every community action program is the involvement of the poor themselves—the residents of the areas and members of the groups to be served—in planning, policy-making, and operation of the program."[80] The goal was "the 'mobilization' of the poor" through "traditional democratic approaches and techniques such as group forums and discussions, nominations, and balloting" and with the use of films, literature, and mobile information centers.[81] Fundamentally, the CAP administrators envisioned the creation of new social and political institutions through which the poor could express their needs and participate in solutions. Thus, they interpreted the ambiguous legislation enacted by Congress as empowering them to democratize the social welfare policy-making process.

Such strategies as those adopted by CAP administrators in the second half of 1964 were not, however, looked upon favorably by either the systems analysts in the Pentagon or the top-level economists of the Bureau of the Budget. The scales fell quickly from the eyes of the assistant director of the BoB, Charles L. Schultze, who lost his enthusiasm for CAPs as a centralizing mechanism and began to perceive them as destabilizing influences under the direction of "a man of little stature and uncertain competence."[82] As Moynihan recalls, "Disillusion with the 'OEO radicals,' as they came to be known in some circles at least of the Executive Office Building, was instantaneous. The Bureau had been promised coordination, and all it has seemingly got was chaos."[83] Also, the adoption of decentralized policy making failed to conceptualize social problems as systems but concentrated instead on local solutions to locally perceived problem sets. Further, the reliance of the CAP management structure on local experience and involvement of the poor prohibited detached, rational analysis of the sort championed by the scientific analysts. They believed that persons deeply enmeshed in the social systems targeted for policy making could neither objectively comprehend those systems nor develop unbiased policy alternatives. Finally, the empowerment of the poor threatened to undermine existing social and institutional structures. In this sense, the destabilizing implications of the CAPs ran contrary to what was a primary objective of the Great Society—the stabilization of American society. As such, by early 1965, BoB officials were looking for a means to reassert centralized control over the new social welfare programs, and their search quickly took them to the Pentagon.

As it became apparent to BoB officials that CAPs would not provide the sort of coordinating mechanism they desired, the application of the PPB system to the civilian agencies as an alternative solution rapidly gained currency. The BoB had been closely involved with Charles Hitch and his staff as they created the PPB system in 1961–1962 in the Department of Defense, and many top-level budget administrators were deeply impressed by the degree of control it afforded.[84] At the same time, PPB's reliance on systems analysis for decision making placed emphasis on quantitative techniques and statistical precision—two qualities held dear by the BoB economists. Further, Sargent Shriver, the director of the OEO, and McNamara were close friends, and the secretary of defense had advised Shriver that he needed a facility like the one that Charles Hitch and Alain Enthoven ran at the Pentagon—a program-planning facility.[85] Taking McNamara's advice, Shriver established a program analysis office reporting directly to him and hired RAND analyst Joseph Kershaw as assistant director for program coordination, planning, evaluation and research. Finally, President Johnson's new special assistant for domestic programs, Joseph A. Califano, Jr., had been a member of McNamara's OSD staff from 1961 until his appointment to head the domestic programs in 1964. By his own admission, Califano's prior experience had been confined to defense, and he "knew nothing about domestic programs, nothing about social programs, and indeed, . . . hadn't even ever thought much about them."[86] However, he had been associated with the implementation of McNamara's centralized management system and could be counted on to champion a broadening of its use.

In October 1964, three members of the BoB's labor and welfare division wrote a lengthy memorandum to the bureau's deputy director, Elmer Staats, which describes the nature of the Defense Department's PPB system, the differences in implementation between the DOD and OEO, and potential problems.[87] The memorandum advises that the application of PPB to the poverty administration would require the recruitment of considerable numbers of social scientists to participate in systems analyses, that the cost-benefit approach to analyses would be more difficult in social welfare policy making due to difficulty in measuring program benefits, and that OEO would have to compensate somehow for "the absence of any Rand Corporation in the poverty area as a starting point."[88] Nevertheless, the BoB staff members emphasized the powerful centralizing effect PPB had had on defense policy making. They commented that before 1961 the OSD had been little more than a "rubber stamp" operation; the PPB system provided the secretary of defense with the information and administrative structure necessary to exercise considerable authority. They found that, "this new type of decision-making ... appears to be the most fundamental innovation of the Hitch-McNamara era. If this is the fundamental essence of the Hitch function, then we believe it is obvious that OEO needs a Hitch-type operation."[89]

Over the winter of 1964–1965 a second source of pressure magnified the desire of OEO and BoB officials to centralize control over the social welfare programs—the increasingly strident complaints of local Democratic politicians concerning the Community Action Programs' destabilizing effects. As was discussed above, an objective of the CAP administrators was to stimulate the creation of new social institutions that would permit the poor a meaningful voice in policy making. If necessary, this meant that existing social and political structures, perceived by the poverty warriors as decrepit and unresponsive to the needs of the poor, would have to be abandoned. Further, the CAP guidelines appeared explicitly to encourage program participants to engage in disruptive behavior. The first CAP program guide, for example, suggested that local organizations provide "meaningful opportunities for residents, either as individuals or in groups, to protest or to propose additions to or changes in the ways which a Community Action Program is being planned or undertaken."[90] In fact, protest and confrontation were viewed by many CAP organizers as at least therapeutic means for the poor to vent their frustrations.

President Johnson had never been favorably inclined toward CAPs and certainly never intended them to mobilize the poor.[91] In fact, he had initially vetoed the idea when it was presented to him by Kermit Gordon and Walter Heller over the Christmas holidays in late 1963.[92] As the programs swang into action in early 1965, Johnson's anxiety promptly turned to antipathy as it became clear to him that the CAPs were being run by "kooks and sociologists."[93] Being a consummate politician, President Johnson recognized that the programs were causing more problems for his friends than for his enemies. The breaking point was reached in June 1965 when the Conference of Mayors was used as a platform by Democratic urban leadership to denounce Johnson's infant social welfare programs. There, Mayors John F. Shelley of San Francisco and Sam Yorty of Los Angeles sponsored a resolution accusing Sargent Shriver of "fostering class struggle."[94] Johnson demanded that his staff find a way to rein in the programs, but, in Moynihan's words, "No one knew how to respond to the President's concerns. Save the Bureau of the Budget."[95]

Charles L. Schultze, one of the frustrated parents of the CAPs, replaced Kermit Gordon as director of the BoB in June 1965 and immediately began working out plans to transfer the PPB system from the Pentagon to the social welfare agencies. Gordon had fortuitously provided Schultze with a knowledgeable and energetic executive officer when, during the previous January, he had hired Henry Rowen away from the OSD to serve as assistant director of the BoB for national security affairs.[96] Rowen, a RAND alumnus and one of McNamara's Whiz Kids, was reassigned by Schultze to begin laying the groundwork for implementation of the defense management system in the civilian agencies—a task Rowen pursued with customary relish. In Rowen's words,

> I had really gotten quite upset about the Great Society program. There I was sitting in the Budget Bureau while all this stuff was coming from across the street—the White House basically.... By then Charles Schultze had become budget director ... and he was very skeptical of a lot of this too. It wasn't that I was the only one who thought that this presented some real problems.[97]

Over the summer of 1965, Schultze and Rowen pulled together a plan for instituting PPB in the civilian agencies of the federal government. According to the design they presented to Joseph Califano, the president's special assistant for domestic programs, each department would draw up a basic multiyear program (a five-year plan was suggested) that established specific goals and objectives.[98] To accomplish this, each department or agency would create a central systems analysis office within the Office of the Secretary to analyze goals and programs on a regular, year-round basis. Operationally, the system would function similarly to the defense PPB arrangement.

On 16 August 1965, three days after Schultze's memorandum was written, Califano passed the document on to President Johnson with his ringing endorsement.[99] Since his appointment to head the domestic programs a year earlier, Califano had been appalled by the "very unsystematic and chaotic and anarchic" state of management in the civilian agencies, particularly in contrast with the OSD. Again, it was the prospect of centralized control that enthused the White House staffer. Califano recommended to the president that, at the next Cabinet meeting, he lend Charles Schultze "overt and substantial support" for the implementation of PPB and that each department or agency head be directed to acquire the personnel necessary to create a central systems analysis office. Califano's note is marked "YES" by the president. Subsequently, at a Cabinet meeting on 25 August that was immediately followed by a press conference, President Johnson directed the implementation of PPB across the federal bureaucratic structure.[100] In his comments, Johnson pointed out that PPB "will improve our ability to control our programs and our budgets rather than having them control us."[101]

Conclusion

Considerable scholarship remains to be done concerning the diffusion of systems analysis and program budgeting to the federal civilian departments and the long-run impacts this had on the American social welfare programs. While this subject goes far beyond the scope of the present chapter, a few observations are warranted. First, as was mentioned above, systems analysis and program budgeting were developed for use in the DOD, and the DOD is uniquely hierarchical among federal agencies in both its structure and its culture. The giving and taking of orders, deference to authority, and the centralization of command are intrinsic to its operation. In this sense, the DOD is a poor model for the democratic ideal many Americans hold for their government institutions. Also, the Department of Defense is remarkably self-contained in the sense that it requires very little coordination with other departments and agencies to prosecute its missions. Its planning-programming-budgeting system program packages could thus identify and pursue objectives with relatively little attention to other agencies. This was not the case in the social welfare arena where an assault on, say, urban decay required the cooperation of myriad federal agencies and departments, none of which could reasonably construct a program package whose output would be urban revitalization.

This situation led to the emergence, during the 1960s, of critiques of systems analysis and PPB based on the argument that because social programs are vastly complex in their political construction, they cannot be addressed by centralized, monolithic hierarchies.[102] Also, as was noted above, the defense problems to which systems analysis was applied were very often quite technical in content and thus lent themselves with relative ease to quantification and precise measurement. This was simply not the case in social welfare issues, where the very "output" of any "weapons system" (e.g., new education curricula) could not be ascertained for years, if at all. This had severe ramifications since, as Ida Hoos argued in her biting critique, "the salient features of systems methodology in theory and practice [are]: (a) it is intolerant of and even antagonistic to unquantifiable variables; (b) it contains a bias against social and human factors; (c) its ethic is that of the marketplace; (d) it is driven by technical optimism."[103]

Furthermore, while systems analysis and PPB had been designed for military applications, their results in the Department of Defense were dubious at best and held even less promise for application to social welfare problems. As McNamara made clear in his 1995 book, *In Retrospect*, the management and decision-making processes adopted by him and his staff contributed directly to the military debacle in Vietnam.[104] McNamara used PPB to separate planning from execution by extracting decision-making authority from lower echelons and concentrating it in the highest levels of management. Systems analysis facilitated this centralization by compensating, apparently, for the alienation of decision makers from the locus of operations. Through systems analysis, McNamara and his staff felt empowered to replace the complexity of real life with simplified models that were lent illusory precision by their quantitative bases. Numerical data became proxies for experience, and the human content of defense problems seemed to fall with disturbing ease out of the scientific analysts' equations. Henry Rowen illustrated this tendency when he reported a conversation with Secretary of the Air Force Harold Brown in 1966:

> First, with regard to tactical air operations, in Vietnam in particular, he thinks it's just very important now that he have a war on for the Air Force, to get just a mite better idea of the effectiveness of tactical air operations. He has had the operations analysis people in the Air Staff drop everything and just work on this problem—it's a serious problem and he thinks we've been doing miserably in understanding it. . . . He knows RAND is doing a lot of work on tactical air in a model-building sense, but what he is talking about is using the Vietnamese data much more.[105]

Indeed, the American public retains no image of the Vietnam War more vivid than the body counts that were broadcast nightly—a grim harvesting of data for the Pentagon decision apparatus.

This fixation on policy research and evaluation accompanied systems analysis and program budgeting as these methods diffused from the Pentagon to the federal social welfare agencies. The intense appetite for policy research created by the implementation of the PPB system can be seen first in the Pentagon where the creation of the defense PPB system in 1961 had dramatically increased RAND's menu of research and analysis projects for the OSD. To supplement these research capabilities, the OSD also created the Logistics Management Institute in 1962 to provide systems analysis and research on logistical problems. Similarly, the post-1965 period witnessed a dramatic increase in federal expenditures for social research as "civilian RANDs" and other "policy research" institutes began to proliferate. As Robert Haveman writes, "Research support was clearly less controversial and risky than, say, community action."[106] In constant 1966 dollars, federal expenditures for poverty-related research and development grew from $2.6 to $6.1 million between 1965 and 1968, more than tripled to $19.0 million in 1971, and then more than quadrupled again by 1976 to $83.9 million.[107] Thus, the research-intensive structure of the Great Society programs virtually created a new and well-funded discipline: policy analysis.

In the Office of Economic Opportunity, for example, the elaboration of PPB was followed by the perception that the poverty administration needed "a basic research organization, much as the Pentagon relied upon the Rand Corporation."[108] As such, in 1966 OEO provided an annual grant to create and sustain the Institute for Research on Poverty at the University of Wisconsin–Madison. Modeled closely on RAND, the Institute supported a staff of about forty economists, sociologists, and other social scientists throughout the late 1960s and 1970s. Most of these scholars also had academic appointments at the university. The creation of the Urban Institute is a further example of the research-intensity of RAND's management methodologies. Indeed, in 1966 the RAND Corporation itself formally diversified from national defense research into social welfare policy research, and by 1972 almost half of RAND's research program was devoted to social welfare.

Finally, the top-down, technocratic allocation of resources created by systems analysis and program budgeting was profoundly undemocratic and, in a nation that prides itself on democratic principles, its use for social policy making raises fundamental issues. Charles Zwick, a RAND analyst who succeeded Charles L. Schultze as budget director in 1968, argued that the "grand strategy" of PPB and systems analysis was "getting more control higher up in the system."[109] The Community Action Programs are a case in point. As Robert Wood notes, "The Community Action Program proceeded on a theory of direct democracy in its purest form, to empower poor people in neighborhoods with authority not only to choose among values but also among techniques of implementation."[110] However, with the construction of OEO's first four-year PPB plan by Joseph Kershaw in September 1965, CAPs were excluded from the program categories or elements. The CAPs' democratizing strategy was an abomination to the defense and BoB officials who championed PPB and systems analysis. These individuals sought systematic, orderly, centrally controlled social welfare programs that would stabilize American society in the face of deepening Cold War tensions. Contrary to democratic traditions, according to Hoos, "The techniques of systems analysis can, if used astutely, remove highly charged political issues from the arena of public debate by relegating them to 'scientific' appraisal."[111] Historians must consider more rigorously how the fate of the Great Society social welfare programs was shaped by the implementation of PPB and the pervasive influences of systems analysis.

Notes

1. Memorandum, L. J. Henderson to F. R. Collbohm, "Discussion with Fred Stephan Concerning the McGraw-Hill Series and RAND Social Science," 2 November 1949, RAND Classified Library, folder "History, Henderson 1949."

2. With regard to the consensus among civilian and military leaders concerning the decisive nature of research and development, see James Phinney Baxter, *Scientists against Time* (Boston: Little, Brown, 1946). For scholarship concerning the formulation of U.S. science policy in the years immediately following World War II, see Vannevar Bush, *Science—The Endless Frontier: A Report to the President* (Washington, D.C.: U.S. Government Printing Office, 1945); U.S. Senate, Committee on Military Affairs, Subcommittee on War Mobilization, *The Government's Wartime Research and Development, 1940–1944* (Washington, D.C.: U.S. Government Printing Office, 1945); Talcott Parsons, "The Science Legislation and the Role of the Social Sciences," *American Sociological Review* 11 (Dec. 1946): 563–666; Perry McCoy Smith, *The Air Force Plans for Peace, 1943–1945* (Baltimore: Johns Hopkins University Press, 1970); Daniel J. Kevles, "The National Science Foundation and the Debate over Postwar Research Policy, 1942–1945: A Political

Interpretation of *Science—The Endless Frontier,*" *Isis* 68 (1977): 5–26; Daniel J. Kevles, *The Physicists: The History of a Scientific Community* (Cambridge: Harvard University Press, 1987; originally published: New York: Knopf, 1977), 340–348; Gregg Herken, *The Winning Weapon: The Atomic Bomb in the Cold War, 1945–1950* (New York: Knopf, 1980); Daniel J. Kevles, "Principles and Politics in Federal R&D Policy, 1945–1990: An Appreciation of the Bush Report," introduction to 1990 reissue of Vannevar Bush, *Science—The Endless Frontier* (Washington, D.C.: National Science Foundation, 1990); and Daniel Lee Kleinman, *Politics on the Endless Frontier: Postwar Research Policy in the United States* (Durham: Duke University Press, 1995). For an analysis of the Soviet experience, see David Holloway, *Stalin and the Bomb: The Soviet Union and Atomic Energy, 1939–1956* (New Haven: Yale University Press, 1994).

3. This historical interpretation is most extensively expressed in the essays collected in Peter Galison and Bruce Hevly, eds., *Big Science: The Growth of Large-Scale Research* (Stanford: Stanford University Press, 1992).

4. See Michael A. Dennis, "A Change of State: The Political Cultures of Technical Practice at the M.I.T. Instrumentation Laboratory and the Johns Hopkins University Applied Physics Laboratory, 1930–1945" (Ph.D. diss., Johns Hopkins University, 1990).

5. On the Manhattan Project see Leslie R. Groves, *Now It Can Be Told: The Story of the Manhattan Project* (New York and Evanston: Harper & Row, 1962); Richard G. Hewlett and Oscar E. Anderson, Jr., *The New World: A History of the United States Atomic Energy Commission, Volume I, 1939–1946* (Berkeley: University of California Press, 1990; originally published University Park: Pennsylvania State University Press, 1962); Richard Rhodes, *The Making of the Atomic Bomb* (New York: Simon and Schuster, 1986); and Barton C. Hacker, *The Dragon's Tail: Radiation Safety in the Manhattan Project, 1942–1946* (Berkeley: University of California Press, 1987).

6. This point is expressed in much of the literature concerning the formation of American science policy in the post–World War II era (see n. 3). See also Herbert F. York, *The Advisors: Oppenheimer, Teller, and the Superbomb* (Stanford: Stanford University Press, 1976); and Michael Sherry, *Preparing for the Next War: American Plans for Postwar Defense* (New Haven: Yale University Press, 1977).

7. Quoted in *40th Year: The RAND Corporation* (Santa Monica: The RAND Corporation, 1989), 3.

8. Fred Kaplan, *The Wizards of Armageddon* (Stanford: Stanford University Press, 1983), 56–58.

9. The following discussion of Project RAND's objectives is from Arthur E. Raymond, "Talking Outline—RAND Briefing of 4-21-47," RAND Document D-99-1, 21 April 1947, RAND Classified Library; L. J. Henderson, "Statement of Purposes, Objectives, and Functions of RAND," RAND Document D-92, 6 March 1947, RAND Classified Library; and, "Commander Brown's First

Report—Origin and Objective of Project RAND, 1st Revision," RAND Document D-138, 18 June 1947, RAND Classified Library.

10. Arthur E. Raymond, "Presentation by Mr. Arthur E. Raymond on Project 'RAND,'" RAND Document D-188, 13 August 1947, RAND Classified Library, 12.

11. L. J. Henderson, "Statement of Purposes, Objectives, and Functions of RAND," RAND Document D-92, 6 March 1947, RAND Classified Library, 1.

12. Raymond, "Presentation by Mr. Arthur E. Raymond on Project 'RAND,'" 1.

13. Memorandum, J. D. Williams to F. R. Collbohm, "Project RAND," RAND Document D-7, RAND Classified Library, 1.

14. On the history of the AMP, see Larry Owens, "Mathematicians at War: Warren Weaver and the Applied Mathematics Panel, 1942–1945," in *The History of Modern Mathematics, Volume II: Institutions and Applications*, ed. David E. Rowe and John McCleary (Boston: Academic Press, 1989), 287–305.

15. For scholarship on RAND's role in the development of computer science and artificial intelligence, see Allen Newell and Herbert A. Simon, "Historical Addendum," *Human Problem Solving* (Englewood Cliffs, N.J.: Prentice-Hall, 1972); F. J. Gruenberger, "History of the Johnniac," *Annals of the History of Computing* 1, no. 1 (1979): 49–64; Pamela McCorduck, *Machines Who Think* (San Francisco: W. H. Freeman and Company, 1979); C. L. Baker, "JOSS Johnniac Open-Shop System," in R. L. Wexelblat, ed., *History of Programming Languages* (New York: Academic Press, 1981); James Fleck, "Development and Establishment in Artificial Intelligence," in Norbert Elias, Herminio Martins, and Richard Whitley, eds., *Scientific Establishments and Hierarchies*, vol. 6 of *Sociology of the Sciences: A Yearbook* (Boston: D. Reidel Publishing Company, 1982); Shirley L. Marks, "JOSS: Conversational Computing for the Nonprogrammer," *Annals of the History of Computing* 4, no. 1 (1982): 35–52; George Bernard Dantzig, *Impact of Linear Programming on Computer Development*, Technical Report SOL 85–7 (Stanford: Systems Optimization Laboratory, 1985); Gwen Bell and Leah Hutten, "A Historical Timeline of Artificial Intelligence and Robotics," *Computer Museum Report* (Boston: The Computer Museum, 1987); Martin Davis, "Mathematical Logic and the Origin of Modern Computers," in E. Phillips, ed., *Studies in the History of Mathematics* (Washington, D.C.: Mathematical Association of America, 1987); Herbert A. Simon, *Models of My Life* (New York: Basic Books, 1991); and Daniel Crevier, *AI: The Tumultuous History of the Search for Artificial Intelligence* (New York: Basic Books, 1993). For discussions of RAND's participation in early space studies, see R. Cargill Hall, "Early U.S. Satellite Proposals," *Technology and Culture* 4 (Autumn 1963): 410–434; Robert E. Bickner, *The Changing Relationship between the Air Force and the Aerospace Industry* (Santa Monica: The RAND Corporation, 1964); Edmund Beard, *Developing the ICBM: A Study in Bureaucratic Politics* (New York: Columbia University Press, 1976); Walter McDougall, . . . *The Heavens and the Earth: A Polit-*

ical History of the Space Age (New York: Basic Books, 1985); David H. DeVorkin, *Science with a Vengeance: How the Military Created the U.S. Space Sciences after World War II* (New York: Springer-Verlag, 1992); and Davis Dyer, "Necessity as the Mother of Convention: Developing the ICBM, 1954–1958," paper presented at the Business History Conference, Boston, Massachusetts, 19 March 1993.

16. Operations research is also known, especially in Great Britain, as operations analysis. The term "operations research" is used here but is understood to be synonymous with operations analysis. For accounts of the origins of operations research, see C. West Churchman, Russell L. Ackoff, and E. Leonard Arnoff, Introduction to *Operations Research* (New York: John Wiley, 1957); Donald W. Mitchell, "Military Operations Research," *Marine Corps Gazette* 41, no. 2 (1957): 47–51; L. Edgar Prina, "The Navy First Used Think Tanks during World War II," *Armed Forces Journal* 106, no. 4 (28 September 1968): 18–23; and William Brett, Emile B. Feldman, and Michael Sentlowitz, *An Introduction to the History of Mathematics, Number Theory, and Operations Research* (New York: MSS Information Corporation, 1974). Robin Rider of Stanford University is currently writing a history of operations research.

17. See Robert Dorfman, Paul A. Samuelson, and Robert M. Solow, *Linear Programming and Economic Analysis* (New York: McGraw-Hill, 1958); George Bernard Dantzig, *Linear Programming and Extensions* (Princeton: Princeton University Press, 1965); Robert McQuie, "Military History and Mathematical Analysis, *Military Review* 50, no. 5 (1970): 8–17; George Bernard Dantzig, "Time-Staged Methods in Linear Programming: Comments, Early History, Future Prospects," *Large Scale Systems*, Studies in Management Science Systems 7 (Amsterdam: North Holland, 1980); George Bernard Dantzig, "Reminiscences about the Origins of Linear Programming," *Operations Research Letters* 1, no. 2 (1981): 43–48; Robert Dorfman, "The Discovery of Linear Programming," *Annals of the History of Computing* 6 (1984): 283–295; Donald J. Albers and Constance Reid, "An Interview with George B. Dantzig: The Father of Linear Programming," *The College Mathematics Journal* 17 (1986): 293–314; and George Bernard Dantzig, *Origins of the Simplex Method*, Technical Report SOL 87–5 (Stanford: Systems Optimization Laboratory, 1987).

18. See Richard Bellman, *Dynamic Programming and Modern Control Theory* (New York: Academic Press, 1965).

19. John von Neumann and Oskar Morgenstern, *Theory of Games and Economic Behavior* (Princeton: Princeton University Press, 1947). For the history of game theory, see R. Duncan Luce and Howard Raiffa, *Games and Decisions: An Introduction and Critical Survey* (New York: John Wiley, 1957); Steven J. Brams, *Superpower Games* (New Haven: Yale University Press, 1985); Eric Rasmusen, *Games and Information: An Introduction to Game Theory* (Oxford: Basil Blackwell, 1989); Ken Binmore, *Essays on the Foundations of Game Theory* (Oxford: Basil Blackwell, 1990); William Poundstone, *Prisoner's Dilemma* (New York: Doubleday, 1992); and E. Roy Weintraub, ed., *Toward a History of Game Theory* (Durham, N.C.: Duke University Press, 1992).

20. On the development of systems analysis, see Olaf Helmer, *Social Technology* (New York: Basic Books, 1966); C. West Churchman, *The Systems Approach* (New York: Delacorte Press, 1968); Guy Black, *The Application of Systems Analysis to Government Operations* (New York: Frederick A. Praeger, 1968); E. S. Quade and W. I. Boucher, eds., *Systems Analysis and Policy Planning* (New York: American Elsevier, 1968); and Seth Bonder, "Systems Analysis: A Purely Intellectual Activity," *Military Review* 51, no. 2 (1971): 14–23. The RAND Corporation archives also appear to contain considerable materials on the origins and development of systems analysis. For example, see James F. Digby, "Operations Research and Systems Analysis at RAND, 1948–1967," RAND Note 2936-RC, April 1989.

21. Charles J. Hitch and Roland N. McKean, *The Economics of Defense in the Nuclear Age* (Cambridge: Harvard University Press, 1960).

22. The use of the term "rational" in this context is from Gregory Palmer, *The McNamara Strategy and the Vietnam War: Program Budgeting in the Pentagon, 1960–1968* (Westport, Conn.: Greenwood Press, 1978), 4. Palmer states, "The rationalist stresses calculated choice between alternatives rather than deference to habit or authority, which, as they might inhibit optimal choice, are always regarded as bad. Favoring calculation, he tends also to favor quantification and thus is often disposed to making his selection from the most easily quantifiable solutions at the neglect of other viable alternatives."

23. Hitch and McKean, *The Economics of Defense*, 2.

24. Ibid., 105.

25. These core recommendations are enumerated in Hitch and McKean, *The Economics of Defense*, 107.

26. The ideas expressed in *The Economics of Defense* reflected the extensive research performed at RAND throughout the 1950s in defense budgeting and cost analysis. For earlier work in this area, see David Novick, *Efficiency and Economy in Government through New Budgeting and Accounting Procedures*, RAND Report R-254 (Santa Monica: The RAND Corporation, 1 February 1954); David Novick, *Weapon-System Cost Methodology*, RAND Report R-287 (Santa Monica: The RAND Corporation, 1 February 1956); David Novick, *A New Approach to the Military Budget*, RAND Research Memorandum RM-1757 (Santa Monica: The RAND Corporation, 12 June 1956); Kenneth J. Arrow and S. S. Arrow, *Methodological Problems in Airframe Cost-Performance Studies*, RAND Research Memorandum RM-456 (Santa Monica: The RAND Corporation, 20 September 1950); Malcolm W. Hoag, *The Relevance of Costs in Operations Research*, RAND Paper P-820 (Santa Monica: The RAND Corporation, 13 April 1956); Gene H. Fisher, *Weapon-System Cost Analysis*, RAND Paper P-823 (Santa Monica: The RAND Corporation, 10 July 1956); David Novick, *Concepts of Cost for Use in Studies of Effectiveness*, RAND Paper P-1182 (Santa Monica: The RAND Corporation, 4 October 1957); and Charles J. Hitch, *Economics and Operations Research*, RAND Paper P-1250 (Santa Monica: The RAND Corporation, 8 January 1958).

27. Charles J. Hitch, *Decision-Making for Defense* (Berkeley and Los Angeles: University of California Press, 1965), 50.

28. Charles Hitch, for example, notes that "to push through the development of the programming system in so short a time and make it work required a Secretary as strong and decisive as Robert S. McNamara." Hitch, *Decision-Making for Defense*, 71.

29. John A. Byrne, *The Whiz Kids: Ten Founding Fathers of American Business and the Legacy They Left Us* (New York: Doubleday, 1993), 36.

30. Gregory Palmer, *The McNamara Strategy and the Vietnam War.*

31. For accounts of the evolution of the Department of Defense from 1947 through 1958, see Alain C. Enthoven and K. Wayne Smith, *How Much Is Enough? Shaping the Defense Program, 1961–1969* (New York: Harper & Row, 1971); Paul Y. Hammond, *The American Military Establishment in the Twentieth Century* (Princeton: Princeton University Press, 1961); Morris Janowitz, ed., *The New Military: Changing Patterns of Organization* (New York: Russell Sage Foundation, 1964); Edward Kolodziej, *The Uncommon Defense and Congress, 1945–1963* (Columbus: Ohio State University Press, 1966); John C. Ries, *The Management of Defense: Organization and Control of the U.S. Armed Forces* (Baltimore: Johns Hopkins University Press, 1964); and James M. Roherty, *Decisions of Robert S. McNamara: A Study of the Role of the Secretary of Defense* (Coral Gables, Fl.: University of Miami Press, 1970). Of particular relevance to RAND and the changing air force position within the defense establishment, see Arnold Mengel to George Tanham, memorandum, "The Air Force—Its Environment and Organization," 23 June 1972, Donald B. Rice Papers, Air Force 1967–79, box 1, RAND Archives.

32. Rockefeller Brothers Fund, *International Security: The Military Aspect* (Garden City, N.Y.: Doubleday, 1958).

33. Ibid., 58.

34. Lt. General John W. O'Neill, interview by Hugh N. Ahmann and James C. Hasdorff, Washington, D.C., 15–18 May 1973, 16–17.

35. The original National Defense Act of 1947 underwent several revisions prior to 1958, most notably in 1949 and 1953. However, these were not as extensive as the 1958 Act.

36. A unified command was defined as a military command with a broad continuing mission, under a single commander and composed of significant assigned components of two or more military services. Unified commands were established and so designated by the president, through the secretary of defense, with the advice and assistance of the joint chiefs of staff. Specified commands were defined as those with a broad continuing mission, also established and so designated by the president through the secretary of defense with the advice and assistance of the JCS, but composed of forces from only one service.

37. Senator John F. Kennedy, press release, 14 September 1960, Roswell Gilpatric Papers, box 4, Symington Committee on the Defense Establishment, Correspondence Aug.–Oct. 1960 folder, John F. Kennedy Library, 3.

38. Senator Stuart Symington to members of the Symington Committee on the Defense Establishment, preliminary draft of committee report, "Modernizing the Defense Establishment," 10 October 1960, Roswell Gilpatric Papers, box 4, Symington Committee on the Defense Establishment, Correspondence Aug.–Oct. 1960 folder, John F. Kennedy Library, 5–6.

39. Symington Committee on the Defense Establishment, final report to President-elect Kennedy, 26 November 1960, Roswell Gilpatric Papers, box 4, Symington Committee on the Defense Establishment, Drafts of Report Part II folder, John F. Kennedy Library.

40. "Proposed Defense Shake-Up Faces Strong Congressional Objections; Puts President-Elect Kennedy on the Spot," *Army Navy Air Force Journal* 98, no. 15 (10 December 1960): 1.

41. Roswell Gilpatric, interview by Dennis J. O'Brien, 5 May 1970, John F. Kennedy Library Oral History Collection, 103.

42. See John Wayne Fuller, "Congress and the Defense Budget: A Study of the McNamara Years" (Ph.D. diss., Princeton University, 1972), 76–77.

43. Roswell Gilpatric, interview by Dennis J. O'Brien, 5 May 1970, John F. Kennedy Library Oral History Collection, 103.

44. Alain C. Enthoven, interview by William W. Moss, 4 June 1971, Washington, D.C., John F. Kennedy Library Oral History Collection, 4.

45. Enthoven and Smith, *How Much Is Enough?* 31–32; Ralph Sanders, *The Politics of Defense Analysis* (New York: Dunellen, 1973), 43–44.

46. Roswell Gilpatric, interview by Dennis J. O'Brien, 5 May 1970, John F. Kennedy Library Oral History Collection, 69.

47. Alain C. Enthoven, interview by William W. Moss, 4 June 1971, Washington, D.C., John F. Kennedy Library Oral History Collection, 5.

48. The Planning-Programming-Budgeting system is referred to in the literature as both PPBS and PPB. The more flexible PPB will be used in the present chapter.

49. For discussions of the conceptual components of the PPB system, see Enthoven and Smith, *How Much Is Enough?* 33–47; Hitch, *Decision-Making for Defense*; Novick, "The Department of Defense," in *idem, Program Budgeting: Program Analysis and the Federal Budget* (Cambridge: Harvard University Press, 1965), 81–117; and Harold A. Hovey, *The Planning-Programming-Budgeting Approach to Government Decision-Making* (New York and Washington: Frederick A. Praeger, 1968).

50. Daniel P. Moynihan, *Maximum Feasible Misunderstanding: Community Action in the War on Poverty* (New York: The Free Press, 1969), 79.

51. Ibid., 80.

52. The following account of the Ford Foundation "Grey Areas" programs is largely from Moynihan, *Maximum Feasible Misunderstanding*; and James L. Sundquist, "Origins of the War on Poverty," in *idem, On Fighting Poverty: Perspectives from Experience* (New York: Basic Books, 1969): 6–33. Interestingly, the term "Grey Areas" was also used during the 1950s in international relations to refer to what is now referred to as the "third world" or underdeveloped world. See Kaufmann, *The McNamara Strategy*, 63.

53. Moynihan, *Maximum Feasible Misunderstanding*, 42.

54. The following discussion of Mobilization for Youth, Inc., is from Moynihan, *Maximum Feasible Misunderstanding*.

55. Richard A. Cloward and Lloyd E. Ohlin, *Delinquency and Opportunity* (Glencoe, Ill.: The Free Press, 1960).

56. For accounts of the PCJD, see James L. Sundquist, "Origins of the War on Poverty"; Richard Blumenthal, "Community Action: Origins of a Government Program" (B.A. thesis, Harvard College, 1967); Moynihan, *Maximum Feasible Misunderstanding*; and Stephen M. Rose, *The Betrayal of the Poor: The Transformation of Community Action* (Cambridge, Mass.: Schenkman Publishing, 1972).

57. Ibid., 11.

58. The "guerrillas" included Frederick O'Reilly Hayes, then assistant commissioner of program planning for urban renewal at the Housing and Home Finance Agency; Lloyd Ohlin; Sanford Kravitz; Richard Boone; and William Lawrence, a PCJD committee staff member.

59. For scholarship on the origins of the Community Action Program, see Peter Marris and Martin Rein, *The Dilemmas of Social Reform* (New York: Atherton Press, 1967); Moynihan, *Maximum Feasible Misunderstanding*; John G. Wofford, "The Politics of Local Responsibility—Administration of the Community Action Program, 1964–1966," in *On Fighting Poverty*, ed. James L. Sundquist (New York: Basic Books, Inc., 1969); Sanford L. Kravitz, "Community Action Programs: Past, Present, Its Future," in *On Fighting Poverty*, ed. James L. Sundquist (New York: Basic Books, Inc., 1969); John C. Donovan, *The Politics of Poverty* (New York: Western Publishing Company, 1967); and Kenneth B. Clark and Jeannette Hopkins, *A Relevant War against Poverty: A Study of Community Action Programs and Observable Social Change* (New York and Evanston, Ill.: Harper and Row, 1969); David A. Grossman, "The CAP: A New Function for Local Government," in *Urban Planning and Social Policy*, ed. Bernard J. Frieden and Robert Morris (New York: Basic Books, 1968); and Sar Levitan, *The Great Society's Poor Law: A New Approach to Poverty* (Baltimore: Johns Hopkins University Press, 1969).

60. Kravitz, "Community Action Programs: Past, Present, Its Future," 56.

61. Conference on the Kennedy Administration Urban Poverty Programs and Policies, transcript, 16–17 June 1973, Brandeis University, John F. Kennedy Library Oral History Collection, 244–245.

62. Ibid., 144–145.

63. Moynihan, *Maximum Feasible Misunderstanding*, 169.

64. William Cannon, Bureau of the Budget, to P. S. Hughes, memorandum, "A Proposed Approach to the Poverty Problem," 12 December 1963, Kermit Gordon Papers, box 12, Poverty folder, John F. Kennedy Library, 1–2.

65. Sundquist, "Origins of the War on Poverty," 22.

66. Quoted in Blumenthal, "Community Action," 58–59.

67. Quoted in Moynihan, *Maximum Feasible Misunderstanding*, 82.

68. Ibid., 83.

69. Conference on the Kennedy Administration Urban Poverty Programs and Policies, transcript, 16–17 June 1973, Brandeis University, John F. Kennedy Library Oral History Collection, 241.

70. William B. Cannon, Bureau of the Budget, to Kermit Gordon, director of the BoB, and Walter Heller, Chairman of the CEA, memorandum, "Shriver Poverty Program," 13 February 1964, Kermit Gordon Papers, box 12, Poverty folder, John F. Kennedy Library, 1.

71. Moynihan, *Maximum Feasible Misunderstanding*, 82.

72. Blumenthal, "Community Action," 94.

73. Adam Yarmolinsky, interviewed by Richard Blumenthal, 9 February 1967. Quoted in Blumenthal, "Community Action," 91.

74. For example, a prospective OEO organization chart from 1964 in Yarmolinsky's papers places Yarmolinsky as deputy director. Anonymous, handwritten organization chart for poverty administration, no date (probably July 1964), Adam Yarmolinsky Papers, box 87, OEO Poverty Task Force folder, John F. Kennedy Library.

75. For accounts of this incident, see Blumenthal, "Community Action," 115–116; Moynihan, *Maximum Feasible Misunderstanding*, 91; and Rowland Evans and Robert Novak, "The Yarmolinsky Affair," *Esquire* 63 (February 1965): 80–82, 122–123.

76. Adam Yarmolinsky to Paul Ylvisaker, 6 October 1965, Adam Yarmolinsky Papers, box 13, Chron File Oct.–Dec. 1965 folder, John F. Kennedy Library, 2.

77. Conference on the Kennedy Administration Urban Poverty Programs and Policies, transcript, 16–17 June 1973, Brandeis University, John F. Kennedy Library Oral History Collection, 254.

78. See Blumenthal, "Community Action," 23–27.

79. Jack Conway, interviewed by Richard Blumenthal, 23 November 1966, quoted in Blumenthal, "Community Action," 122.

80. Office of Economic Opportunity, *Community Action Program Guide*, vol. 1 (Washington, D.C.: G.P.O., February 1965), 7.

81. Ibid.

82. William B. Cannon, Bureau of the Budget, to Charles L. Schultze, director of the Bureau of the Budget, memorandum, "Community Action Program," 25 September 1965, NARA, Records of the Bureau of the Budget, Entry 7N, Series 61.1a, Director's Office—Correspondence 1964, box 161, folder R1-2 Office of Economic Opportunity, folder no. 1965, RG 51, 2.

83. Moynihan, *Maximum Feasible Misunderstanding*, 144.

84. See David E. Bell, interview by Robert C. Turner, 11 July 1964, John F. Kennedy Library Oral History Collection, 76–77.

85. Robert A. Levine, interview by Stephen Goodell, 26 February 1969, Santa Monica, Calif., Lyndon B. Johnson Library Oral History Collection, tape 1, p. 2.

86. Joseph A. Califano, Jr., interview by Robert Hawkinson, 11 June 1973, Washington, D.C., Lyndon B. Johnson Library Oral History Collection, 4.

87. Labor and Welfare Division, Bureau of the Budget (Forrer, Turen, and Sutton), to Elmer B. Staats, deputy director of the Bureau of the Budget, memorandum, "Is There a Hitch in OEO's Future?" 8 October 1964, NARA, Entry 8B, Directors, Deputy Directors, and Assistant Directors Office Files, 1961–1967, Series 61.1b, box 22, Office of Economic Opportunity folder, RG 51.

88. Ibid., 3–4.

89. Ibid., 2.

90. Office of Economic Opportunity, *Community Action Program Guide*, 16.

91. William Selover recalls, "The President had never intended the program to mobilize the poor and tap its spokesmen for the purpose of opposing either local or federal political organizations, or both. But that is, in fact, what happened." William C. Selover, "The View from Capitol Hill: Harassment and Survival," in *On Fighting Poverty: Perspectives from Experience*, ed. James L. Sundquist (New York: Basic Books, 1969), 169.

92. Conference on the Kennedy Administration Urban Poverty Programs and Policies, transcript, 16–17 June 1973, Brandeis University, John F. Kennedy Library Oral History Collection, 79.

93. Johnson's depiction of the CAP staff members is from Robert A. Levine, interview by Stephen Goodell, 26 February 1969, Santa Monica, Calif., Lyndon B. Johnson Library Oral History Collection, tape 1, p. 29.

94. Selover, "The View from Capitol Hill," 181; and Robert L. Lineberry and Ira Sharkansky, *Urban Politics and Public Policy* (New York: Harper & Row, 1971), 293–294.

95. Moynihan, *Maximum Feasible Misunderstanding*, 144.

96. Kermit Gordon, director of Bureau of the Budget, to President Lyndon Johnson, memorandum, "Appointment of Assistant Director," 26 January 1965, Ex FG 11-1/A, box 55, White House Central Files, Lyndon B. Johnson Library.

97. Henry S. Rowen, interview by David A. Hounshell and David Jardini, Stanford, Calif., 5 December 1994, 12.

98. Charles L. Schultze, director of the Bureau of the Budget, to President Lyndon Johnson, memorandum, "National Goals," 13 August 1965, Ex FI, box 1, FI Finance 12/1/64–8/25/65 folder, White House Central Files, Lyndon B. Johnson Library, 3–4.

99. Joseph Califano to President Lyndon Johnson, memorandum, 16 August 1965, Ex FI, box 1, FI Finance 12/1/64–8/25/65 folder, White House Central Files, Lyndon B. Johnson Library.

100. The adoption of PPB was required in the following departments and agencies: Agriculture, Commerce, Defense, HEW, Housing and Urban Development, Interior, Justice, Labor, the Post Office, State, Treasury, Agency for International Development, Atomic Energy Commission, Central Intelligence Agency, Federal Aviation Agency, General Services Administration, NASA, National Science Foundation, Office of Economic Opportunity, Peace Corps, United States Information Agency, and Veterans Administration. Additionally, a formal PPB system was encouraged in the Civil Aeronautics Board, Civil Service Commission, Export-Import Bank of Washington, Federal Communications Commission, Federal Home Loan Board, Federal Power Commission, Federal Trade Commission, Interstate Commerce Commission, National Capital Transportation Agency, National Labor Relations Board, Railroad Retirement Board, Securities and Exchange Commission, Selective Service System, Small Business Administration, Smithsonian Institution, Tennessee Valley Authority, and United States Arms Control and Disarmament Agency.

101. Quoted in Willard Fazar, presentation to the Institute for Applied Technology, U.S. Department of Commerce, "The Planning-Programming-Budgeting System," 20 December 1965, NARA, Records of the Bureau of the Budget, Entry 7N, Series 61.1a, Director's Office—Correspondence 1964, box 134, Folder F3-1 Planning-Programming-Budgeting, 1965, RG 51, 2.

102. This body of criticism is dominated by the work of Charles E. Lindblom and Aaron Wildavsky. See Charles E. Lindblom, "The Science of 'Muddling Through,'" *Public Administration Review* 19, no. 2 (Spring 1959): 79–88; "Decision-Making in Taxation and Expenditures," in National Bureau of Economic Research, *Public Finances: Needs, Sources, and Utilization* (Princeton: Princeton University Press for NBER, 1961); David Braybrooke and Charles Lindblom, *A Strategy of*

Decision: Policy Evaluation as a Social Process (New York: Macmillan, 1963); Charles Lindblom, *The Intelligence of Democracy* (New York: Macmillan, 1965); and Aaron Wildavsky, "The Political Economy of Efficiency: Cost-Benefit Analysis, Systems Analysis, and Program Budgeting," *Public Administration Review* 26, no. 4 (December 1966): 292–310.

103. Ida R. Hoos, *Systems Analysis in Public Policy: A Critique*, revised ed. (Berkeley: University of California Press, 1983), 248–249.

104. Robert S. McNamara, *In Retrospect: The Tragedy and Lessons of Vietnam* (New York: Random House, 1995).

105. Henry S. Rowen, memorandum of conversation with Harold Brown, 16 August 1966, J. R. Goldstein Papers, Henry Rowen Correspondence Before Arrival box, RAND Archives.

106. Robert H. Haveman, *Poverty Policy and Poverty Research: The Great Society and the Social Sciences* (Madison: University of Wisconsin Press, 1987), 32.

107. Ibid., 31, table 3.1.

108. Ibid., 166.

109. Charles J. Zwick, interview by David McComb, 1 August 1969, Miami, Fl., Lyndon B. Johnson Library Oral History Collection, 21.

110. Robert Wood, "The Great Society in 1984: Relic or Reality?" in *The Great Society and Its Legacy*, Marshall Kaplan and Peggy L. Cuciti, eds. (Durham: Duke University Press, 1986), 21–22.

111. Hoos, *Systems Analysis in Public Policy*, 96.

References

Blumenthal, Richard. "Community Action: Origins of a Government Program." B.A. thesis, Harvard College, 1967.

Braybrooke, David, and Charles Lindblom. *A Strategy of Decision: Policy Evaluation as a Social Process*. New York: Macmillan, 1963.

Byrne, John A. *The Whiz Kids: Ten Founding Fathers of American Business and the Legacy They Left Us*. New York: Doubleday, 1993.

Clark, Kenneth B., and Jeannette Hopkins. *A Relevant War against Poverty: A Study of Community Action Programs and Observable Social Change*. New York and Evanston, Ill.: Harper and Row, 1969.

Cloward, Richard A., and Lloyd E. Ohlin. *Delinquency and Opportunity*. Glencoe, Ill.: The Free Press, 1960.

Donovan, John C. *The Politics of Poverty*. New York: Western Publishing Company, 1967.

Enthoven, Alain C., and K. Wayne Smith, *How Much Is Enough? Shaping the Defense Program, 1961–1969*. New York: Harper & Row, 1971.

Evans, Rowland, and Robert Novak. "The Yarmolinsky Affair." *Esquire* 63 (February 1965): 80–82, 122–123.

Fuller, John Wayne. "Congress and the Defense Budget: A Study of the McNamara Years." Ph.D. diss., Princeton University, 1972.

Grossman, David A. "The CAP: A New Function for Local Government." In *Urban Planning and Social Policy*, edited by Bernard J. Frieden and Robert Morris. New York: Basic Books, 1968.

Hammond, Paul Y. *The American Military Establishment in the Twentieth Century*. Princeton: Princeton University Press, 1961.

Haveman, Robert H. *Poverty Policy and Poverty Research: The Great Society and the Social Sciences*. Madison: University of Wisconsin Press, 1987.

Hitch, Charles J. *Decision-Making for Defense*. Berkeley and Los Angeles: University of California Press, 1965.

Hitch, Charles J., and Roland N. McKean. *The Economics of Defense in the Nuclear Age*. Cambridge: Harvard University Press, 1960.

Hoos, Ida R. *Systems Analysis in Public Policy: A Critique*, revised edition. Berkeley: University of California Press, 1983. Originally published Berkeley: University of California Press, 1972.

Hovey, Harold A. *The Planning-Programming-Budgeting Approach to Government Decision-Making*. New York and Washington: Frederick A. Praeger, 1968.

Janowitz, Morris, ed. *The New Military: Changing Patterns of Organization*. New York: Russell Sage Foundation, 1964.

Kolodziej, Edward. *The Uncommon Defense and Congress, 1945–1963*. Columbus: Ohio State University Press, 1966.

Kravitz, Sanford L. "Community Action Programs: Past, Present, Its Future." In *On Fighting Poverty*, edited by James L. Sundquist. New York: Basic Books, 1969.

Levitan, Sar. *The Great Society's Poor Law: A New Approach to Poverty*. Baltimore: Johns Hopkins University Press, 1969.

Lindblom, Charles E. "Decision-Making in Taxation and Expenditures." In National Bureau of Economic Research, *Public Finances: Needs, Sources, and Utilization*. Princeton: Princeton University Press for NBER, 1961.

Lindblom, Charles E. *The Intelligence of Democracy*. New York: Macmillan, 1965.

Lindblom, Charles E. "The Science of 'Muddling Through,' " *Public Administration Review* 19, no. 2 (Spring 1959): 79–88.

Lineberry, Robert L., and Ira Sharkansky. *Urban Politics and Public Policy*. New York: Harper & Row, 1971.

Marris, Peter, and Martin Rein. *The Dilemmas of Social Reform*. New York: Atherton Press, 1967.

McNamara, Robert S. *In Retrospect: The Tragedy and Lessons of Vietnam*. New York: Random House, 1995.

Moynihan, Daniel P. *Maximum Feasible Misunderstanding: Community Action in the War on Poverty*. New York: The Free Press, 1969.

Novick, David, ed. *Program Budgeting: Program Analysis and the Federal Budget*. Cambridge: Harvard University Press, 1965.

Office of Economic Opportunity. *Community Action Program Guide*, vol. 1. Washington, D.C.: U.S. Government Printing Office, February 1965.

Ries, John C. *The Management of Defense: Organization and Control of the U.S. Armed Forces*. Baltimore: Johns Hopkins University Press, 1964.

Roherty, James M. *Decisions of Robert S. McNamara: A Study of the Role of the Secretary of Defense*. Coral Gables, Fl.: University of Miami Press, 1970.

Rose, Stephen M. *The Betrayal of the Poor: The Transformation of Community Action*. Cambridge, Mass.: Schenkman Publishing, 1972.

Sanders, Ralph. *The Politics of Defense Analysis*. New York: Dunellen, 1973.

Selover, William C. "The View from Capitol Hill: Harassment and Survival." In *On Fighting Poverty: Perspectives from Experience*, edited by James L. Sundquist, 169. New York: Basic Books, 1969.

Sundquist, James L. "Origins of the War on Poverty." In *On Fighting Poverty: Perspectives from Experience*, edited by James L. Sundquist. New York: Basic Books, 1969.

Wildavsky, Aaron. "The Political Economy of Efficiency: Cost-Benefit Analysis, Systems Analysis, and Program Budgeting." *Public Administration Review* 26, no. 4 (December 1966): 292–310.

Wofford, John G. "The Politics of Local Responsibility—Administration of the Community Action Program, 1964–1966." In *On Fighting Poverty*, edited by James L. Sundquist. New York: Basic Books, 1969.

Wood, Robert. "The Great Society in 1984: Relic or Reality?" In *The Great Society and Its Legacy*, edited by Marshall Kaplan and Peggy L. Cuciti, 21–22. Durham: Duke University Press, 1986.

11

The Limits of Technology Transfer: Civil Systems at TRW, 1965–1975

Davis Dyer

In the late 1990s, top executives at TRW Inc., the Cleveland, Ohio–based automotive supplier and aerospace contractor, spoke enthusiastically, but also cautiously, about the company's new ventures in "civil systems." The term refers to a fast-growing business based on applying management expertise, software systems, and technological capabilities originally developed in the company's work on defense programs in wholly different areas. In 1996, for example, TRW engineered a comprehensive public safety 911 response system in Atlanta for the Olympic Games. The company manages similar big, complex programs for the Federal Aviation Administration, the Internal Revenue Service, the Department of Energy, and other civilian government customers. As a result of such efforts, between 1986 and 1996 TRW's aerospace businesses dramatically lessened their dependence on the U.S. Department of Defense—a vital necessity in an era of declining and flattening defense expenditures—from about 95 percent to about 65 percent of sales.

Hence TRW's enthusiasm for civil systems. The company's caution reflects earlier and less happy experiences in a similar vein: a strand of new initiatives that TRW and other aerospace companies pursued during a particularly active period of diversification that started in the mid-1960s and lasted about a decade. In TRW's case, leaders of the company believed that it had developed valuable management capabilities while serving as contractor for systems engineering and technical direction on the U.S. Air Force ballistic missile programs and performing similar services for other government-funded projects. Company leaders also believed that these capabilities would transfer readily to other sorts of technological, social, and economic challenges: managing large collections of information, coordinating traffic in urban areas, cleaning up the environment, solving the housing crisis, improving the delivery of medicine, and overcoming energy shortages. The company undertook many small, civil systems projects for federal non-defense and local agencies and commercial customers. These initiatives multiplied during an aerospace industry downturn in the early 1970s, when company leaders believed that civil systems might represent a counter-cyclical business and facilitate steady growth and employment. Many other aerospace companies followed suit, invading commercial and civilian government markets, hoping and expecting that the latest management methods and mastery of the most advanced technologies would ensure success.

Across the aerospace industry the returns from these efforts were dismal. Most of TRW's civil systems ventures lost money or eked out meager returns during brief life spans. The company came to recall its first experiences in civil systems painfully, as a time in which its most optimistic assumptions were shattered and its capabilities exceeded their limits. In the aftermath, TRW excercised much greater discipline and caution in seeking to apply systems management tools and techniques in new contexts. When again facing an imperative to diversify with the sustained aerospace industry downturn that began in the mid-1980s, TRW defined a narrow scope of projects that it would pursue. As a result, TRW's civil systems ventures of the late 1990s differed markedly from its first attempts two decades before.

This chapter deals with why and how TRW first branched into civil systems, expanded its efforts, and then shut them down. It also considers the lessons and conclusions TRW drew from the experience and the legacies that lingered on into the late 1990s.

Push and Pull

TRW entered the business of civil systems in the mid-1960s for reasons related to the corporation's strategy of diversification, the specific circumstances and personalities of its aerospace businesses, and its (partly self-promoted) image as pioneers of the latest management practices. The company both pushed and was pulled into pursuing new opportunities outside its traditional markets.

TRW was formed in 1958 through a merger between Cleveland-based Thompson Products, Inc., a precision manufacturer of parts for automobiles and aircraft, and Los Angeles–based Ramo-Wooldridge Corporation, an advanced scientific and engineering services company that supported the U.S. Air Force ballistic missile program.[1] The merger and a subsequent divestiture of former Ramo-Wooldridge activities dedicated to long-range planning for the air force had far-reaching effects on both constituent parts of TRW. The changes brought Thompson Products new capabilities in aerospace engineering and electronics to offset the decline of its biggest business in fabricating parts for jet engines, as well as a new source of management talent. For its part, Ramo-Wooldridge gained access to Thompson Products' capital, corporate management systems, and manufacuring expertise, while also freeing itself to become a producer of aerospace and electronic equipment and systems.[2]

The merger created an imperative for TRW to strike a new balance in its sales mix. In 1959, about three-fourths of its revenues originated in defense contracts. To new Chairman and CEO J. David Wright, this dependence on the defense budget seemed highly risky. At the same time, Wright championed rapid growth—a goal that typified many big American corporations during "the go-go years" of the 1960s. In 1961, he challenged TRW to achieve $1 billion in sales—about three times its current total revenues—before his scheduled retirement in 1970. Acquisitions became the means to satisfy the objectives of both growth and diversification. In the mid- to late-1960s, a trickle of deals became a flood, as TRW bought several dozen companies, including some that were themselves already diversified businesses. TRW also aggressively employed other means of growth: improving operations, developing new products, opening new markets and segments, and increasing foreign sales. The company crossed the $1 billion sales threshold in 1967, and by the end of the decade, it managed more than fifty divisions organized in six groups.

Under the company's decentralized organization structure and management policies, the heads of each operating group and division pursued the corporate policies of growth and diversification. These objectives presented special challenges to TRW's Space Technology Laboratories (STL; after 1965, TRW Systems Group), which had provided systems engineering services to the air force. In 1960, virtually all of STL's revenues were tied to air force missile programs that were shifting from development, in which TRW had played a prominent part, to an operational phase in which its role diminished significantly. Led by Ruben F. Mettler, STL rapidly attempted to reposition itself. It established new businesses in building spacecraft, rocket motors and thrusters, and associated electronics and communications equipment, and it moved into a new campuslike facility in Redondo Beach (not far from Los Angeles International Airport) called Space Park.

During the early 1960s, STL scored notable successes, including big, long-term contracts to build spacecraft for NASA and other customers, as well as supporting electronic and communications systems. In these instances, the elaborate and formal management methods, policies, and systems that Ramo-Wooldridge had developed for the missile programs proved a key advantage. STL also formed a new unit to explore commercial development of some proprietary electronic products such as test equipment and high-speed cameras. This organization later became the foundation of a diverse group of acquisitions in electronic instrumentation, process control devices, and products and equipment for the energy industry. STL ran these operations as commercial businesses, and there was little overlap of personnel or management methods between them and the major activities at Space Park.

In 1963, STL moved farther afield after Mettler recruited James P. Brown, a marketing executive from Litton Industries, to lead additional attempts to diversify. In pursuit of new customers for existing capabilities, Brown supported initiatives in space hardware and electronic systems and helped land new business in providing systems engineering services to the U.S. Navy's Anti-Submarine Warfare Program. He also sought new markets for STL's skills in systems engineering and program and project management in wholly new areas, including nondefense government work and commercial business. Along the way, Brown received strong encouragement from Simon Ramo, cofounder of Ramo-Wooldridge.

In the mid-1960s, Ramo enjoyed a special status in TRW. The merger had made him a multimillionaire as well as one of the company's biggest stockholders. Although he served atop TRW for two decades (as executive vice-president from 1958 to 1961 and as vice-chairman from then until his retirement at age sixty-five in 1978), he had no direct operating responsibilities and seldom took part in routine management meetings. He spoke frequently with Wright and his successor as chairman, Horace Shepard, and, in avuncular fashion, took special interest in the former Ramo-Wooldridge businesses. But his involvement with the company was episodic, punctuated by commitments elsewhere. "Not with total facetiousness," Ramo told an interviewer in 1984, "I really did my retiring in 1958, because I shifted to a pattern of activity which continues to this very moment." This pattern (which still held in the late 1990s) consisted of a whirlwind schedule of meetings, consulting jobs, writing projects, teaching assignments, speaking engagements, and cultural and philanthropic endeavors. For a time, Ramo kept an office at Space Park, but he worked mostly out of a small office about ten miles away near his home in Beverly Hills. In addition to his duties at TRW, between 1963 and 1967, he was president of Bunker-Ramo, a computer business that had originated at Ramo-Wooldridge and that TRW sold to Martin Marietta. He also served on many national advisory boards and councils, on several corporate boards, and as an adjunct professor at Caltech.[3]

During the 1960s, Ramo spent most of his time looking outward from TRW. A reporter described his role in the company as "a kind of self-monitored radar to anticipate coming perils or opportunities." In this context, he became increasingly preoccupied with a subject he would speak and write about for decades. A fundamental problem facing society, he believed, was "the imbalance between accelerating technology and lagging social maturity"—a major theme of his speeches and writings from the early 1960s on, as well as of two books written for the general public: *Cure for Chaos* (1969) and *Century of Mismatch* (1970). He pointed to the major scientific breakthroughs of the mid-twentieth century—the harnessing of atomic energy, the cracking of the genetic code, the exploration of space, the coming of the computer—as signs of extraordinary technological progress. Yet at the same time problems abounded: unchecked population growth, crowded cities, an increasingly polluted environment, the apparent depletion of nonrenewable energy sources, and a fearsome struggle between decentralized and centralized economies for world dominance. The combination of these contrary trends added up to a "paradoxical hodgepodge of sophisticated technological progress and social primitivism."[4]

For Ramo, this paradox offered both social and political challenges as well as business opportunities. There was a prominent—and profitable—role, he declared, for companies that applied "an intellectual discipline for mobilizing science and technology to attack complex, large-scale problems in an objective, logical, complete, and thoroughly professional way." The "systems approach," he further explained,

> depends upon use of a team of cooperating experts in both the technological and nontechnological aspects of the problem to be analyzed. It starts by definition of goals and ends with a description or a design of a harmonious, optimum ensemble of the required men and machines, with such a corollary network of flow of information and materials as will cause this system to operate to solve the problem or to fill the need. The approach includes use of sophisticated techniques for assembling and processing the necessary data,

> comparing alternative approaches to evaluate their relative benefits and shortcomings, providing compromises, making quantitative analyses and predictions where they are appropriate, and seeking out judgments from experience of the past and creative innovations where, in turn, they are indicated. Resting in part on the computer to assist in weighing and relating facts and relationship, the systems approach is an extended, a somewhat automated, common sense. It is more especially a reasoned and total, rather than a fragmentary, look at problems, seeking to push confusion and hit-or-miss decision-making into the background and leaning heavily on rational, concrete judgments.

Ramo prodded TRW to apply the skills, experience, and techniques proven on missile and space programs to a new order of business. He even popularized a term, "civil systems," to refer to systems engineering and large-scale, technical project management approaches to challenges of natural resource development, environmental cleanup, mass transportation, housing, urban planning, and health care delivery—all of which he urged TRW to pursue.[5]

In 1965, notions that Brown and Ramo were pushing won public endorsement when California Governor Edmund G. "Pat" Brown encouraged members of the aerospace industry to conduct "the California studies," "an examination and evaluation of the potential for the effective expansion of systems engineering methods and techniques to other government program [*sic*] encompassing socio-techno-economic areas." These areas included waste management, crime and delinquency prevention, statewide information systems, and transportation. Needless to say, the studies concluded that techniques of systems engineering and program and project management were readily applicable outside the aerospace industry. These techniques included analyses to establish systems requirements and to break down complex problems into subsystems with interfaces, use of visual and graphic displays, and a host of specific measures for organizing, budgeting, monitoring, and reviewing performance.[6]

At the same time, TRW was pulled into civil systems as a consequence of its own historical successes. During the late 1950s and early 1960s, as details of the missile programs became public—partly through Ramo's speaking and writing—the techniques of management and organization of large-scale projects rippled across to other aerospace companies and then into the economy generally. (TRW was by no means the only management pioneer in large-scale systems projects, and the ripples moved in all directions. The U.S. Navy team that developed the Polaris missile, for example, also became renowned for its Program Review and Evaluation Technique [PERT] for scheduling, monitoring, and reviewing progress in complex projects.)

Government agencies and private corporations frequently invited Ramo, Mettler, and other TRW missile pioneers to speak about systems engineering and program management and specific applications of computer modeling and graphical communications aids for planning, scheduling, and control. Inevitably, some of these invitations led to business opportunities, which ranged from pro bono work on community projects, to paid consulting assignments, to formal engagements to provide systems engineering services. During the site activation of a Minuteman wing near Grand Forks, North Dakota, for example, TRW had developed graphical techniques for scheduling and tracking projects. These techniques attracted notice in government and industry circles and led TRW to contract with a consortium of oil companies to provide scheduling and task identification services and to help with automating some management tasks.[7]

TRW's first true civil systems projects were direct outgrowths of missile management. In 1966, the company won a $200,000 contract from the state of California's Office of Planning to develop a regional land-use plan for portions of Santa Clara county. The premise for this project was a supposed parallel "between the creation or revitalization of a city and planning for the installation of a missile site and the provisioning for its manpower, housing, transportation, communications, and security." As described by a TRW publicist, the California land-use plan began with

> an analysis of the situation today and an educated estimate of the needs for tomorrow. Out of those analyses has come an information system suitable for determining where industries and shopping centers should be located to best serve the community. Time-phased plans for installation of public utilities can be developed. Routing of streets and highways and the construction of schools are more efficiently determined. Zoning decisions yet to be made are forecasted. Flood control, recreation facilities, health and welfare programs, fire and police protection, and the range of service industries may be identified well in advance of their actual introduction into community life.[8]

President Lyndon Johnson's "War on Poverty" presented TRW with more civil systems opportunities. In 1966, the federal Office of Economic Opportunity engaged the company to provide training sessions on systems management for local governments and colleges and universities. TRW volunteered to test the approach in the San Bernardino office. From there, prospects began to multiply: with the Federal Water Pollution Control Authority on systems approaches to water resource management; with the U.S. Department of Commerce, to investigate a new high-speed ground transportation system; and with the city of Redondo Beach, to analyze and improve the flow of automobile traffic in the South Bay area. Through one of Ramo's connections, the provincial government in Alberta, Canada, retained TRW to help design and plan a new $100 million medical center in Edmonton.[9]

TRW's initial civil systems contracts were small, although in 1969 they added up to about $10 million in revenues. The move into civil systems also initiated a modest shift in hiring practices in TRW Systems Group, as it recruited a handful of sociologists, economists, statisticians, and political scientists into its technical staff.[10]

EXPANDING HORIZONS

In the early 1970s, TRW's eagerness to find new markets for its management capabilities swelled as a result of a recession in the aerospace industry. Although the war in Vietnam would linger on until the mid-1970s, Richard Nixon had won the presidency in 1968 with a pledge to end it, and spending on defense had already begun to turn down. At the same time, despite the triumphs of the Apollo Program, NASA's budget began to shrink, and future funding levels were uncertain. Finally and unluckily for TRW, these factors coincided with the conclusion of several big spacecraft programs and some classified work that did not lead to follow-on work.[11]

The aerospace recession devastated TRW's operations at Space Park in Los Angeles. Between February 1970 and October 1971, employment plunged from 15,500 to 9,500. (One of the casualties, parts application engineer Jerry Buss, landed on his feet. A shrewd investor in real estate, he accumulated a fortune and became famous as the owner of the Los Angeles Lakers professional basketball team.) At the same time, TRW cut back sharply on internally financed activities and slashed overhead. Although the company eventually climbed back on the basis of core capabilities in spacecraft and electronic systems, scoring notable wins with NASA (Pioneers 10 and 11 and the High Energy Astronomical Observatories) and the Department of Defense (the Fleet Satellite Communications System), it also redoubled efforts to diversify. By the mid-1970s, led by Richard DeLauer and George Solomon (Mettler had since moved to Cleveland as president of the corporation), TRW Systems Group was engaged in businesses as diverse as charge authorization systems for consumer credit, housing and real estate development, minerals exploration, construction engineering, and urban transportation systems (see tables 11.1 and 11.2).

The applied R&D organization inside TRW Systems Group provided a particularly fertile source of ideas. One line of business emerged from propulsion technology originally developed for the Lunar Excursion Module Descent Engine (LEMDE), an innovative, variable-thrust rocket motor TRW had developed for NASA's Apollo program. Once the moon landings were over, however, the company lost out to competitors for the next big rocket motor program, the Space Shuttle. That touched off a furious scramble for new opportunities in related areas, including high-temperature, high-pressure combustion. The most important application was high-energy chemical lasers, which had many potential uses, most obviously in weaponry, but also in industrial operations such as cutting, drilling, and welding. A combustor based on LEMDE held promise for reducing nitrogen oxide emissions from industrial processes by a factor of five or more; another device called a "charged droplet scrubber" used a technology derived from electrostatic colloid thrusters to control particulate emissions from industrial processes. By early 1972, development was far enough along for TRW to form with Mitsubishi Corporation a joint venture called Civiltech Corporation to begin marketing this technology in Japan.[12]

TRW also pushed into civil systems through acquisitions and joint ventures. In January 1970, DeLauer formed a new organization called the Systems Application Center and charged it to explore ways to "make systems technology available to various commercial markets." The foundation of the center consisted of two late 1960s acquisitions, Credit Data Corporation, a consumer credit reporting operation, and the much smaller Credifier Corporation, a Santa Monica, California–based company that produced charge authorization hardware (for credit card approvals) compatible with NCR cash registers.[13]

TRW's immediate task was to make sense of these operations, install new management systems and cost controls, and formulate plans for long-term growth. At Credit Data, TRW retained existing

Table 11.1
TRW's major ventures in civil systems, 1968–1975

Name	Description
Canada Systems, Ltd.	Joint venture in 1971 between TRW, London Life Insurance, the T. Eaton Company, and the Steel Company of Canada, Ltd., to pool data processing capabilities and pursue civil systems opportunities in Canada. TRW sold its interest in the mid-1970s.
Community Technology Corporation	Start-up in 1971 to promote community planning and develop manufactured housing. Created two subsidiaries, Colorado Land Development, a joint venture with Mitsubishi, and Community Shelter Corporation. Built prototype housing in Sacramento and other locations and acquired land in Littleton, Colorado, for a model community. Sold at a loss in 1974.
Credit Data Corporation	1968 acquisition of credit bureau. TRW provided data processing expertise in plan to grow business dramatically. The foundation of TRW Information Systems & Services, a major unit of the company eventually put up for sale in 1996.
Credifier Corporation	1969 acquisition of small maker of credit authorization systems. Later part of a major TRW initiative in commercial electronics involving point-of-sale terminals. Divested through a joint venture with Fujitsu in 1980.
International Decision Techniques	Joint venture between TRW and J.P. Morgan, Inc., to develop software to aid decision analysis for financial investment companies. Formed in 1969; discontinued in early 1970s.
OMAR Explorations, Ltd.	Joint venture between TRW, Barringer's Ltd., and two other parties to use TRW technology in airborne sensors for mineral exploration. Formed in 1968; TRW sold its interest in 1975.

Table 11.2
TRW's major ventures in transportation and environmental operations, 1972–1977

Name	Description
Civiltech Corporation	Joint venture (50/50) in 1972 with Mitsubishi to market TRW pollution control technology and research capabilities in Japan. Transferred to Energy Systems Group in 1976. Sold in 1985.
The DeLeuw, Cather Company	Acquisition in 1972 of multinational architectural and engineering firm. Sold in 1977.
National Distribution Systems	Joint venture formed in 1971 between TRW, Eastern Airlines (majority stake), and Ralph M. Parsons Co. to establish a nationwide network of warehouses for use by a wide variety of customers. Dissolved soon after start-up because of disagreements among the partners.
Traffic Control	Start-up at Space Park to apply systems engineering to problems of urban traffic control. Built traffic control systems south of Los Angeles, in Baltimore, Denver, and other communities. Discontinued in mid-1970s.
Transit Systems	Start-up at Space Park to apply systems engineering techniques to rail networks and public transportation systems. Established the computer control systems for the Washington, D.C., METRO and other customers. Discontinued in mid-1970s.

management, including the president, Bud Jordan. At TRW's direction, he pruned unprofitable activities and sought to increase national coverage by direct investment, acquisition of existing credit bureaus, and cooperative arrangements. With the help of TRW software specialists and engineers, he instituted programs to improve the credit information itself, including tighter controls on data entry, standardization of formats, and new techniques for data compression, storage, transmission, and access. Jordan also tapped TRW's financial resources to build a new national computer center in Anaheim, California.[14]

At Credifier, TRW changed management and installed a team of aerospace engineers. The new team moved swiftly to upgrade the company's original product line, which used a primitive tape storage system that was slow in use and had to be replicated at each location. In

November 1970, TRW veteran Donald Kovar led a crash program to develop a new programmable charge authorization system. Two months later, the company announced its System 4000, which included digital display terminals connected over telephone lines to a central mini-computer. The product proved an instant technical and marketing success, although problems in pricing and service and maintenance postponed crossover in profitability for several years. Nonetheless, System 4000 spawned several other similar—and marginal—businesses, including "Validata," a charge authorization service sold to the airlines, and a security access control system for governmental agencies and defense contractors.[15]

The Systems Application Center also assumed responsibility for the civil systems ventures begun in the 1960s (see table 11.1). Many of these led short and troubled lives. TRW's efforts to apply principles of systems engineering in hospitals and health care planning, for example, at first seemed promising. The initial project in Edmonton, Alberta, led to another at Walter Reed General Hospital near Washington, D.C. TRW soon found, however, that the architects with whom it typically partnered in these ventures could provide systems analysis and planning much cheaper than it could. In addition, several of the TRW personnel engaged in the work left the company to form their own lower-cost consulting practice. In the early 1970s, TRW simply abandoned the business.[16]

Other early civil systems ventures also struggled. In 1968, TRW had agreed to form a joint venture with Barringer Research Limited, which took a playful name: OMAR Explorations, Inc.—"Ramo" spelled backwards. The business organized to explore for minerals in western Canada using airborne sensors and other technology developed by TRW. Delays and disputes in finalizing the agreement postponed the startup until the fall of 1972. TRW agreed to commit $1 million per year to OMAR, but the venture soured almost immediately. In October 1975, TRW sold its stake to Barringer Research for just $15,000. Another joint venture with Eastern Airlines and The Ralph M. Parsons Company called National Distribution Services aimed to "operate all or a portion of any client's physical distribution in any part of the United States." This entity also fell apart quickly due to differences among the partners.[17]

Still another problematical civil systems venture was Community Technology Corporation, which TRW formed in June 1971 to develop manufactured housing and pursue land development. This entity spawned several subsidiaries, including Colorado Land Development, which, in a legacy from earlier land-use studies, partnered with Mitsubishi Corporation to acquire real estate for development in Littleton, Colorado, and Community Shelter Corporation (CSC), which planned to produce modular housing. In the summer of 1972, CSC obtained a contract from the U.S. Department of Housing and Urban Development to develop modular housing under a program called "Operation Breakthrough." The contract called for CSC to build twenty prototype single-family houses near Sacramento. CSC licensed technology that Aerojet General had developed to wrap rocket motor cases from composite materials, announced a new product called Fiber-Shell, and poured about $5 million in equity and loans to build a small factory in Sacramento.

TRW's initial plan was to use the wrapping technology to fabricate cubical shapes that would serve as frames for entire rooms and that could be bolted together into housing units. TRW failed to get the process to work efficiently, however, and faced significant cost penalties. The government had set a target cost for manufactured housing at $9 per square foot; TRW's cubes exceeded the target by a factor of more than ten. Eventually, TRW engineers abandoned the approach and got the costs down—but still far above the government target—by making prefabricated wall panels that could be stapled together. Once the prototype houses were finished, TRW scrambled for orders and built a handful of houses elsewhere in California and in New Mexico. The venture never came close to paying a return, however, and it was finally sold at a loss in the mid-1970s, when TRW also exited the community development business.[18]

TRW also pursued civil systems work through another Space Park organization called, confusingly, the Civil Systems Center—another unit that encountered difficulties. This unit supervised projects in such areas as mass transportation systems, environmental systems, and general engineering services (see table 11.2). One project, for example, involved an automated, computerized system for controlling the flow of automobile traffic in urban areas. The initial installation was in the South Bay area (including Redondo Beach, home of Space Park) near Los Angeles, where TRW obtained a $645,000 contract to develop a system called SAFER (Systematic Aid to Flow on Existing Roadways). The system employed a central computer system to monitor sensors placed at key intersections, but unfortunately, according to one of the project engineers, it had a lot of bugs in it. SAFER "actually screwed up traffic" for a while, he recalls, and it sparked "a lot of complaints from our own employees." Eventually, TRW ironed out the problems, and the system remained in service for many years. The company also bid successfully for installations on the East Coast between Washington, D.C., and Philadelphia.[19]

TRW bid low on most of these projects, accepting losses to gain the business and experience. The losses soon piled up to nearly $20 million, however, amid fierce price competition from low-cost rivals. In addition, unhappy memories of clashes with a multitude of local governments—hearings, delays, community protests, grandstanding, and contract discussions played out on the evening news and in local newspapers—lingered with Mettler. "Whatever our difficulties in dealing with the Defense Department or the Air Force," he recalled, "they were nothing compared to dealing with the city of Philadelphia."[20]

The Civil Systems Center enjoyed better luck in related work. Early in 1972, the Washington Metropolitan Area Transit Authority awarded a $4.9 million contract to TRW to develop the Automatic Train Supervision System for the city's new subway system, the METRO. The METRO operated a ninety-eight-mile transit network, and the Transit Authority engaged TRW to design and build a dual computer complex, a display subsystem, and a manual control subsystem to be housed in a central control facility. The system included not only computerized control, but also digital communications. TRW not only made money on the deal but also acquired a partner on the project: DeLeuw, Cather, Inc., an architectural and engineering consulting firm with headquarters in Chicago and about 1,400 employees in offices around the world. TRW planned to keep DeLeuw, Cather as an autonomous unit whose normal operations might lead to new systems engineering opportunities.[21]

The Civil Systems Center also oversaw several ventures in Canada, where a number of big companies and provincial governments had evinced interest in Ramo's ideas and had noted TRW's contributions to the Edmonton hospital project. Early in 1971, for example, TRW joined with three big Canadian companies to establish The Canada Systems Group, Inc. The partners pooled their data processing requirements into a common computer center and planned to offer general computer-based information services. Canada Systems also offered for the Canadian market civil systems services related to air and water pollution, urban development, mass transportation and traffic control, medical services, and other public problems. As in the United

States, however, the response was underwhelming, and the principal source of revenue at Canada Systems was the original agreement to provide data processing services to the founding partners.[22]

During the early 1970s, TRW's ventures in civil systems made slow headway as aerospace engineers scurried to transfer advanced technology and management practices into markets generally unreceptive to such innovation. Most of these ventures drained financial resources and management time—a typical fate of startups. The questions most of them faced were the same: could they compete on cost? Could TRW develop the requisite marketing skills and understanding of customers? How long would it take before they made significant contributions to TRW's business? Would TRW have staying power, or would senior management pull the plug? By the mid-1970s the last of these questions loomed increasingly large.

RECKONING AND REEVALUATION

Early in 1975, TRW Chairman Horace Shepard began his annual report for the preceding year by noting that it had been "the most successful year in TRW's history. Sales, net earnings and earnings per share all set records." Yet Shepard also noted that TRW's performance had persisted through "a most difficult business environment" including a recession brought on by the energy crisis. What he did not mention, although it was an enormous concern to him and other leaders of the company, was TRW's plunging stock price. After trading in the mid-$30 range for several years and hitting a high of $43.25 during the first quarter of 1971, the stock tumbled to just over $10 per share during the latter half of 1974. The company's price/earnings ratio fell to within a very low range of 3.6 to 5.3. Amid these circumstances, Standard & Poor's and Moody's threatened to downgrade TRW's historical "A" rating, a measure that would have significantly increased the company's cost of financing during an already difficult period.[23]

Clearly, strong measures were necessary to help restore investor confidence in the company, and top management took some quick and decisive actions. TRW trimmed its proposed capital spending budget by $250 million over a three-year period and implemented cost control measures across every business. It backed out of a pending deal to acquire a sizable manufacturer of valves and flow control systems. It also became clear to Mettler that "one of the things we had to do was make some choices. We could do a few things such as data communications, rack and pinion steering and international automotive, and we could support the growth in space and defense. But not all the children could go to college." Among other things, that meant "limiting further diversification to exceptional circumstances; evaluating acquisition proposals and new product development in the context of the new, higher [financial return] goal; [and] developing a higher competence in divestment and in product line selection."[24]

Among the casualties of the new corporate discipline were most of TRW's Space Park–based diversification efforts. Within months, the company sold or liquidated most of its ventures in civil systems, including OMAR Explorations; Canada Systems; Community Technology Corporation; DeLeuw, Cather; and the initiatives in transit and traffic control. TRW also reassigned responsibility for its credit reporting and authorization systems outside of Space Park to its commercial electronics group. By the mid-1970s, few aerospace personnel remained in these ventures, which were led by managers experienced in financial services or in the particular markets in which TRW competed.[25]

The financial crisis of the mid-1970s permanently altered TRW's thinking about the applicability of its management technologies and capabilities in areas outside its experience. For the most part, the company was content to rely on its core businesses as the source of future growth. Another big spacecraft win—NASA's Tracking and Data Relay Satellite System—and new opportunities in avionics, electronic warfare, classified programs, and large-scale software systems for the Department of Defense more than compensated for the loss of employment and revenues in civil systems.

There was one exception to the stick-to-the-knitting pattern, however, and it, too, met a sad fate: TRW Energy Systems Group. TRW's initiatives in energy harked back to several acquisitions, including Mission Manufacturing and Reda Pump, that proved successful in the late 1960s. To many observers, especially those who looked at trends in U.S. production and consumption of oil, moreover, the need to find alternative sources to petroleum seemed obvious. In September 1973, TRW scored a coup by recruiting John S. Foster—known as "Johnny"—as vice president of energy research and development. A University of California, Berkeley Ph.D. in physics, Foster was an eminent scientist with high-level contacts in Washington. At Lawrence Radiation Laboratory in the 1950s, he had worked closely with Edward Teller on the development of the hydrogen bomb. Since 1965 he had served as director of defense research and engineering, the top scientific and technical position in the Department of Defense.[26]

Foster's hiring proved timely, to say the least. A month after he joined TRW, war broke out between Israel and its Arab neighbors, and during the next year and a half, oil prices tripled, triggering a worldwide energy crisis and a deep recession in most of the developed world. In Washington, federal officials responded by stepping up funding of energy-related research and development and a series of organizational changes that culminated in 1977 in the formation of a cabinet-level Department of Energy.

Although TRW could hardly afford to invest hundreds of millions of dollars in a major energy development initiative—the strategy that several big oil companies pursued, for example—it could follow the tried and true path of performing customer-funded R&D work on specific energy-related problems, with reasonable expectation that commercial opportunities would eventually develop. Given Foster's connections with the Atomic Energy Commission and Ramo's prominent role in national science and technology affairs—early in 1974 he was named to the White House Advisory Council on Energy Research and Development, and he later worked closely with Vice President Nelson Rockefeller on the Ford administration's science and technology policies—such a course seemed natural. Such a course also avoided the stretch that many of the civil systems projects had represented.[27]

Foster soon took charge of a new Energy Systems Group that included three principal operations: the first focused on government-funded planning and analysis activities; the second performed similar work for the U.S. Environmental Protection Agency; and the third sought to commercialize technology developed in the other units. TRW veterans directed most of the group's activities, although in 1976 Foster hired an executive from Shell Oil to assist with commercial development. Energy Systems Group devoted most of its attention to R&D projects, and it carried out many small studies and mounted several major projects, including some derived from earlier TRW initiatives[28] (see table 11.3).

By the end of the 1970s, Energy Systems Group was reporting annual sales of about $50 million derived from a variety of projects. Virtually all of these started with small steps and proceeded slowly. Unfortunately, all of them also ended in frustration during the 1980s. The problem was not so much trouble in transferring aerospace technology

Table 11.3
Principal TRW ventures in energy systems, 1973–1979

Name	Description
Coal Combustion Systems	High-pressure coal combustion process using technology derived from Lunar Excursion Module Descent Engine. Product a retrofit device to utility or industrial boilers that burn oil or gas.
Low NOx Burner	Clean-burning combustors for utilities and industry also using technology derived from Lunar Excursion Module Descent Engine. Combined with Civiltech Corporation (table 11.2).
Energy Engineering Services	Studies, planning, and analysis for the U.S. Department of Energy.
Environmental Services	Miscellaneous research and testing services for U.S. Environmental Protection Agency.
BEACON Process	New process for gasification of coal after mining. Marketed in partnership with SOHIO (forerunner of BP America).
TRW Carbon	Offshoot of the BEACON Process resulting in a unique carbonaceous material for use as an additive in various industrial products.
Coalbed Methane Extraction	Proprietary process for gasifying coal *in situ*.
Gravimelt	Proprietary process for reducing the sulfur content of coal.
Hot Spot	Extraction of oil from tar sands.
Isotope Separation	New process for enriching uranium for nuclear power generation.
Ocean Thermal Energy Conversion	Tapping ocean thermal energy to generate electricity.

and methods to this market as difficulties in the market itself. The sustained decline of oil prices that began in 1981 rendered many of TRW's energy ventures uneconomical. At the same time, the new Reagan administration proved unenthusiastic about supporting big energy projects and curtailed spending for energy-related R&D. As a result, TRW had little choice but to wind its energy programs down. Foster moved up to become a corporate vice president of technology, in effect succeeding Ramo in the position. For nearly ten years, TRW mounted no further efforts to develop new outlets for its capabilities in systems engineering and program management.

LESSONS AND LEGACIES

During the late 1980s and early 1990s, the changing face of world politics and mounting pressure to balance the federal budget once again portended a gloomy future for TRW's operations in defense and space markets. Recognizing the need to diversify and determined to avoid previous misadventures, several company planners undertook an informal review of earlier initiatives to transfer aerospace technology and methods of management to commercial markets. The study reached few definite conclusions, noting that TRW had fared better when entering commercial markets through acquisition rather than through startups and when these ventures had been led by experienced commercial managers rather than by aerospace engineers.[29]

From the distance of several decades and an outsider's perspective, however, many other factors also contributed to TRW's troubles in civil systems. In many instances, the company's strategy was so muddled and organizational responsibilities so ill-defined that no one, no matter how well equipped with powerful analytical tools and techniques, could succeed. There were also issues of incentives and motivation for key personnel. As is evident from the number of civil systems startups that coincided with the aerospace recession of the early 1970s, many TRW employees gravitated toward civil systems ventures as an alternative to being laid off, rather than because they were champions of new opportunities. At the same time, the company protected its most valuable scientists and engineers in core aerospace activities, where they continued to be attracted by more exciting technical and managerial challenges.

Beyond these considerations, however, the assumptions behind TRW's push into civil systems were flawed in significant ways. Claims for the transferability of the systems approach, for example, were too bold and broad. There was little to disagree with in Ramo's assertion that commercial and nondefense organizations could benefit from "an extended, a somewhat automated common sense." On the other hand, such thinking could be—and was—used to justify entry into virtually any kind of market or activity, including straightforward technology transfers (such as propulsion to combustion) or diversification via acquisition (such as credit reporting). And once there, aerospace management tools and techniques offered few tangible advantages and many drawbacks. Indeed, in many "civil systems" ventures TRW personnel quickly abandoned the systems approach and embraced ways of managing appropriate to the industry. Unfortunately, this realization often came too late and created handicaps that the company proved unable to overcome.

Another flawed assumption was the belief that the systems approach was inevitably superior to traditional management techniques. In some cases, of course, it was, proving an effective way to oversee development of missiles, spacecraft, and other complex products and systems featuring fast-changing technologies and many uncertainties. But for most of the commercial and local government opportunities TRW pursued, the systems approach was too elaborate and costly. It was also a poor substitute for experience in specialized markets. The company's expertise in systems design and engineering, for example, offered no advantage against entrenched competitors in housing, land development, and hospital administration. These competitors possessed far better understanding of customer needs and pricing and cost issues in their respective businesses. In many instances, competitors thwarted TRW with entirely conventional technologies and management methods.

In any event, TRW drew a large dose of humility from its experiences in civil systems. In the 1990s the company approached the problem of transferring technology and capabilities from aerospace to nondefense markets in ways consciously and conspicuously different from those of earlier days. One stratagem called for the company to focus on markets such as automotive supply that it already knew well. In 1990, TRW Chairman and CEO Joseph Gorman established the Center for Automotive Technology (CAT) at Space Park. The center was led initially by Peter Staudhammer, a storied veteran of such programs as the Lunar Excursion Module Descent Engine and the Viking biological instrument package, as well as many classified projects.

Staudhammer set up CAT with corporate funding to provide support to automotive projects already in progress or on the drawing board. In contrast to past technology transfer efforts, CAT did not seek to invent new products or systems and push them onto its customers; rather it focused on meeting the specific requirements of TRW's automotive project engineers. The new center operated with a small staff and a total annual budget of about $3 million, not including contributions from the divisions on specific projects. CAT personnel participated in work long under development, such as side-impact restraint systems, and wholly new products, such as collision-avoidance systems that used advanced sensors and diagnostics based on radar systems that will anticipate and help avoid crashes. Other projects in development included "intelligent" cruise control to automatically adjust speeds and maintain proper spacing between vehicles on highways; "smart" struts that employed the piezoceramic wafers used to help dampen vibration in spacecraft; new shock absorbers using electrorheological magnetic fluids to yield dramatic improvements in the smoothness of ride; electronic lane change aids to compensate for the blind spots not covered by rearview mirrors; and advanced keyless entry systems that remember the preferences of individual drivers and automatically adjust seat position and angle, mirror angles, climate control, and radio station.

TRW's aerospace units also resumed the search for business outside of their traditional markets. In contrast to the diversification efforts of the civil systems era, the approach this time was judicious: new programs played to TRW's traditional strengths in managing large-scale, complex problems and appeared to provide a foundation on which it could continue to build. In the late 1980s, the company moved its Systems Integration Group—the units responsible for selling systems engineering and program management capabilities—from Los Angeles to a new headquarters complex near Washington, D.C. The site was chosen both to emphasize a focus on the federal government as TRW's principal customer and to escape the perceived "technology-driven" mentality of Space Park. According to general manager John Stenbit, the group focused on "nationally important and technically challenging problems." That meant, he said, "Congress sets up a subcommittee, worries about the problem continuously, and funds money." Stenbit noted two important differences between TRW's current approach to such business and that of the 1960s. First, the company no longer concerned itself primarily with technical problems of optimization. Rather, it dealt with "enormous, complex problems where there's a highly litigious and emotional set of characters who are influencing the particular problem." Second, "the percentage of the actual solution that comes from TRW" declined dramatically. "This is a much less vertically integrated business than it used to be," Stenbit said, "because the domains are getting broader and more complex."[30]

In the early and mid-1990s, TRW landed several big systems projects with nonmilitary customers: a contract potentially worth $1 billion over ten years to support the Department of Energy to help manage the problem of nuclear waste disposal; a $138 million contract from the Federal Aviation Administration to assist in improving the nation's air traffic control systems; and an agreement worth potentially $400 million to help modernize the computer systems at the Internal Revenue Service.[31] In winning these and other projects, the company placed less emphasis on its general ability to manage complex problems than its specific expertise in information systems and information management. This shift became still more evident late in 1997, when TRW merged with BDM International, Inc., a company specializing in information technology and services for commercial and government customers. The combined entity, called TRW Systems & Information Technology, is one of the five biggest providers of information technology and services to the federal government.

TRW's experiences in civil systems during the late 1960s and early 1970s constitute an interesting case of organizational learning. Through such organizations as the CAT and the Systems Integration Group, TRW learned to apply its skills and capabilities in systems engineering and technical management in new areas more prudently and patiently than ever before. Whether these new efforts will pan out, of course, cannot be known with certainty. TRW is well aware of the challenges ahead and it will surely encounter problems and difficulties. But it also remains convinced that its systems management capabilities can, when properly applied, provide a foundation for new business and that it will make progress by avoiding the mistakes of the past.

Acknowledgments

This chapter is partly based on research funded by TRW for the author's book, *TRW: Pioneering Technology and Innovation, 1900–1995* (Boston: Harvard Business School Press, 1998). The author is grateful for comments on an earlier version delivered at the Dibner Institute conference in May 1996. Comments from Hans Klein and Thomas P. Hughes were especially helpful in revising the chapter.

NOTES

Many of the notes refer to records that are the property of TRW Inc. The author consulted these documents, most of which were maintained in an archive at Redondo Beach, California. Copies of other internal TRW documents, including minutes of the board of directors and news releases, were consulted at TRW Inc.'s corporate headquarters in Cleveland, Ohio.

1. After the merger, the official name of the company was Thompson Ramo Wooldridge, Inc., which was abbreviated in 1965 to TRW Inc., the name used here.

2. In 1954, Ramo-Wooldridge had agreed to forego production of hardware as part of its contract to provide systems engineering and technical direction to the air force. This hardware ban proved a constraint on the company's growth and starting in 1957, it sought to have it overturned. Competitors in the aerospace industry were not eager for TRW to be freed, however, and the issue proved a source of controversy for several years. TRW was released from the hardware ban in stages. The process was finally completed in 1960, when TRW divested its long-range planning activities for the air force as the basis for a new nonprofit entity, Aerospace Corporation.

3. Simon Ramo, interview with Christian G. Pease, Entrepreneurs of the West Oral History Program, University of California at Los Angeles, 1984.

4. Walter McQuade, "Caution! Si Ramo at Work," *Fortune*, November 1970, 105–107 and 185–192; Simon Ramo, "Mobilizing Science for Society," The Charles M. Schwab Memorial Lecture, delivered at New York City, 22 May 1968.

5. Simon Ramo, *Cure for Chaos: Fresh Solutions to Social Problems through the Systems Approach* (New York, 1969), v–vi and 6–7. Although this work appeared at the end of the 1960s, it is representative of Ramo's thinking and writing throughout the decade. See, for example, "The New Pervasiveness of Engineering," *Journal of Engineering Education* 53, no. 2 (October 1962): 1237–1241; "Management in an Information-Automation Era," *Proceedings of the International Management Congress*, CIOS 13 (September 1963): 198–203; "Putting Technology to Work—the Social/Political Barrier," *Space Digest*, October 1964, 56–62; and "The Coming Technological Society," *Engineering and Science* 28, no. 3 (December 1964): 9–13.

6. Robert W. Hovey, "History and Trends of Systems Engineering," unpaginated typescript, April 1970; K. H. Borchers, C. S. Lightfoot, and R. W. Hovey, "Translation and Application of Aerospace Management Technology to Socio-Economic Problems," *Journal of Spacecraft and Rockets* 5, no. 4 (April 1968): 467–471.

7. Glenn E. Bugos, "Programming the American Aerospace Industry, 1954–1964: The Business Structures of Technical Transactions," *Business and Economic History* 22, no. 1 (Fall 1993): 210–222; "Civil Systems Grow in SEID Management Systems Section," *Sentinel*, 7 October 1966, 2. (*Sentinel* is a TRW internal employee publication.)

8. "TRW to Study Land-Use Planning," *Technology Week*, 6 June 1966, unpaginated reprint; Herbert H. Rosen, "Technology Transfer in the Service of Mankind" (TRW Inc., n.d., 1971?), 1–2.

9. "TRW's Civil Systems Work Discussed on 'Today' Show," *Sentinel*, 4 March 1967, 3.

10. "Civil Systems Grow," 2; McQuade, "Caution! Si Ramo at Work," 184.

11. George Solomon, interview with the author, 13 February 1991.

12. "STD Develops Combustor to Help Combat Air Pollution," *Sentinel*, 18 December 1970, 2; "Japanese, TRW in Joint Venture," *Sentinel*, 4 February 1972, 2; "Space Science Invention Fights Industrial Pollution," *Sentinel*, 13 December 1974, 2.

13. TRW memo to managers, 12 January 1970; "New TRW Operating Unit Formed for Commercial Market Systems," *Sentinel*, 6 February 1970, 1; "Sommer, Burnett Named to New Top Posts," *Sentinel*, 2 October 1970, 1; Donald Kovar, interview with the author, 16 November 1990; TRW Inc., "TRW Five Year Plan," 6 February 1970, Section 8.

14. TRW Inc., "TRW Five Year Plan," 6 February 1970, Section 8; TRW Inc., "1975 Plan," April 1971, Section 8; John E. Davis, "Commercial Electronic Equipment at TRW: An Historical Analysis," TRW Inc., March 1984; Ruben F. Mettler, interview with the author, 11 May 1994; J. Sidney Webb, interview with the author, 14 February 1991; and Richard Campbell, interview with the author, 12 February 1991.

15. "New Credit Authorization System Introduced by TRW Data Systems," *Sentinel*, 4 June 1971, 1 and 5; Webb and Campbell interviews, as well as Kovar interview, 16 November 1990; Melvin Shader, interview with the author, 18 June 1991; and Davis, "Commercial Electronic Equipment," passim.

16. See the recollections of TRW manager Tom Harrington in Peter D. Stenzel, "Lessons Learned: A Study of TRW Space & Defense Sector's Civil and Energy Businesses," TRW Inc., Space & Defense Planning Department, 1990, 32–33.

17. "Summary of Joint Venture Agreement, TRW Inc.–Barringer Research Limited," 24 September 1968; board minutes, 26 April and 1 December 1972 and 15 October 1975; Ted Wilson interview in Stenzel, "Lessons Learned," 19; "Joint Venture Forms New Service Company," *Sentinel*, 1 October 1971, 2.

18. TRW Inc., board minutes, 21 July 1971 and 12 April, 19 July, and 13 December 1972; TRW Inc., news release, 21 June 1971 and 4 August 1972; Harold Hirsch, interview with the author, 19 June 1991; TRW Inc., annual reports for 1972 and 1974; "TRW Selected for Operation Breakthrough," *Sentinel*, 6 March 1970, 8. According to Tom Harrington, whom Peter Stenzel interviewed in 1990, the entrepreneur who bought TRW's assets in Sacramento eventually did quite well by selling modular housing in Saudi Arabia. Stenzel, "Lessons Learned," 34. See also Stenzel's interview with Dean Lowrey in "Lessons Learned," 53–56.

19. "County Supervisors Select TRW for Computerized Traffic Control," *Sentinel*, 2 July 1971, 2; Tom Harrington interview in Stenzel, "Lessons Learned," 35–38.

20. Mettler interview, 11 May 1994.

21. "TRW Will Provide Supervision System for Washington, D.C. Transit Authority," *Sentinel*, 7 April 1972, 6; board minutes, 26 April 1972; Tom Harrington interview in Stenzel, "Lessons Learned," 46–47; "DeLeuw, Cather—It's Part of TRW," *Sentinel*, 2 April 1976, 5; *Wall Street Journal*, 11 October 1977, 21.

22. TRW Systems Group, interoffice correspondence, 20 January and 24 May 1971; board minutes, 21 October 1970 and 17 February 1971; TRW news release, undated [1971?].

23. TRW Inc., annual report for 1974; TRW Inc., Data Book, 1970–1974. Taking a longer view of the company's performance, the value of TRW's common stock had been subject to a sustained decline since the middle of 1968.

24. TRW memo to managers, 8 October 1974; Mettler interview, 11 May 1994; TRW interoffice correspondence, 7 August 1975, "Some Highlights of the 1975 Corporate Planning Conference."

25. In 1980, the company spun off the credit authorization business through a joint venture with Fujitsu. Five years later, shortly before divesting most of its commercial electronics operations, it made the credit reporting business the foundation of a separate business group called TRW Information Services. This unit grew rapidly for several years until it ran into a series of legal and regulatory barriers and competitive inroads. Following a big loss in 1991, TRW put the business on the block and finally found a buyer in 1996.

26. TRW memo to managers, 10 September 1973; "Foster Joins TRW Energy Research and Development," *Sentinel*, 14 September 1973, 1; John Foster, interview with the author, 12 February 1991.

27. Foster interview, 12 February 1991; TRW interoffice correspondence, 12 November 1973; Simon Ramo, *The Business of Science: Winning and Losing in the High-Tech Age* (New York: Hill & Wang, 1988), chaps. 5 and 6, esp. 212–217.

28. TRW organizational announcement, 29 December 1975.

29. TRW Inc., Space & Defense Planning Department, "Summary of TRW Ventures," 18 May 1990.

30. John P. Stenbit, Remarks from the Annual Analysts Meeting, 25 May 1994, La Jolla, California, 55.

31. TRW Inc., news releases, 2 March 1990, and 13 February, 20 May, and 12 December 1991. Early in 1997, the new head of the IRS decried the state of the agency's information systems but specifically exempted TRW from responsibility. The company remains the agency's principal information technology contractor.

12

From Operations Research to Futures Studies: The Establishment, Diffusion, and Transformation of the Systems Approach in Sweden, 1945–1980

Arne Kaijser and Joar Tiberg

Introduction

In 1948 the Nobel Prize in Physics was awarded to the British physicist Patrick Blackett for his discoveries in cosmic radiation. Blackett, a fifty-year-old professor from the University of Manchester, was already something of a hero in his home country, not because of his scientific work, but because of his contributions during World War II to the development of operations research (OR).

At the outbreak of the war, Blackett was asked to head a group of scientists at the Anti-Aircraft Command working on a new air defense system based on radar technology. The group improved the functioning of this system considerably, a factor of great importance during the Battle of Britain. The success of this group led to the establishment of similar groups of scientists at other Commands in the British army, navy, and air force. Their activities were referred to as Operations Research, and "Blackett's circus," as the groups were sometimes called, played an important role throughout the war in improving the efficiency of a variety of military operations.

The Nobel Prize awarded to Blackett was an important event in the early history of operations research in Sweden. Blackett stayed in Stockholm for ten days, and he gave several lectures and had many informal discussions with leading Swedish scientists from universities and from the recently founded National Defense Research Institute (FOA). He probably also met high-ranking military officers. Blackett's wartime experience and his personal charisma created a strong interest in OR and was one impetus for establishing OR in Sweden.[1]

The purpose of this chapter is to describe and analyze the development of the systems approach in Sweden from 1945 to 1980. We use the term "systems approach" to signify a specific intellectual tradition, which started with operations research in the 1940s, but later developed in a variety of ways into systems analysis, policy analysis, and futures studies. Our demarcation is mainly sociological in character; that is, we start with operations research and the people and organizations in this particular field, and then we move on to the activities that grew out of operations research, which to a great extent involved the same people and organizations. The boundaries of this tradition are somewhat arbitrary. First, there were activities similar to operations research long before World War II. In particular, scientific management, developed by F. W. Taylor and others at the turn of the century, had a number of similarities with OR. Second, there were other activities in the postwar period that resembled the systems approach (for example, economic and regional planning) but did not involve operations researchers.

One area the chapter focuses on is the gradual transformation of the systems approach. Operations research was primarily tactical, aimed at improving existing weapon systems or production systems, and the methods employed had a scientific and mathematical bias. In the late 1950s, operations researchers, or operations *analysts* as they were often called, gradually turned to broader problems of finding the best designs for future systems. Handling uncertainty became a key issue, and the OR toolbox was complemented with economic methods and models and with concepts such as flexibility and adaptivity. To emphasize this wider approach, the term "systems analysis" was introduced. In the late 1960s, the approach was broadened once again to more general societal problems, and methods from the political and social sciences supplemented the earlier approaches. In the United States, this new approach was labeled policy analysis, but in Sweden the emphasis was more on what was called futures studies.

Another focus of this chapter is on the diffusion of the systems approach into different sectors of Swedish society. In the early 1950s, OR was carried out by four academic groups in Stockholm. They worked with both military and civilian applications, often on a consultancy basis. In the late 1950s, OR was established in the defense sector on a permanent basis, and there has been a fairly coherent and organizationally unified development of the systems approach in this sector since then. There were also attempts to establish OR more permanently in industry, in the academic world, and in the public sector, but in these civilian sectors the systems approach has been more scattered, with no strong organizational center. The Swedish Operations Research Association, established in 1959, played an important role as a meeting place for operations analysts and also as a lobbying organization for the systems approach, but it had no permanent staff and limited economic resources.

The second section describes the development of the systems approach in the Swedish defense establishment, and the third section deals with the systems approach for civilian applications. This means that the article is not strictly chronological, as these two developments occurred in parallel. In the concluding section some characteristic patterns and factors in the development of the systems approach in Sweden are outlined.

The sources for the article are of four kinds: archive material;[2] contemporary official publications and journals; interviews with eleven persons involved in or knowledgeable about OR activities in the 1950s, 1960s, and 1970s;[3] and secondary literature.[4]

The Systems Approach in the Swedish Defense Establishment

At the outbreak of World War II, Sweden had a weak defense establishment with equipment that was largely obsolete. After the German invasions of Denmark and Norway in the spring of 1940, acute fear of a German invasion arose in Sweden, and the country started a rapid military buildup. The country was blocked from most of its foreign trade, and the war economy focused on establishing and expanding a domestic industry for all kinds of military equipment. Thanks to a strong and diversified engineering industry, within a few years Swedish industry was able to produce everything from airplanes and tanks to ammunition and gas masks. This domestic industry for military equipment was in large part retained after the war, although the levels of production were lowered.

The Swedish government succeeded in its ambition to keep the country out of World War II. In the late 1940s, Sweden tried to establish a Nordic defense treaty with Denmark and Norway. However, these countries chose to join NATO in 1949, and Sweden decided to return to its traditional neutrality policy, based on the formula: "non-alignment in peace, aiming at neutrality in case of war." Even if the Swedish neutrality was not as strict as officially maintained,[5] this decision meant that Sweden would have to rely largely on its own military strength. There has been a broad political consensus ever since that a strong military defense is necessary in order to gain confidence and respect for the country's neutrality policy. Sweden has been unique in this respect. Apart from Israel, no other nation of Sweden's size has built such a strong defense establishment and such a diversified domestic industry for military equipment.[6]

The First OR Activities

Blackett's visit to Stockholm in 1948 was not the first time that operations research was discussed in Sweden, but it spread interest in OR to wider circles. Furthermore, Blackett arranged for a British OR specialist, C. E. G. Bailey, to come to Stockholm in February 1951 to give three lectures on operations research to an audience of officers, academics, and civil servants. Bailey gave an account of British experiences concerning the kind of problems that had been solved, the techniques that had been used, and the forms of organization that had been most effective. Based on the impressions he had received during his stay in Sweden, he also drew some conclusions, emphasizing that

> O.R. methods have still much to offer to Sweden, and in particular to Swedish Defense Forces. I believe those forces are still in the stage where small groups of men could cause an increase in the effectiveness of 2 or 3 times in the units they are attached to. First class men are needed and they will have to be paid for.[7]

However, in an official letter from the Defense Staff in September 1951, it was argued that the Swedish defense establishment could not afford to employ full-time civilian personnel for OR work. It was recommended instead that scientists be employed on a part-time consultancy basis, and that young scientists should be encouraged to do their compulsory military service as operations analysts.[8] Thus, in the early 1950s the first OR activities were started in the Swedish military forces on a small scale. Each of the services established cooperation with a specific academic environment in Stockholm.

The air force started cooperating with Professor Lamek Hulthén at the department of mathematical physics at the Royal Institute of Technology (KTH). Hulthén had been a consultant to FOA in matters related to atomic weapons since 1945,[9] and moreover he knew Blackett personally.[10] Hulthén was teaching the students of technical physics, which was one of the most prestigious academic careers in Sweden at the time. Each year he would suggest that one or two of his best graduating students should do their military service as so called "research technicians" for the air force.

Hulthén was a scientific advisor for these "research technicians" on a consultancy basis, advising them about suitable problems and appropriate methods for solving them. Furthermore, a career officer was assigned to provide them with military expertise. In the early years the OR work was focused on aircraft warning systems, and it was concluded that the present warning system at military air bases would not give sufficient time for airplanes to take off in case of a sudden air attack. When the commander in chief of the air force was informed about these findings, he was quite disturbed about this deficiency.[11]

The Navy Staff established cooperation with Professor Harald Cramér from the department of mathematical statistics at the University of Stockholm. Cramér was a man of high standing in the academic world; in 1951 he became vice chancellor of the University of Stockholm, and in 1959 he became chancellor of all Swedish universities. In 1948, Cramér and some of his colleagues had established the Statistical Research Group (SGF) to do statistical studies and OR studies on a consultancy basis for industry (this will be outlined in the second part of this chapter). Now the group extended its work to include the military sector. It focused on mathematical models of naval artillery. In particular the probabilities for hitting targets using different firing modes were analyzed as a basis for developing more efficient battle tactics.[12]

The Army Staff chose FOA as its OR partner. FOA had been founded in 1945 as a joint research agency for all the defense authorities and was based on wartime experience of defense research. In the 1950s it became the center for Swedish military R&D and its personnel grew from 200 in 1946 to more than 1,600 in the late 1960s.[13] The OR work in the Army Staff was headed by a young mathematician at FOA, Lars-Erik Zachrisson. He developed a special kind of game theory (Markov games) for analyzing battlefield behavior. This theory was applied in the analysis of duels between tanks, or similar situations, as a basis for developing better battle tactics.[14]

The OR work in the Swedish military in the first half of the 1950s was thus of limited scope and was mainly carried out by young "research technicians" doing their military service. This way of organizing OR work was cheap, but prevented the buildup of a more permanent competence base. Most of the OR work was focused on finding the most efficient use for existing weaponry and other resources.

The purpose was to identify weak spots and to propose better fighting methods and battlefield tactics. This focus was similar to that of the allied OR groups during World War II.

From Operations Research to Systems Analysis

In the early 1950s, Sweden's military situation changed with the growing tensions between East and West, and the rapid armaments race on both sides. There was a growing conviction both in the military and among leading politicians that Sweden's military equipment was becoming obsolete and that new, modern weapon systems would have to be developed. Fighter aircraft and nuclear weapons were seen as two key technological areas.

The director general of FOA, Hugo Larsson, became an outspoken proponent for using operations research as a key instrument in this technological modernization. In an article entitled "Which Weapons Shall We Choose?" Larsson warned that the traditions and prestige of the three services could have a dangerously restrictive effect. He argued that it was imperative to make a thorough analysis of the kind of weapons an enemy might use against Sweden in the future, and that choices of future weapon systems must be based on this analysis. "Everywhere abroad Operations Research is used for this end," he asserted, and argued that Sweden should do the same.[15]

Larsson convinced the minister of defense of the importance of OR, and in 1955 the minister commissioned the Supreme Commander to investigate the future organization of military OR. A special committee with representatives from the services was appointed to this end. In its final report it recommended the creation of one central OR office, which was to coordinate and support the work of permanently established OR groups in the army, navy, and air force staffs. A tug-of-war developed between the Defense Staff and FOA about the location of this central office. The final outcome, in 1958, was that this office was placed at FOA, within a new planning department, called FOA P.[16]

At the same time, FOA became more and more involved in the overall military planning process. In 1959, the director general of FOA, Martin Fehrm, became a permanent member of the Military Council under the supreme commander, which had previously consisted only of the commanders in chief of the services. This was a recognition on the part of the military of the increasing importance of R&D in the development of Swedish defense. It was largely the task of the personnel of FOA P to support Fehrm's work in the council.[17]

Thus, in the late 1950s, military OR work was organized on a permanent footing. At the same time a gradual transition of the content of OR work toward more strategic issues began. The role of the analyst was changing from a "trouble-shooter" and efficiency improver of existing weapon systems to a planner and designer of future weapon systems. It is interesting to note that the term "weapon system" was increasingly used at this time. A large part of the inspiration for this development came from the United States. In a 1957 article, an OR researcher from FOA very clearly acknowledged this American influence when he pointed out that "the Americans have coined the term Systems Analysis for this kind of OR."[18]

OR for Military Planning

When FOA P was created it established permanent OR groups within the services and Defense Staff. In the first years, there was only one permanently employed analyst on each staff, but gradually the groups were expanded to a full size of about five persons each. The central OR office at FOA P was kept rather small and was primarily responsible for methods development, coordination of the work of the OR groups, and the recruitment of analysts. Between 1958 and 1970 the number of people on the OR staff grew from six to more than forty.[19]

Carl-Gustav Jennergren, an expert on explosives from FOA's physics department with no prior OR experience, was appointed director of FOA P. Jennergren became deeply involved in the military planning process as Fehrm's adviser, and this influenced the work of his staff. In the early years of FOA P, a great deal of energy was devoted to the development of new methods for long-range planning: evaluation principles for comparing the effectiveness of different weapon system designs and defense compositions, methodology for large-scale war games carried out by military officers for simulating combat situations, and computer simulations of large-scale military operations.[20]

At the outset, the methodologies employed had a quantitative bias. For example, many analysts thought that measures of effectiveness of future weapon systems could be quantitatively determined. However, when the analysts were confronted with broader and more future-oriented issues, the limitations of pure quantitative methods became obvious. Instead, more qualitative questions about the flexibility and adaptability of weapon systems to future changes in enemy weaponry were seen as vitally important. The participation in the military planning process thus stimulated a reevaluation of methods and approaches. Jennergren and his staff also found that actual planning and implementation was not as rational as they had assumed, and they became increasingly interested in the planning and implementation process as such and in how it could be improved.[21]

Important impulses for new methods and approaches came from abroad, and in particular from the United States. In 1961, a group of FOA P analysts visited the RAND Corporation, which had extensive experience in military long-term planning. In the following year some prominent RAND researchers, among them the OR pioneer E. S. Quade, came to Stockholm to give lectures and seminars on systems analysis and long-term planning. The RAND researchers advocated a broad multidisciplinary approach, and they stressed the importance of including economists. This had an impact on FOA P's subsequent development, according to Jennergren. As a direct result, FOA P recruited some economists, who came to play an important role in the development of more sophisticated methodology and organization for military long-term planning.[22]

In the mid 1960s, analysts at FOA P played a vital role in the development of a new planning system for the defense sector. It had four major elements. The first involved broad studies of international developments within the political, military, economic, technical and social domains as a basis for constructing possible cases of aggression in the form of "crisis scenarios." The second element, a prospective planning process with a fifteen-year perspective, included the design of different future defense structures for different budgetary constraints. The third element was directed toward the planning and design of major weapon systems, such as new fighter aircraft. The fourth element, based on the first three, was the formulation of five-year program plans, which were adapted—"rolled"—every year.[23]

FOA P: The Spider in the Military Web

In the 1960s FOA P gradually became a very influential "spider" in Sweden's military web. Through its OR groups it had close ties with and extensive knowledge of each of the three services and of the properties and efficiency of existing weapon systems. FOA provided an access to the expertise of the other FOA departments doing military research on new weapon systems. And the involvement in the military planning process offered a bird's-eye view of the military sector. This broad expertise gave FOA P considerable influence.

This influence is clearly demonstrated in the nuclear weapons issue, which was the overarching question for military planning in Sweden from the mid-1950s to the mid-1960s. FOA was the center for research on nuclear weapons, and around 1960 a large number of FOA's personnel were involved in different aspects of this research. In the mid-1950s most of the emphasis was on how to construct nuclear weapons as quickly as possible. There was a fairly broad consensus that acquisition of nuclear weapons was of critical importance for Sweden and would have a significant deterrent effect.

In the late 1950s opposition to nuclear weapons grew, within both the opposition and the governing Social Democratic Party. As a result, nuclear weapons research was partly reoriented toward assessing the effects of nuclear bombs on military installations, military personnel, and the civilian population. These studies indicated that the effects were much more pervasive than had previously been imagined. Based on these results, operations analysts initiated studies and war games together with officers to clarify the tactical, strategic, and political aspects of nuclear warfare. Two important results emerged from these studies. First, the previous conception of limited nuclear warfare restricted to the battlefield was abandoned; once nuclear weapons were introduced in a battle, it would be very difficult to prevent an escalation. Second, the lack of limits on nuclear warfare meant that the effects of such a war would be a total catastrophe for Sweden. These studies played a decisive role in showing leading military figures limitations of the nuclear strategy and led to a new consensus in the mid-1960s not to develop nuclear weapons in Sweden.[24]

Analysts, Officers, and Politicians

From the late 1950s onwards, operations analysts acquired a growing influence in the military and on the military planning process. This "civilian invasion" into the military hierarchy was in a way the most essential feature of the systems approach in defense. It is hardly surprising that a number of prominent officers felt threatened in their professional role. An illustrative example of this occurred when the head of the military academy wrote an article in one of the major Swedish newspapers entitled "Strategists and Counting Experts," questioning whether military decisions should be based on "seemingly mathematical calculations instead of applying military judgment, intuition, and experience."[25]

Serving as an operations analyst in the military demanded a certain diplomatic skill, and on several occasions analysts lacking this skill were frozen out. In most cases, however, the operations analysts were able to establish a successful cooperation with their military counterparts. One reason for this was that the analysts were seen as a valuable resource in the eternal battle for economic resources between the services.[26]

At the Defense Staff, representing the highest level in the military hierarchy, the OR researchers were considered valuable in preparing decisions about how to optimize allocation of resources between the services. When Jennergren met the supreme commander for the first time, after having been appointed head of FOA P, the latter exclaimed: "Very good, write up your damned formulas so that we can finish all these discussions about resource allocation!"[27]

In the mid-1960s the politicians and administrators at the Ministry of Defense also began to see operations analysts as a valuable resource. At this time the general political attitude to defense became more restrictive, and politicians were worried about the cost escalations in the development of a new aircraft, Viggen. As a result, there was growing antagonism between government and the military authorities.[28] The operations analysts were highly knowledgeable about the services and new weapon systems, and yet did not belong to the military hierarchy. So, in 1965, a handful of analysts were employed by the ministry to develop a planning system which would give government a stronger position vis-à-vis the military.[29]

The Swedish historian Wilhelm Agrell has described the major changes in the Swedish defense in the postwar era in terms of three concepts or trends: technology orientation, professionalization, and politicization. The development of new, technically sophisticated weapon systems has been a key issue in defense policy and has led to allocation conflicts between the services. The growing complexity of these systems has led to a change in the military profession, with higher demands on theoretical knowledge and management skills. And the high costs and long lead times of these weapon systems have forced the politicians to try to increase their control over the long-term development of the defense sector.[30]

The role of the operations analysts can be seen against these three trends. Their background and training, in combination with their organizational location at FOA and the military establishments, gave them a unique knowledge of weapon systems. This knowledge made them a valuable resource both for professional officers in their fight for resources and for politicians in their efforts to curb excessive military ambitions.

Above, we have tried to explain the rapidly growing influence of operation analysts in Swedish defense in the 1960s. Another salient feature is the persistence of FOA P as an organization and of the systems approach as an intellectual tradition within the defense sector. In fact, the whole organizational structure of OR work and the military planning process built up in the 1960s is more or less intact today.

Civil OR in Sweden

The civil operations research that developed in Sweden after World War II derived its inspiration from several sources. In the initial stages, as we will see, military OR activity was the most important of these sources. The military's OR work became the mold for the civilian approaches.[31] But early civilian OR also had clear roots in the intellectual environments arising around the classic administrative sciences. One such was the rationalization movement dating from the 1910s.[32] Inspired by F. W. Taylor's concepts of scientific management, many ideas of rationalization found support in influential industrial and academic environments at that time. The Federation of Swedish Industries (Industriförbundet), formed in 1910, stated in its first annual report, for example, that industrial management was to be a subject of key importance for the organization. In 1913, two years after its original publication, the Federation published Taylor's book *The Principles of Scientific Management* in Swedish. The Royal Academy for the Engineering Sciences (IVA)—which was established in 1919 to support technical research and development—also became an important institution for the spread of the rationalization movements ideas.[33] During the interwar period this movement, with its echoes of Taylor, swept over Swedish industry, and gradually concepts such as time studies and production planning became standard prose among top Swedish industrialists. New statistical tools for the optimization of stockholding and for the planning of long-time investments also appeared at this time. In 1934, a big conference organized by IVA and the Swedish Association of Engineers (Teknologföreningen) dealt with time studies and in 1945 the Swedish omnibus book *Handbook of Industrial Productions Economy and Organization* was published.[34]

The core ideas of the rationalization movement were developed further and the theories found an institutionalized form when education and research programs in industrial economy and organization were established at the universities. In 1940, the first professorship in industrial economy and organization was set up at the Royal Institute of Technology in Stockholm.[35] Courses for engineers were offered in production planning, among other things. The material used in that course later appeared in revised form, as a Swedish textbook on operations research.[36] Thus, even if few of the OR pioneers in Sweden after 1945 stressed the fact, the OR techniques that they proposed had methodological roots in the Swedish academic world independent of any military influence.

Early Civilian OR

In 1947 Professor Harald Cramér delivered a speech at IVA on statistical methods for industry. Most of Cramér's speech consisted of an exposition of the acute need for more resources in this research area. He mentioned the American efforts to raise money for this purpose and pointed to the work done at the Statistical Research Group at Columbia University as worthy of imitation. At the end of the lecture, Cramér mentioned OR. He stressed its wartime successes and its continuing military relevance, and argued for its future industrial use:

> ... according to experts in the matter, military OR in time of peace will be followed by industrial OR, where the efficiency of different industrial processes of production will be studied in the same way, with mathematical/statistical methodology.[37]

In 1948 the Statistical Research Group (SFG) was formed at Cramér's department for mathematical statistics at Stockholm University. The group carried out commissioned research in OR for industries in Sweden and abroad. Most of their work in the 1950s was "clean" OR: empirical, and aimed at finding solutions to well-defined and isolated industrial problems. In other words, studies similar to the work done in the armed forces during the 1950s. These early OR studies, industrial as well as military, thus resembled earlier Taylor-influenced work; they aimed at making existing processes and technologies more effective, rather than replacing them with new ones.

Among other things, the analysts at SFG worked for Swedish newspapers on distribution issues, for the oil industry on prospecting methods, and for the Swedish state mines and steel plants on transportation and location models. Around the corner from SFG in downtown Stockholm was the School of Economics where the Group for Model Conception in Business Administration had its office. This group, under Professor Paulsson Frenckner, became the platform for the introduction of OR into managerial economics in Sweden. The two groups did not compete for commissions. The demand for the new services was far higher than the group's combined capacity.

Three kinds of arguments frequently recurred in the rhetoric of the Swedish OR propagandists. One was simple: it sold OR as tried-and-true for military applications.[38] Now, it was argued, the techniques that had helped to win the war could also help corporations hit their business competitors. A second argument arose from the idea that in modern industry, labor would gradually be replaced by machines and other new technologies. These technologies were expensive, locking production in set patterns for long periods of time. Analysts argued that inflexibility, characteristic of highly technology-dependent industries, required a broad general knowledge about processes of change and about future trends. OR teams, experienced in applying new mathematical/statistical methods, were ready to provide this knowledge. A third argument used by the OR propagandists was to point to a new class of managerial problems in modern industry which could only be solved with OR techniques. This new class of problems resulted from the multidivisional character of modern industry and led to inconsistent and sometimes conflicting goals within corporations. With the tools of OR, the managers were promised a way to re-create harmony and regain an overview and control of their organization. To obtain that overview, however, the management of the corporation had to allow the OR teams considerable freedom in their work. This was because the OR methodology, according to analysts, was organizationwide in scope, which meant that no matter what specific problem the analyst was to examine, it had to be examined as part of the whole. This inherent characteristic of OR, analysts claimed, required the teams to have access to all kinds of information, including that which might appear irrelevant to the uninitiated.

The young scientists selling OR to Swedish industry were clearly not modest in their ambitions. One of them, at the IVA meeting in 1953, outlined three levels on which OR should be applied: the lowest level (compared with weapons in the military) was concerned with effectiveness in use and arrangement of machines. A tactical level was for handling production efficiency and optimization of the resources.

Finally, there was the strategic level, where problems concerning the organization as a whole, location issues, market analysis, and so forth were discussed.[39] Another analyst presented a survey of articles in the British and American OR journals. He showed that industrial subjects outnumbered military and theoretical/philosophical and historical topics, and that this trend was on the increase. Using this survey, he argued that OR was soon going to be an indispensable tool for any industrialist seriously wanting to compete in the market.[40]

Shortly after the IVA meeting, an ad hoc committee for operations research was created within IVA. In 1955, the committee sponsored a study tour for Professor Paulsson Frenckner to the United States. During the six-week trip, Frenckner visited twelve universities and a countless number of corporations, think tanks, and state agencies where OR was practiced. In 1956, IVA invited top Swedish industrialists, academics, and military figures to an OR conference. Frenckner started by presenting his experiences from the study tour in the United States. Other papers were presented as well, on OR for industrial management, OR methods for optimization of quality in the processing industry, and the optimization of route design for public transport. The conference resulted in the creation of a permanent committee within IVA for OR issues.

The Swedish Operations Research Association

In February 1959 Professor Edwards H. Bowman from MIT in Boston delivered a speech at the Stockholm School of Economics entitled "Some Difficulties in Doing Operations Research." Invitations were sent out to forty-five men—academics, military figures, and industrialists. The letter of invitation informed them that an OR association was going to be formed, and that Bowman's speech was to be the starting signal for the club's activities.[41] Contacts had been made in advance with the sister associations in America and Canada and their statutes and membership regulations had been copied. On this occasion, then, the Swedish Operations Research Association (SORA) was formed, and it soon became part of the International Federation of Operations Research Societies (IFORS). Harald Cramér was appointed president and Lamek Hulthén and Paulsson Frenckner were elected to the board. From the start, analysts from FOA P also occupied high offices within the association. A year later, a group of twenty SORA members attended the second annual IFORS conference in Aix en Provence.[42]

The association started up with a lecture series on classic OR problems such as queue theory and location problems. By the mid-1960s, a wider variety of subjects was seen at SORA's meetings. Planning and security issues in road traffic, transport planning analysis for the switch to righthand traffic, peace research, and more were dealt with at recurrent "OR days" arranged by SORA.[43] The association grew from around 100 original members to about 400 in the early 1970s. Throughout the 1960s the majority of the members were engineers and mathematicians/statisticians. Many of the most active members of SORA were employees at FOA P. Courses in OR for civilians were given by FOA P analysts during the 1960s, and measures for the establishment of institutionalized education for operations analysts were discussed within the association. A step toward this kind of program was taken in the mid-1960s when FOA P agreed to start courses in classic OR for students in engineering. The FOA mathematician Lars-Erik Zachrisson, who developed game theory for tank duels in the early 1950s, became a teacher and researcher at KTH with his salary paid by FOA. Later, in 1969, KTH managed to get government funding for establishing a new department named "Optimization and Systems Theory," and Sachrisson became permanent professor.[44] An important step for the institutionalization of OR in Sweden had been taken.

Early Swedish OR: Consensus

The first period in Swedish OR activity is characterized by consensus among its different practitioners. There was enough work for everyone at this time, and there was no explicit antagonism between the different spokesmen for OR. The members of the IVA committee agreed, not only about the benefits of the new techniques, but also about the best course of action in the launching process. No voices critical of OR thinking were raised at this time. Consequently, there were no discussions about the possible dangers of applying the new techniques, or about overall aims.

The situation in the United States at this time was rather similar. Giandomenico Majone explains the lack of conflicting interests in early American OR by the fact that it was a young and developing profession.[45] As such, it was missing a mechanism for quality control that is one of the distinguishing features of more established professions. Ida Hoos sees another factor behind the consensus situation, that is to say, the "width" of OR. She claims that the myriad forms and manifestations that constituted the new "science" made it almost immune to attack and critical reading.[46]

The situation in Sweden and the United States differs from that in Britain, where efforts were made by left-wing intellectuals, among whom were prominent OR heroes from the war, to create alternative and more radical OR after the war.[47] Rosenhead has shown how conflicting British interests involving consumers, scientific workers, nationalized industries, and government ministries fought over control of OR in the late 1940s. The defeat of the Labour party in the 1951 elections meant an end to efforts to develop an OR methodology aimed at increasing the power of consumers at the expense of that of private capital.[48]

There were no groups trying to establish a more radical OR in Sweden after the war. Why not? Many explanations are possible. The left-wing views of the scientific community creating OR in Britain were not prevalent among their Swedish counterparts. Furthermore, the atmosphere that had characterized the British OR groups during the war—hard and innovative work for a common aim, to win the war—never prevailed in the Swedish OR environment. In addition, Swedish industry in 1945 stood intact and ready to provide the country with economic wealth. Nationalization of industry was therefore not such a lively idea in Sweden as it was in Britain, where industry was in ruins and private capital was scarce.[49] It was not until the late 1960s that conflicting interests around the systems approach arose in Sweden.

OR in the Public Sector

By the early 1960s, OR had changed substantially, compared to the work of the early 1950s. Many teams had left the easily isolated tactical problems, and the studies had broadened. Operations analysts, at FOA and elsewhere, had by the early 1960s understood that many of the problems they examined were too complex to be understood only with the help of their square tools of mathematics and mathematical statistics. Through meetings with RAND representatives, FOA analysts became convinced of the need to incorporate, among other things, economic considerations in their analyses. Consequently, a group of economists entered the Swedish OR scene in the early 1960s. From this time on, OR changed its focus from industrial applications to increasing number of applications in the public sector, such as health, housing, and social welfare.

In 1958, a paper by Professor Tore Dalenius on queue theory in the health sector was published.[50] Dalenius was a statistician from SFG and had been prominent in the creation of SORA. In the mid-1960s he was engaged by the Council for the Rationalization of the Health Sector (SJURA) to make an inventory of the possibilities of using OR methods to solve problems arising in the health care system. The report describes how questions concerning everything from the optimization of hospital size and location to personal administration and optimal diet for patients could be answered with the help of quite rudimentary OR methods. Two organizations for the planning of the rapidly growing health sector appeared at this time, and a commission for OR in the health sector was formed in the mid-1960s on the initiative of IVA and the State Technical Research Council. In 1968, the State Planning and Rationalization Institute for Health Issues (SPRI) was established by the government.[51]

The activities within SORA partly reflect the change in Swedish OR methodology from the early 1960s. The proportion of speeches at SORA's OR days concerned with public administration increased while topics concerning military applications diminished. OR days in the mid-1960s were also explicitly focused on systems analysis. The RAND connection was emphasized, as several of the speeches were delivered by FOA P analysts with close contacts with the United States.[52] The association's membership structure did not, however, change correspondingly. The preponderance of natural scientists continued into the 1970s. SORA also retained its industry-oriented position and remained an environment where OR in its more original mathematical form was practiced. This continued to be true even when the reputation of that kind of OR fell dramatically among planners and the general public in the 1970s. In 1974, SORA published a booklet containing a selection of recent applications of OR. They were all very traditional studies concerning optimization of industrial processes and similar topics. From the mid-1970s onward the membership and activities of SORA decreased. Today the association is not very active.

Systems Analysis in Sweden

From around the mid-1960s the term systems analysis (SA) started to be used in Swedish OR circles. The first systems analytical studies initiated in the 1960s differed in a number of ways from traditional OR. Per Molander has written a survey of systems studies in Sweden.[53] He holds that SA dealt largely with synthesizing new systems or part systems, something that OR did not do. SA problems, Molander says, were more dynamic and their time perspectives were longer. Furthermore, the goals for SA studies were, unlike in OR, only rarely well defined. SA studies also differed from early OR in that they often tended to avoid fixed systems demarcations. Moreover, the systems analysts worked with a higher general uncertainty than did the OR teams in the 1950s. These characteristics result partly from the extended time horizons in SA studies.

New academic disciplines were incorporated when OR was extended to systems analysis. Theoretical fields such as organizations theory and decision theory were included in the SA work of the 1960s. This was because the importance of considering the organizational frames within which a decision should be made was recognized by systems analysts both at FOA and in civilian sectors at this time. This, in turn, was a result of the growing concern that the results of OR work often appeared to have no impact on decision making within the organization where the studies had been performed. In other words, the results from OR work in some way "got lost" on the way up to the decision-making layers of the organization.

FOA P felt the need to establish courses and research in systems analysis and planning methodology. To this end a research group in planning theory was set up by FOA at the mathematics department at KTH in 1969. Lars Ingelstam, a young mathematics professor with previous experience of OR work at FOA P, was appointed head of this group. Ingelstam and his staff came to play an important role in the introduction of futures studies in Sweden, as will be outlined below.

In the United States, the more executive oriented policy analysis appeared at this time. Attempts to create an academic environment with such policy analytical features were made by FOA representatives in the early 1970s, but failed. The executive-oriented American policy studies do not appear to have fit into the Swedish academic organization, at least not at this time. Policy analysis, of the kind argued for by Yehezkel Dror and others, never became institutionalized in Sweden.[54] Instead a series of studies oriented toward a more general public appeared in the early 1970s.

Swedish Futures Studies

In the early 1970s a new process of change in the systems approach began in Sweden. Political, domestic, and international situations changed dramatically. Economic stagnation and the energy crisis replaced the unquestioned optimism of the 1960s, and the awareness of environmental problems and the situation in the developing countries grew, especially among young citizens. People with a past in the OR environment at FOA P, and with close ties to the Social Democratic Party, now started to argue for a broadening of the systems approach to long-term analysis of entire sectors of society, suggestions not very different from the studies that British OR radicals had proposed in the 1940s.

Swedish industry also showed a growing interest in long-range planning and futures studies. As early as 1969, IVA had presented a report suggesting a joint institute for futures studies between private industry and the government. The report was inspired by IVA's experiences as a client at the Hudson Institute under Herman Kahn. With the IVA report, a period of intense debate over the aims of, and control over, futures studies started in Sweden. The conflict reflects the general antagonistic political climate of the time.

The IVA proposal for a joint institute was rejected by the Social Democratic government. Instead, Prime Minister Olof Palme appointed a committee—with Alva Myrdal, a well-known promoter of the welfare state, as chairman. Lars Ingelstam, head of the FOA-financed group for planning theory, became secretary. The committee was given the task of investigating possible forms for Swedish futures research. It presented its report in the summer of 1972, suggesting state-financed forms for futures research and stressing that the complexity of the issue should not be neglected.[55] The committee stressed that "odd" and less well-established competence be utilized as part of the effort to avoid the creation of an isolated "science" with sole rights in futures studies, and opposed any form of technological determinism. It also emphasized the need for broad public participation and involvement in the work of futures studies. In 1973 a secretariat for futures studies was founded and was housed in the prime minister's office.

Lars Ingelstam was appointed to lead the secretariat. He recruited young scientists, some of whom were analysts from FOA P, to carry out futures studies. A general goal of the secretariat's work, Ingelstam writes, was "the raising of the forecast consciousness" of political decision makers, civil servants, and the general public.[56] Four initial studies were begun at the secretariat in the early 1970s: "Sweden's International Conditions," "Resources and Raw Materials," "Energy and Society," and "Working Life in the Future." One of the aims was to stimulate general public debate on these issues.

Many of the reports from this first generation of futures studies questioned the growth paradigm and led to heated public debates. Systems studies were for the first time in Sweden not only a concern for a small planning elite. The final report from the energy study was particularly influential. It outlined two contrasting energy scenarios, one based on renewable energy sources and one on nuclear energy. The report, entitled *Solar versus Nuclear,* became an important element in the debate preceding the 1980 Swedish referendum on nuclear power. It was published in numerous editions and also reached an international audience when translated into English.[57]

A second generation of studies started in 1978 on the themes "Care in Society," and "Sweden in a New World Order."[58] The third generation in the early 1980s concentrated on regional issues, human communication, and values and changes in values.

The reports from the secretariat were often controversial, not least for leading politicians. In particular, the energy report questioning the decision that Sweden should be highly dependent on nuclear power led to a weakening of the secretariat's political position. Gradually the secretariat was removed from the political center. This process started in 1975, when the secretariat was transferred from the prime minister's office to a much less prestigious address at the Ministry of Education.[59] Political support continued to weaken, and by the early 1980s the secretariat was nothing but a subdivision within one of the state research councils.[60]

The epoch when Swedish futures studies were conducted at the prime minister's office and were of top political interest was thus short. But the studies undoubtedly had contributed to a growing societal interest in planning issues.

Concluding Discussion

In the early 1960s, some dozen years after Blackett and Cramér first introduced the concept of operations research in Sweden, an organizational basis for this activity had been created. OR work was carried out on a regular basis in the defense sector, and also in some academic environments consulting for industry. A new paradigm for the study of industrial, military, and social issues was developing. Highly skilled persons had been recruited, people who soon occupied positions in various sectors in society and thereby gave the systems approach momentum.

The mid-1970s were the high-water mark of the systems approach in Sweden. Futures studies were carried out at a secretariat placed at the prime minister's office. The studies were largely conducted by people fostered in the systems approach tradition, and some of them had a considerable political impact. In the defense sector an elaborate planning system, developed by the analysts at FOA P, had been firmly established.

In this concluding discussion we will try to discern some characteristic patterns and factors in the development of the systems approach in Sweden. We will argue that three factors in Swedish society have been important for the way in which this approach was established and diffused in Sweden. Furthermore, we will underline two ways in which this approach was transformed from the 1940s to the 1970s. However, we want to emphasize the tentative character of our analysis. To be able to conclude what has been characteristic for the development in Sweden one would have to compare it with developments in other countries, but this has been outside the scope of this chapter.

The firm and persistent position of the systems approach in the defense sector is one salient pattern. Sweden's foreign policy of neutrality and nonalignment is an important factor for explaining this. When Sweden in the late 1940s failed to create a Nordic Defense Treaty and chose to stay outside NATO, there was a consensus among the political parties that the country had to have a strong military defense, including a domestic capacity for developing new, sophisticated weapon systems. In the mid-1950s, the director general of FOA argued that operations research was essential as a basis for choosing which future weapon systems to develop. This led to the creation of a permanent OR organization, with OR groups located at the Military Staffs. This OR organization became an important element in the country's military planning process in the 1960s and has remained so ever since.

Another characteristic is the small size of and the proximity within the Swedish "systems approach community." A close planning elite of operations analysts was formed in the 1950s and 1960s, and many of these met regularly at meetings of the Swedish Operations Research Association. This enabled a fast diffusion of new ideas and approaches within this community. An important reason for this proximity has been the geographical concentration of OR activities to the Stockholm area, which in turn has to do with Stockholm's role as capital and the localization of all military staffs and organizations like

FOA and IVA in the city. Furthermore, there were academic groups at the Stockholm University, the Royal Institute of Technology, and the Stockholm School of Economics that were able to respond to the growing demand for operations analysts.

A third characteristic in the Swedish history of the systems approach is the penetration of, and the political importance of, systems-based futures studies in the 1970s. This, we argue, was due to the Social Democratic Party's leading role in the steering of Swedish society at this time (in the early 1970s the social democrats could look back to some forty continuous years in office) together with the party's ambition to reconstruct society and build a solid welfare state. The social democrats' welfare ambitions and the systems approaches developed at FOA found each other. A result of that meeting was the broad, interdisciplinary futures studies of the 1970s.

The content of the systems approach has changed substantially from early OR in the 1950s, over systems analysis in the 1960s, and to futures studies in the 1970s and early 1980s. This process of change has been decisively influenced from abroad. In the first stage, these influences came from the United Kingdom, and the OR experiences there during World War II. Later, from the late 1950s onward, the inspiration came almost exclusively from the United States, primarily from the RAND Corporation, to which FOA had created contacts. Another reason for the broadening of the systems approach was that since the early 1960s FOA had gotten more deeply involved in the military planning process, which "forced" the analysts, in order to be credible, to broaden their approaches.

Not only the methodology but also the mode of work among analysts has gradually changed character. OR originally was focused on guidance for military executives, in other words, a secret activity taking place behind locked doors. Also on the civilian side, in industry, the analysts directed their energy toward the high-ranking executives, and their advice was, as a rule, secret.

In the early 1960s, an increasing openness was palpable among analysts, strangely enough starting in the military sector. This was partly due to the fact that OR had been incorporated into the military planning process, which was a political process, and as such had to be made public, at least partly. Furthermore, the nuclear weapons debates created a broader interest in defense issues within the electorate. Some analysts at this time chose to make results public, and got involved in the political debate over the issue. As a result, a series of booklets and paperbacks on military planning and on nuclear arming written by OR analysts appeared in the bookstores around the mid-1960s.

With the futures studies in the 1970s, an era of more programmatic openness started in Swedish systems thinking. This is not surprising, as the explicit purpose of these studies was to try to create a public debate on planning- and future-oriented issues. This aim was attained, and the systems approach had firmly taken the step out of the sealed rooms and into the Swedish general public.

The broadening in content, and the growing openness in the 1960s, together seem to have contributed to a gradual "rubbing out" of the specific character of the systems approach. In the 1980s and 1990s, conceptions such as OR, systems analysis, and futures studies are no longer widely used. It is true that within the military sector the OR organization and the whole planning system still exists, but it seems to have become more set in a fixed mold than was the case in earlier times. Against this background one could argue that the systems approach in Sweden is fading out. However, the trains of thought it contained have been incorporated into new academic disciplines and into computer software. Furthermore, in modern forms of organization much of the systems thinking elaborated during the 1950s and 1960s is taken for granted. What has happened though is that the systems approach community has dispersed and lost its identity, which is illustrated by the only slumbering existence of SORA and other similar associations.

However, as history often seems to be more circular than linear, perhaps it is soon the time for him or her to appear again: the young scientific genius, ready to provide the world with unimpeachable formulas of everything's future development.

Acknowledgments

We are grateful to the people we have interviewed for having given of their time. Furthermore, we want to thank the participants at the conference "The Spread of the Systems Approach" at the Dibner Institute in Cambridge in May 1996—and in particular our commentator Professor John Staudenmaier—as well as our colleagues at the Department for History of Science and Technology, at KTH, for valuable comments on earlier versions of the chapter. Research for this chapter was made possible by funding from the Swedish Research Council (FRN).

NOTES

1. According to interviews with C. G. Jennergren and G. Tidner. See also Dagens Nyheter, 8, 9, 10, and 11 December 1948.

2. At the War Archive in Stockholm C. G. Jennergren has left extensive material about OR in the military sector (Krigsarkivet. FOA: Ö.IV.d, box 1–12.) In the following this is referred to as "the Jennergren file." Furthermore, the archive of the Swedish Operations Research Association, SORA, has been used.

3. Rolf Björnerstedt, Jan-Robert Eklind, Nicolai Herlofsson, Lars Ingelstam, Carl Gustav Jennergren, Mårten Lagergren, Bengt Nagel, Brita Schwarz, Peter Steen, Torbjörn Thedéen, Gunnar Tidner.

4. Publications dealing specifically with OR introduction in Sweden are C. G. Jennergren, S. Schwarz, and O. Alvfeldt, eds., *Trends in Planning: A Collection of Essays from the Planning Department of the Swedish National Defence Research Institute* (Stockholm: FOA, 1977); B. Rapp, "Operations Research in Sweden," *OMEGA International Journal of Management Science* vol. 16, no. 3: 189–195, 1988; Per Molander, *Systemanalys i Sverige—en översikt* (Stockholm: Swedish Council for Systems Analysis, 1981); and Björn Wittrock, *Choosing Futures—Evaluating the Secretariat for Futures Studies* (Stockholm: Swedish Council for Planning and Coordination of Research, 1985).

5. A recent government commission has made clear that all Swedish military planning in the 1950s and 1960s concentrated solely on defense against attack from the Soviet Union, and there was considerable secret cooperation with NATO to enable support in case of an attack from the Soviet Union. See *Om kriget kommit . . . Förberedelser för mottagande av militärt bistånd 1949–1969*, SOU 1994:11.

6. Wilhelm Agrell, *Vetenskapen i försvarets tjänst* (Lund: Lund University Press, 1989), 240.

7. Bailey's lecture notes are kept at the War Archive, in the Jennergren file, box 11.

8. Ibid.

9. *FOA och kärnvapen*, FOA VET om försvarsforskning nr 8 (Stockholm: National Defense Research Institute, 1995), 78f.

10. Hulthén visited Blackett in Manchester in the summer of 1948 to discuss meson theory, according to an interview with Professor Nicolai Herlofsson. Herlofsson was a guest researcher at Blackett's department in Manchester in 1948, and later became a colleague of Hulthén at the Royal Institute in Stockholm.

11. Interview with Professor Bengt Nagel, who was the first "research technician" recruited by Hulthén. See also Kungl. Krigsvetenskapsakademiens Årsberättelse 1955, 224.

12. Interview with Thedéen. See also Kungl. Krigsvetenskapsakademiens Årsberättelse 1955, 223.

13. See Wilhelm Agrell, *Vetenskapen i försvarets tjänst* (Lund: Lund University Press, 1989); Hans Weinberger, "Physics in Uniform: The Swedish Institute of Military Physics, 1939–1945" in Svante Lindqvist, ed., *Center on the Periphery: Historical Aspects of 20th-Century Swedish Physics* (Canton, Mass.: Science History Publications, 1993), 141–163; and Ann Kathrine Littke and Olle Sundström, eds., *Försvarets forskningsanstalt 1945–1995* (Stockholm: National Defense Research Institute, 1995).

14. Kungl. Krigsvetenskapsakademiens Årsberättelse 1955, 222. In the 1950s and 1960s Sweden concentrated solely on defense against attack from the Soviet Union, and there was considerable secret cooperation with NATO to enable support in case of an attack from the Soviet Union. See *Om kriget kommit . . . Förberedelser för mottagande av militärt bistånd 1949–1969*, SOU 1994:11.

15. William Pansar, *Militärteknisk tidskrift*, no. 11–12, 1955, 1–4.

16. M. Lagergren, "Operations Research in Swedish Defence," in *Trends in Planning*, 110f. See also O. Krokstedt. *Förberedande utredning angående organ för operationsanalys vid försvaret*, mimeo, Stockholm, 1957.

17. Interview with C. G. Jennergren. See also C. G. Jennergren, "Operationsanalys och den rådgivande funktionen," in *Försvarets forskningsanstalt, 1945–1995* (Stockholm: FOA, 1995).

18. T. Magnusson, "Operationsanalys för försvaret" in *Militär teknisk tidskrift*, no. 2, 1957, 1–6.

19. Lagergren, 112ff.

20. Ibid.

21. Interview with Jennergren. See also Lagergren, 113f.

22. Interviews with Jennergren and Tidner.

23. B. Schwarz, "Programme Budgeting and/or Long-Range Planning," in *Trends in Planning*, 40ff. See also L. Grape & B-C. Ysander, *Säkerhetspolitik och försvarsplanering* (Stockholm: SNS, 1967).

24. See for example Agrell, 175; R. Björnerstedt, "Sverige i kärnvapenfrågan," *Försvar i nutid*, no. 5, 1965; *Svenska kärnvapenproblem*, R. Björnerstedt & L. Grape, eds. (Stockholm: Aldus/Bonnier, 1965); and *FOAs kärnvapenforskning*, 87ff.

25. G. A. Westring, "Strateger och räknemästare," Svenska Dagbladet 23/9 1960.

26. Interviews with Jennergren, Tidner, Schwarz, and Lagergren.

27. Interview with Jennergren.

28. Agrell, 243.

29. Hellman, 91f., and Agrell, 243.

30. Agrell, 108 and 133.

31. This is not only the case for OR. Industry, in Sweden and abroad, has a long tradition of copying the military in management issues.

32. An in-depth study of the Swedish rationalizations movement is found in Hans de Geer, *The Rationalization Movement in Sweden: Efficiency Programs and Social Responsibility in the Interwar Years* (Stockholm: SNS, 1978).

33. Bo Sundin writes that one of the issues that led to the formation of IVA was the conflict between scientifically oriented engineers with roots in the civil service tradition and the "modern" engineers who emphasized the role of the engineer as an industrial manager trained in economics, etc. Bo Sundin, *Ingenjörsvetenskapens tidevarv* (Umeå: Almqvist & Wiksell International, 1981), 63 ff.

34. Carl Tarass Sällfors, ed., *Handbok i industriell driftsekonomi och organisation* (Stockholm, 1945).

35. The first professor was the engineer Carl Tarras Sällfors.

36. Berglund et al., *Vad är operationsanalys?* (Stockholm: Aldus/Bonnier, 1965).

37. "Statistiska metoder i tekniken," lecture by Harald Cramér at IVA, 4 December 1947. Printed in IVA 1948:3.

38. In 1953, Harald Cramér gave a second lecture at IVA in which he stated that wartime OR had obtained industrial relevance "simply by exchanging terms such as weapons, attack and defense methods to machines and production methods." "Operationsforskning," lecture by Harald Cramér held at IVA, 27 November 1952. Printed in IVA 1953:2. The same arguments were used in both Britain and the United States. See Stephen P. Waring, "Management by Numbers: Operations Research and Management Science," in *Taylorism Transformed: Scientific Management Theory since 1945* (Chapel Hill, University of North Carolina Press, 1991), Rosenhead (1989), and M. Fortun and S. S. Schweber, "Scientists and the Legacy of World War II: The Case of Operations Research (OR)," *Social Studies of Science*, vol. 23 (1993): 595–642.

39. Bengt Magnusson, lecture at IVA, 27 November 1952. Printed in IVA 1953:2.

40. Nils Blomqvist, lecture at IVA, 27 November 1952. Printed in IVA 1953:2. The journals he referred to were *JORSA* (*Journal for Operations Research of America*) and the publication from *The English Operations Research Club*.

41. The invitation was signed by employees both from SFG and from Frenckner's group at the Stockholm School of Economics.

42. The first international OR conference was held in Oxford in 1957. Sweden sent the fourth biggest group, after England, the United States, and France.

43. Svenska operationsanalysföreningens jubileumsskrift 1979–02–26. SORA's definition of OR reads "the preparation of support for rational decisions by means of systematic, scientific methodology and, when feasible and purposeful, quantitative models."

44. Interviews with Jennergren and Thedéen.

45. Giandomenico Majone, "Systems Analysis: A Genetic Approach," in H. J. Miser and E. S. Quade, eds., *Handbook of Systems Analysis* (Chichester: John Wiley, 1985).

46. Ida R. Hoos, *Systems Analysis in Public Policy: A Critique* (Berkeley: University of California Press, 1972).

47. In 1945 the Association of Scientific Workers (AScW) presented the report *Research into Consumers Needs*. Several of the authors of the report were operational researchers.

48. J. Rosenhead, "Operational Research at the Crossroads: Cecil Gordon and the Development of Post-War OR," *Journal of the Operational Research Society* (*JORSA*), vol. 40 (1989): 3–28.

49. Twenty percent of British industry was nationalized after the war. In Sweden, attempts in the late 1940s to nationalize only small industrial sectors failed for political reasons. See Thomas Jonter, *Socialiseringen som kom av sig: Sverige oljan och USA:s planer på en ny ekonomisk världsordning 1945–1949* (Stockholm: Carlsson, 1995).

50. T. Dalenius, *Operationsanalys inom sjukvården* (Stockholm: SJURA, 1966).

51. One of its emloyees in the early 1970s was Mårten Lagergren, operations analyst and head of FOA P's OR division in the 1960s. In the late 1970s and early 1980s Lagergren led the study "Care in Society" at the Secretariat for Futures Studies.

52. SORA, newsletter 29, 1969.

53. Per Molander, *Systemanalys i Sverige—en översikt* (Stockholm: FRN, 1981).

54. In 1982 Dror was in Sweden and attended a SORA meeting where he presented the paper "Policy Analysis for Top Executives."

55. *Att välja framtid: Ett underlag för diskussion och överväganden om framtidsstudier i Sverige*, Stockholm, SOU 1972:59.

56. Lars Ingelstam, "Forecast for Political Decisions," in Peter R. Baehr and Björn Wittrock, eds., *Policy Analysis and Policy Innovation* (London: Sage, 1981).

57. J. Lönnroth et al., *Solar versus Nuclear: Choosing Energy Futures* (Oxford: Pergamon Press, 1980).

58. Björn Wittrock, "Futures Studies without a Planning Subject: The Swedish Secretariat for Futures Studies," in Peter R. Baehr and Björn Wittrock, eds., *Policy Analysis and Policy Innovation* (London: Sage, 1981).

59. H. Glimmel and S. Laestadius, "Swedish Futures Studies in Transition," *Futures*, July 1987, 635–680.

60. Wittrock, 1981.

13

The International Institute for Applied Systems Analysis, the TAP Project, and the RAINS Model

Harvey Brooks and Alan McDonald

The symposium on which this volume is based included a panel discussion on IIASA, the International Institute for Applied Systems Analysis. The panel was chaired by Harvey Brooks, who has been involved with IIASA since the institute's inception in 1972 and chaired the U.S. Committee for IIASA from 1982 to 1989. Other panelists were Howard Raiffa and Roger Levien (IIASA's first two directors), William C. Clark (former IIASA project leader and current member of the U.S. Committee for IIASA), and Alan McDonald (executive director of the U.S. Committee for IIASA). In chapter 14 Roger Levien describes his experience in managing systems analysis at RAND and IIASA. Here we will offer highlights from the other panel presentations.

Early History: 1966–1972

What we know today as the International Institute for Applied Systems Analysis came out of an initiative by President Lyndon Johnson in 1966. In December of that year Johnson asked McGeorge Bundy to explore with the Soviet Union, U.S. allies, and Eastern Europeans the possibility of establishing an institute to study problems common to advanced industrial societies. Bundy had recently resigned as Johnson's special assistant for international security affairs to become president of the Ford Foundation. Johnson's initiative had two motivations. First, it was one of a number of "bridge-building" initiatives intended to reduce East-West tensions by increasing contacts through cooperative constructive enterprises of common interest to the United States, the Soviet Union, and their respective allies. But it was also intended to benefit all countries by expanding knowledge and use of the new techniques of systems analysis. Hitherto these had been used mostly, with considerable success, for military purposes. The institute would thus both advance systems analysis research and help disseminate systems analysis expertise throughout the industrialized world to the general benefit of all societies. In the decision memo for the president that triggered Bundy's assignment, Deputy Special Assistant for National Security Affairs Francis Bator wrote,

> Those of us who have worked on the idea have in mind an institution based on the proposition that *all* advanced economies—capitalist, socialist, communist—share the problem of efficiently managing large programs and enterprises: factories and cities, subway systems and air traffic, hospitals and water pollution. There is great demand—in Russia and Yugoslavia as well as the UK and Germany —for the new techniques of management designed to cope with these problems.

Bundy consulted initially with U.S. and Western academics, business leaders, executive branch officials, congressmen, and ambassadors from the Soviet Union and several U.S. allies, and he commissioned a study by RAND on the feasibility of such an institute.[1] The results of the consultations and of the study proved sufficiently promising that in May he traveled to London, Moscow, Paris, Bonn, and Rome, together with Eugene Staples of the Ford Foundation, and, in Moscow, Carl Kaysen, director of the Institute for Advanced Study at Princeton University. On all their stops they found considerable interest in increasing international cooperation in the study of large common problems and strong support for Johnson's initiative.

In the Soviet Union Bundy's opposite number was Jermen Gvishiani, deputy chairman of the State Committee for Science and Technology and son-in-law of Soviet Premier Aleksei Kosygin. In the United Kingdom and Italy the key players were, respectively, Sir Solly Zuckerman, chief scientific adviser to Prime Minister Harold Wilson, and Aurelio Peccei, founder of the Club of Rome. In France it was Pierre Massé, president of Electricité de France. All shared the U.S. interests of reducing East-West tension and spreading the techniques of systems analysis, but each negotiator also had separate interests of his own. The Soviets had a particular interest in incorporating modern techniques of forecasting in their planning process. They were enamored of large-scale systems, believing that systems analytic successes in military applications would be transferable to the increasingly complex and interdependent problems of modern civil societies. As IIASA's first director, Howard Raiffa, later put it, the Soviet leaders hoped that with these methods and a big enough computer they could finally make Marxist economics work as effectively as envisioned by its original proponents. Nonetheless, they also preferred that the institute focus on theoretical research rather than on applications that might examine Soviet experiences in too cold and unfiltered a light. Solly Zuckerman, on the other hand, preferred an institute focused wholly on applications. He was also very skeptical of the global modeling championed by Aurelio Peccei's Club of Rome and featured in *Limits to Growth*.[2] In addition to Peccei, both Gvishiani and Rennie Whitehead, who later represented Canada in the negotiations, were members of the Club of Rome. But Zuckerman had an ally in the United States, where the National Academy of Sciences (NAS) shared his dim view of global modeling.

In February 1968 Zuckerman invited representatives from the United States, the Soviet Union, France, the Federal Republic of Germany, and Italy to a meeting at the University of Sussex. As had been the case in the initial discussions, participants would take part as individuals rather than as representatives of their respective governments, although it was generally understood that all were in close contact with their governments. However, the day before the meeting Gvishiani pulled out. His formal reason was a flare-up of tensions in Berlin, but the real reason, as he conveyed several months later to Peccei and Zuckerman, was that Zuckerman had not matched his invitation to the Federal Republic of Germany with one to the German Democratic Republic.

Gvishiani and the Soviets remained seriously interested, but the "German problem" was equally serious. The U.S. and FRG governments would simply not provide funding for an institute in which the German Democratic Republic was a member. Conversely, the Soviet Union had a vested interest in equal treatment for the German Democratic Republic as a sovereign nation in its own right. The eventual solution was to make the institute nongovernmental. Rather than the United States and the Soviet Union being members, for example, the National Academy of Sciences and the Soviet Academy of Sciences would be members. More importantly the FRG and GDR members could be the Max Planck Institute and the Academy of Sciences of the GDR. Neither German government would have to recognize the other.

The next major step came in July 1969 at a meeting of Bundy, Peccei, Zuckerman, and Gvishiani in Moscow. In addition to a written aide-mémoire outlining a governance structure for the institute, the meeting produced a key oral understanding. The institute's council chairman would be from the Soviet Union, the director from the

United States, the location would be in the United Kingdom, and, at the suggestion of Gvishiani, the sole official language would be English. The provision that the institute be located in the United Kingdom did not survive the subsequent British expulsion of hundreds of Soviet diplomats for spying. But the other three provisions held substantially longer. English is still IIASA's sole official language. The director has always been from the United States, except for 1981–1984, when a Canadian, Buzz Holling, held the post. And it was not until June 1997, as IIASA neared its twenty-fifth anniversary, that the IIASA Council was chaired by anyone other than the Soviet or, after 1990, Russian representative.

The year 1969 also marked the beginning of Richard Nixon's presidency. Philip Handler, president of the NAS, took over the lead U.S. role from McGeorge Bundy. Through the next three years until IIASA's formal founding in October 1972 Handler and Gvishiani wrestled with three main disagreements—location, the relative authority of the council and director, and the relative importance of methodological and applied research. The Soviets wanted to locate the institute in Vienna. The Americans preferred Fontainebleau in France. The Soviets wanted a relatively powerful council. The Americans were concerned that the director have the flexibility and authority needed to run a first-class research operation. The Soviets were hesitant about research moving beyond methodology. Westerners believed that, to be of value, methodological research on systems analysis had to be continually tested by experience in actual application, and therefore applications were essential if the institute was to be worthwhile.

The location issue was formally resolved in favor of Vienna, by a split vote, less than a month before IIASA's founding. The deciding factor against France was that the French required an intergovernmental agreement. This the United States and FRG considered impossible as long as there was a member organization from the GDR. On the second issue, ultimate authority for hiring and firing ended up with the director, although a member organization could require that any recruits from its country come from a list it prepared. The only quotas in the final charter are that two-thirds of research scholars must come from countries with member organizations and that "[e]ach member institution shall have the right to have at least one research scholar selected from among its nominees." The third issue, the balance between methodological and applied research, has been one of continuing discussion and evolution. As the Soviet national member organization (NMO), and indeed other members, gained experience with IIASA, the range of acceptable applications expanded. A project on demography, for example, was originally vetoed by the Soviet NMO. A year later, however, a Soviet review criticized IIASA for lacking just such a project, and the Soviet NMO became more accommodating. In the end, one of IIASA's considerable success stories resulted from a collaboration between an American demographer, James Vaupel, and a Soviet mathematician, Anatoli Yashin. A second example concerned research on negotiations. Initially there was real reluctance to touch topics close to existing international negotiations. In 1974 IIASA's Director Howard Raiffa proposed a summer workshop on the Law of the Sea negotiations in which workshop participants could explore models analyzing issues addressed by the negotiations and generally share ideas and analyses in an informal setting. The proposal was rejected by IIASA's council as too "political." In contrast, twenty years later the council's strong support for research addressing international negotiations on population and development, global warming, and other environmental issues was almost taken for granted.

During the final two years of negotiations the parties also finally settled on the eventual name. Many had been proposed and used at various stages in the discussions. Each word held the potential for disagreement. Should it be the Center for Study of the Common Problems of Industrialized Societies, or would it be better to refer to advanced societies? Should the word "international" be included? What constitutes a society? Should it be a center or an institute? Should center be spelled "center" or "centre"? Particularly problematic were words like "cybernetics" or "operations research" that carried connotations that one or another of the parties considered at odds with its view of the institute. "Applied systems analysis" was a phrase Howard Raiffa plagiarized from the title of a book he had recently written. No one had preconceived notions of its meaning, and as a new phrase for a unique institute it held the right mix of specificity and ambiguity to be acceptable to all.

The formal founding of IIASA finally came on 3–4 October 1972 in London. There were founding members from twelve countries: Bulgaria, Canada, Czechoslovakia, France, FRG, GDR, Italy, Japan, Poland, the United Kingdom, the United States, and the Soviet Union. Jermen Gvishiani was elected the first chairman of the governing council, and Howard Raiffa was appointed the first director. The founding twelve NMOs were joined within a few years by members from Austria, Finland, Hungary, the Netherlands, and Sweden.

IIASA's Research

It is now a quarter century since IIASA's founding and more than thirty years since Johnson's original assignment to McGeorge Bundy. To date the institute has succeeded against both its objectives: bridge-building across the East-West political divide and advancing both the theory and application of systems analysis.[3]

IIASA's research success has come more from how the institute assembles from the bottom up the components of a successful analysis than from any definitive top-down formula for carrying out such an analysis—or even for defining what "applied systems analysis" really means. According to IIASA's strategic plan,

> The Institute's strategic goal will be to conduct international and interdisciplinary scientific studies to provide timely and relevant information and options, addressing critical issues of global environmental, economic, and social change, for the benefit of the public, the scientific community, and national and international institutions.

But even IIASA's staff would be hard pressed to provide a concise, consistent definition of "applied systems analysis" beyond defining it simply as "what IIASA does." So what does IIASA do?

First, IIASA brings people from different disciplines and countries together for sustained research under one roof. In this it is almost unique and distinctly different from most international exchanges and workshops that bring groups together for a short time. This fosters specialized interdisciplinary and international connections that are more likely to create and consolidate real conceptual breakthroughs. William Clark cited several examples in his presentation at the conference. One such unexpected successful connection was between ecologists intent on descriptive analyses of pest infestations and operations researchers working on optimization and dynamic programming techniques. The result was both new analytic methods and fundamentally new ways of looking at and reformulating practical real-world policies, including a shift in policy focus from spraying bugs to cutting trees. A second example is connections prompted by Tjalling Koopmans, a Nobel laureate and researcher at IIASA in the 1970s. Koopmans noticed commonalties in the instabilities encountered by IIASA researchers studying populations, climate, organic chemistry reaction patterns, and macroeconomics. He brought them all together in several workshops that eventually led to significant advances in nonlinear dynamics. A third example is one we will return to in more detail below, IIASA's RAINS (Regional Acidification Information and Simulation) model.

Second, IIASA seeks to increase the comprehensiveness of existing analyses. In the early 1980s for example, nearly every country was busy producing national models of its energy supplies and demands. Gaps between the two indicated what a country could expect to import. IIASA's study was the first to ask if all these expected imports might not add up to more than the available exports. What were the prospective *global* gaps between energy supply and demand, and what might be done to close them? Twenty-five years later IIASA is exploring the possibility of similar transnational gaps relating to social security. Population aging in industrialized countries is prompting numerous national studies analyzing current and alternative national pension policies. National policies, however, will have international repercussions through their effects on multinational employers, migration, and monetary flows. IIASA is exploring how best to provide a global perspective that contributes today to national pension studies what its energy research contributed in the 1970s to national energy studies. Increasing the comprehensiveness of existing analyses does not always mean adding a global perspective. In IIASA's current energy research, for example, it means linking greenhouse gas emissions calculated by energy models to climate models, RAINS, and models of international agricultural production, consumption, and trade.

Third, IIASA's most effective projects have simplified complexity while maintaining credibility with both scientists and decision makers. For an analysis to be used it must be understandable, and it helps greatly if it can be presented in simple, clear terms. The objective of simplification can be at odds with the second feature of IIASA's approach listed above, increasing comprehensiveness. The challenge is to do both, thereby providing relatively simple aggregated models in the end that rest on underlying detailed comprehensiveness analyses. Sometimes the detailed models are developed at IIASA. Sometimes they are adapted or incorporated whole from other institutions. In both cases, it is important that the institute's research and applications maintain credibility and support from the scientific community through peer-reviewed publications and direct reviews at workshops and conferences.

Fourth, IIASA engages decision makers throughout its analyses. This has two effects. It increases the decision makers' understanding and ownership of the analysis, and that makes it more likely that the analytic results will influence eventual decisions. Second, it allows IIASA to respond early and often to the needs and preferences of decision makers. The result is an analysis more likely to address the issues that matter most to decision makers, and to offer answers to the questions they are most likely to ask.

Fifth, IIASA builds networks extending beyond its walls that incorporate new mixtures of disciplines. The international and interdisciplinary connections fostered under the first feature of IIASA's approach listed above are now complemented by a network of alumni approaching 2,000 researchers. And IIASA actively builds networks well beyond its own alumni. In the 1970s Howard Raiffa navigated between global modeling's strong advocates (e.g., Aurelio Peccei) and its strong detractors (e.g., Solly Zuckerman) by establishing IIASA as a site for regular presentations and reviews of global models. IIASA would not build its own global model but would host regular conferences to review and document the research of others. Global modelers liked the forum this provided. The critics liked the requirement for documentation and the opportunity for skeptical review. IIASA's network involved both. Since the 1981 publication of IIASA's initial global energy study, its subsequent research in the field has always included annual meetings of the International Energy Workshop (IEW). The IEW is run jointly by IIASA and Stanford University and annually assembles approximately fifty energy research groups from around the world to compare their models, their results, and their plans. In the field of natural resources and the environment essentially every international science program in the past two decades has been influenced by the research and networks begun at IIASA in 1972. Ideas and connections initiated at that time through IIASA have persisted and expanded and now thoroughly permeate international efforts such as the International Geosphere-Biosphere Programme (IGBP), the International Human Dimensions of Global Change Programme (IHDP), and the Intergovernmental Panel on Climate Change (IPCC).

The Transboundary Air Pollution Project and RAINS

Our presentation of these five characteristics of successful systems analyses at IIASA does not mean every project in IIASA's first quarter century has scored an A-plus on all five features. Rather, those that have been most successful have better incorporated these features in their approaches than have less successful projects. One of IIASA's most successful projects and best examples of all five characteristics is the Transboundary Air Pollution (TAP) Project, which began in 1983 and developed the RAINS model of the impact of acidification in Europe. The TAP Project and RAINS exhibit each of the characteristics mentioned above, particularly simplification and maintenance of credibility in the eyes of both scientists and political decision makers; early and continual involvement of policy makers and scientists together in the formulation of key questions and the periodic review of progress; and the building of scientific and political networks beyond IIASA's walls capable of taking on a life of their own, continuing the work, and providing critical data inputs to the continued improvement and updating of the models developed at IIASA.

The origins of the TAP Project lie in the Convention on Long-Range Transboundary Air Pollution (LRTAP), which was adopted in Geneva on 13 November 1979 and entered into force on 16 March 1983.[4] By May 1993 its membership had grown to thirty-nine parties, including the European Union. In addition, EMEP (Cooperative Programme for the Monitoring and Evaluation of Air Pollution in Europe) had been established as early as 1977 under the auspices of UN/ECE (Economic Commission for Europe) with the cooperation of WMO (World Meteorological Organization) and, initially, UNEP (United Nations Environment Program). The creation of the IIASA RAINS model[5] began shortly after LRTAP came into force as a means for providing scientific support for negotiations under the LRTAP Convention. By the late 1980s it began to play a central role in the negotiations. As stated in a 1989 report from a task force under the Executive Body of the Convention:[6]

> An integrated assessment model that can assist in cost-effectiveness analysis is now available.... The Task Force ... recommends that the RAINS model be used by the Parties to the Convention, the Executive Body, and the various subsidiary bodies.

The model was to estimate environmental impacts for all of Europe, including the European part of the Soviet Union, with a resolution of 150 km × 150 km. Simulations with the model were to extend back to 1960 for historical perspective, and forward to 2040 to insure that long-term consequences of different control policies would be adequately taken into account. Owing to the large spatial coverage and time horizons involved, the time steps used in the calculations had to be rather large (seasonal or annual). Both the long time horizon and the large spatial grid were necessary because of the sensitivity to short-term local variability in meteorological conditions of the atmospheric transport model used to convert pollutant emissions into environmental impacts. The validity of this averaging was subsequently checked by using the model to reproduce historical trends in pollutant deposition from 1960 to the time of application of the model, with sufficient agreement to provide reasonable confidence in using the model to make projections into the future with the same assumptions.

The RAINS model was subdivided into six submodels for each pollutant to be studied, as follows:[7]

1. Pollutant emissions and costs of reducing them
2. Atmospheric transport and deposition
3. Soil acidification
4. Lake acidification
5. Groundwater sensitivity to acidification
6. Forest impact of pollutants (indirect impact of soil acidification on forest health, and the direct effect on trees from exposure to gaseous forms of pollutants)

In the initial use of the model the only pollutant considered was SO_2. Later, after 1986, NO_x and NH_3 (ammonia) were added in order to take into account active forms of nitrogen as well as sulfur. More recently tropospheric ozone and emissions of its precursor, volatile organic compounds (VOCs), have also been added.

For (1) it was necessary to have energy projections for each country and their emissions under various assumptions regarding installation of technologies for reducing emissions. For this purpose so-called energy pathways were developed from official projections of each of the participating governments for twelve energy sources and six economic sectors from 1980 to 2000. These projections were taken as givens for most of the scenarios computed. However, the project later developed two alternate energy "pathways," one a "high conservation" pathway and the other a high natural gas pathway. The pollution mitigation technologies could be applied to these pathways as well as to the original ones. The conservation pathway generally contemplated phasing out of nuclear power as well as emphasis on the maximum feasible use of renewable energy sources in addition to hydropower. However, it was carried out for only ten of the twenty-seven countries, which accounted for only one sixth of the total energy production of the twenty-seven countries.

In the case of sulfur, five "options" were considered in principle for reducing emissions. These were energy conservation, fuel substitution, use of low sulfur fuels or desulferized fuels, desulferization during the combustion process, and desulferization of the flue gases after combustion.[8] Only the rates of installation of the last three of these were taken as variables that could be adjusted to control emissions in the future. A set of costs for installing these technologies that was compatible across countries was developed. However, as will be discussed later, the first two options were not treated as variables, but were simply taken from the official projections provided by governments.

The second set of submodels after energy pathways and emission reduction options and their costs dealt with atmospheric transport processes and resulted in transfer matrices based on the EMEP model.[9] The emissions derived from the first set of submodels were used as input and transformed into deposition patterns for all the countries including the European parts of Russia. The output of these submodels was then offered as a set of options for the user, including maps with isolines for sulfur and nitrogen deposition, colored maps with deposition patterns subdivided into classes of deposition, and tables showing the source of the calculated deposition at a particular point in Europe designated by latitude and longitude. With these output displays comparisons among scenarios and emission abatement strategies could be carried out.

The remaining four submodels for soils, lakes, groundwater, and forests were heavily dependent on knowledge and data provided by a number of collaborating institutions and experts in various parts of Europe. For example, one of the most complicated impact questions arose in the case of forests. Here the "SO_2 forest impact model is based on empirical data of forest dieback from Czechoslovakia's Erzgebirge Mountain region ... damage to trees is assumed to occur if the accumulated dose exceeds a threshold level, which depends strongly on climate ... Temperature is used in the model as an indicator of climate stress ... threshold levels and doses in the model are decreased as climate stress increases ... This method provides a consistent way to take into account the different levels of climate stress experienced by trees at different elevations and latitudes in Europe."[10]

The scenario evaluation was done within the framework of the LRTAP Convention, and has been established as an official international collaboration among the parties to the convention. This collaboration has also included financial support from a dozen parties.

Moreover, through review meetings in which scientists from all over the world participated, it was possible to convince all parties of the quality of the EMEP model. This enhanced RAINS's credibility as did further regular review meetings at IIASA. The meetings at IIASA covered each of the submodels one by one, and three times, larger meetings were convened to review the model as a whole, particularly the linkages between the submodels, and to discuss the problems with these linkages. These integrated review meetings also provided a forum in which policy advisers to governments could appraise the state of the model and assess its limitations and help formulate questions they deemed most important for the models to address from a policy standpoint.[11] Assessment was generally done in a nonconfrontational, cooperative fashion where the credibility of information derived from self-reporting by individual participating governments was taken for granted.[12]

The original goal in designing the RAINS model had been that a policy maker would sit at a computer monitor and operate the model during actual international negotiations. Although this idea was abandoned as impractical at an early stage, user friendliness remained a very important design criterion, and the models were implemented for personal computers so that alternative scenarios could be tested in the presence of policy makers more or less in real time so as to influence discussions. At general review meetings in 1985 and 1986 representatives from many ministries in Europe and North America were briefed in depth on RAINS and its potential uses, and they in turn offered valuable advice about the kind of questions they wanted the model to address. It was because of their input that the decision was taken to extend considerably the first RAINS submodel (dealing with energy scenarios and abatement strategies). The participants also pointed out that the RAINS model should include all the relevant pollutants. This advice led IIASA to implement submodels for NO_x and NH_3 (emissions and cost calculations). More recently emissions of VOC and the formation of tropospheric ozone has also been included in the model in response to the evolving agenda of governments and their negotiators.

Two basic ways of using the RAINS model have been developed: scenario analysis and optimization analysis.[13] To conduct a scenario analysis, a user begins by specifying an energy pathway and a control strategy. The implications of these inputs can then be examined. The user has the option of examining output from any of the submodels, for example, sulfur emissions in a particular country or group of countries, costs of sulfur control for each country, sulfur deposition or SO_2 concentration at different locations in Europe or mapped for all of Europe, or maps showing the time history of soil acidification, lake acidification, or risk to forest area from exposure to SO_2. This can be an iterative process, in which the user normally examines an output implied by one set of inputs and then, based on a subjective judgment, selects an alternative control strategy for comparison with the previously selected strategies.

Alternatively, in optimization analysis, the user can start with goals of environmental protection and/or economic constraints and work the model backward as it were to determine a country by country strategy in Europe to accomplish these goals, using a single objective, linear programming approach which can accomplish one of the following tasks:

- Given a fixed upper limit on expenditures for emissions reductions, determine where the most sulfur can be removed from emissions.
- Given an environmental target (sulfur concentration or deposition) in a specified region of Europe, determine where the minimum amount of emissions should be removed to meet the target.
- Given an environmental target, determine where emissions should be reduced to minimize the cost of removal and still meet the target.

As indicated previously, fuel switching and conservation have not yet been treated as variables in the optimization process. This has meant that trade-offs between, say, energy conservation and switching to nuclear from coal for electricity generation as opposed to installing desulferizing equipment were not considered as "options" on a country by country basis in the optimization process. In the long run, of course, this is a serious limitation in the applications of the model so far used. The justification for this restriction was given in the following terms:[14]

> Although fuel switching is often an effective way to reduce emissions, it may be limited for historical, institutional, or political reasons (e.g., N. E. Hay, ed., *Natural Gas Application for Air Pollution Control*, American Gas Association, Liburn, GA, 1986). Because of these limitations, which are specific to a country or a site, it is extremely difficult to derive reliable cost estimates for substitution strategies at the European level. In addition, even a rough estimate must account for the costs of changing the infrastructure to the new fuel demand conditions; this would require extensive information on the capacities and age structure of the European energy system ...

> because the costs of fuel substitution policies might have a large effect on the economy (e.g., the trade balance), a factor that is not included within ENEM, the costs of fuel substitution have been excluded from ENEM.

Furthermore, these costs and intangible considerations will be heavily dependent on future political developments, such as the formation of and membership in the European Union, that might make trade considerations less salient, and the adoption of a common currency. This could prove to be a very serious limitation on the use of RAINS, since the cost variables that are omitted are potentially larger in their impact on emissions and environmental impact than those that are considered in the other three mitigation options. This is particularly heavily affected by nuclear power projections (and to a lesser extent renewables), since in some cases the phasing down of nuclear power will considerably increase the amount of sulfur control that has to be used as well as affecting how it needs to be allocated among countries. On the other hand, with greater integration of the European economies, and improved data collection on power plants, there is no reason that fuel switching and conservation cannot eventually be incorporated into the optimization process, and this is currently being considered.

Even with this serious limitation, however, RAINS can show that dramatic improvements in cost-effectiveness can be achieved by allowing the percentage sulfur emission reduction required to vary among countries. As one specific example, we can consider the following:

In 1991 the estimated cost of Current Reduction Plans of European parties to the LRTAP Convention was about DM 12 billion per year. One possibility was that each country would invest its committed portion of this sum within its own borders. Alternatively each could put its share into a common fund to be redistributed among the countries in such a way as to achieve the greatest level of environmental improvement throughout Europe (defined as the lowest attainable level of sulfur deposition in Europe as a whole for a given financial investment). It was estimated that for the same financial investment this alternative would reduce the peak deposition in Central Europe by roughly 40 percent in comparison with Current Reduction Plans by directing money to those countries that are the major contributors to the high deposition levels and can reduce their emissions most cheaply. The result would be a decrease in the area of forest soil in Central Europe with a $pH < 4.0$ from 17 percent to 4 percent between 1995 and 2000.[15]

These principles involved in the optimization procedure developed in RAINS have now been embodied in a formal agreement. The most recent regulatory step is the revised sulfur protocol that was signed in Oslo in June 1994. It is both binding and specific, including both specified targets and timetables. The key feature of the agreement is that these elements are combined with flexibility on the basis of so-called critical loads.[16] The crux of this approach is that emission reductions should be set based on the varying effects of air pollutants, and the variations in abatement costs among countries, rather than by choosing an equal percentage reduction target for all countries involved.[17] It is probably still too early to foresee how well the goals of the 1994 protocol will be met. In this respect, it is well to keep in mind how difficult it is in practice to separate the effects of formal regulatory agreements from other factors, such as current political mindsets and well-publicized anecdotes of dubious scientific validity but dramatic public impact such as the "Waldsterben" crisis of 1981–1982 in Germany, which served to focus public attention and political energy on the control of acid rain.[18] The situation has been nicely summarized by Wettestad in the following terms:

> Summing up: a regulatory development may be discerned, with regulations gradually becoming more binding, specific and more fine-tuned to ecological and economic variations among the countries. However, the impact implications of this regulatory development are quite tricky. It seems that substantial emissions reduction took place at a time before regulations became more specific and binding, and the main part of the more recent emissions reductions probably has nothing to do with the international regulations and/or their design. The possible "softer," more diffuse "tote-board"[19] effects remain to be substantiated.

Nonetheless, however well the parties to the second sulfur protocol do eventually in meeting their commitments, and however much of any success is attributed by future historians to the protocol itself, the TAP Project is a good example of how IIASA can carry out systems analysis first to develop new valuable policy insights, and second to encourage the incorporation of these insights in actual policies.

EVOLUTION AND RESILIENCY

Again, the presentation above of the five characteristics of successful systems analyses at IIASA does not mean every project in IIASA's first quarter century has scored an A-plus on all five features. Rather, those that have been most successful, such as the TAP Project, have better incorporated these features in their approaches than have less successful projects. This, rather than any top-down recipe, we believe has been responsible for IIASA's continuing success over twenty-five years, even as issues and expectations in the field of systems analysis have changed.

Two examples of such changes were presented by the panel. Clark cited IIASA's ecological research. At the time of IIASA's founding the focus was on methods and applications to maximize yields from various natural resource populations. IIASA had a number of successes in this area, advancing the field by introducing both optimization techniques as discussed above and adaptive resource management methods.[20] By the mid 1980s attention had turned to issues of global environmental change and transboundary pollution, of which the RAINS model described above was the centerpiece. Another example is the institute's work on water pollution and climate change.[21] In the 1990s IIASA's ecological research has expanded to include human behavior as an integral part of the ecosystem, and this has led to the incorporation of a much larger social science component with recruitment of more social scientists and the extension of the institute's networks into the international social science community. IIASA's recent project on the Implementation and Effectiveness of International Environmental Commitments is a good example of this effort.

For his example of changes in systems analysis and IIASA over twenty-five years Alan McDonald turned to the field of energy. In 1975 as a young engineer at General Electric's Fast Breeder Reactor Department in Sunnyvale, California, McDonald had been given the task of determining safety criteria for a design known as the prototype large breeder reactor (PLBR). He would calculate how safe the reactors needed to be, and design engineers would then design the PLBR to these specifications. The idea that systems analysis could provide a definitive, unique answer to the question of how safe a nuclear reactor should be seems hopelessly naïve today, but it was not discouraged as naïve at the time. Although the search for such a definitive safety standard was ultimately unsuccessful (as was the PLBR), what is significant is that the best and most useful work at the time was coming out of IIASA. The institute was leading the world in exactly the research for which McDonald had turned to systems analysis. At the time IIASA was only three years old.

Today setting safety standards for LULUs (locally undesirable land uses) such as nuclear reactors is more often seen as a negotiation among multiple parties with different information, expectations, and preferences. In such a negotiation there is no single decisive fairness principle that is theoretically and practically perfect for all situations. Rather there are a number of fairness principles that in practice can serve as focal points around which negotiators can coordinate their expectations.[22] In their simplest forms fairness principles fall into three categories: parity, proportionality, and priority. Many can be characterized by common phrases such as "split the difference," "first come, first served," "take turns," and "returns proportional to investments." As parties gain familiarity and trust with each other, they can overlay combinations of simple fairness principles to create joint gains that can make all parties better off. The process of calculating and agreeing to such mutually beneficial but more complicated combinations of fairness principles can be greatly facilitated by joint cooperative research. Many of IIASA's more recent projects play exactly this role. The TAP Project, and its role in shifting negotiators from an agreement based on equal-percentage reductions by all countries to a less symmetrical but more cost-effective agreement, is an excellent example.

EPILOGUE

Even as systems analysis and the nature of IIASA's contributions have evolved over the last twenty-five years, so have the politics surrounding the institute. With President Reagan U.S. policy toward the Soviet Union shifted from bridge-building to isolation and pressure. The White House terminated government support for U.S. membership in IIASA at the end of 1982 and pressured the NAS into withdrawing as the U.S. NMO. The U.S. membership was transferred to the American Academy of Arts and Sciences in Cambridge, Massachusetts. Through the remainder of the 1980s the American Academy raised private funds, gradually built congressional support, restored partial National Science Foundation (NSF) funding, and, by the end of 1989, through the crucial intervention of Allan Bromley as well as several other members of the Bush administration, backed by support in the Congress, persuaded the president to restore White House backing. Nonetheless, the weakened U.S. participation halted and slightly reversed what had been a steady increase in IIASA's research budget through 1982.

With the end of the Cold War, IIASA's members negotiated a new strategic plan for the institute, focused on global change including technological change, economic change, and environmental change. As noted above the institute is very much involved in particularly the

IPCC, IGBP, and IHDP as well as providing research leadership in the areas of transboundary air pollution, demography, and energy policy. Political challenges remain, although they have changed since the 1970s and 1980s. Where the United States was an unreliable financial partner during most of the 1980s, the economic difficulties in Russia now make its annual contributions the largest recurring financial question mark. Where support for IIASA in the 1970s went against Cold War conventions, funding for IIASA must now contend with the fashion for cutting international research budgets not only in the United States, but in some other previously strongly supportive countries as well. Throughout IIASA's basic strategy has remained unchanged—to provide the best that systems analysis has to offer even as the field and our expectations continue to evolve.

Notes

1. Roger Levien and Sid Winter, "International Center Proposal," The RAND Corporation, internal memorandum, 11 April 1967.

2. Donella H. Meadows, Dennis L. Meadows, Jørgen Randers, and William W. Behrens III, *The Limits to Growth* (New York: Universe Books, 1972).

3. Alan McDonald, "The International Institute for Applied Systems Analysis," New York Academy of Sciences, forthcoming; Hugh J. Miser and Edward S. Quade, eds., *Handbook of Systems Analysis: Craft Issues and Procedural Choices* (Chichester, England: Wiley, 1988);. Hugh J. Miser, ed., *Handbook of Systems Analysis: Cases* (Chichester, England: Wiley, 1995).

4. Juan Carlos di Primio, "Monitoring and Verification in the European Air Pollution Regime," IIASA Working Paper WP-96-47, June 1996, 7.

5. IIASA's development of RAINS was led first by Eliodoro Runca (Italy) and subsequently by Leen Hordijk (Netherlands), Roderick Shaw (Canada), and Markus Amann (Austria).

6. Joseph Alcamo, Roderick Shaw, and Leen Hordijk, *The RAINS Model of Acidification: Science and Strategies at Europe* (Dordrecht: Kluwer Academic Publishers, 1990), 2.

7. Ibid., 2–6.

8. Ibid., 81.

9. Ibid., chap. 4, 115–178; Leen Hordijk (National Institute of Public Health and Environmental Protection, The Netherlands), "The Use of the RAINS Model in Acid Rain Negotiations in Europe," *Environmental Science and Technology*, 25:4 (1991), 598.

10. Alcamo et al., *The RAINS Model of Acidification*, 7; also chap. 7, 263–296. For Czechoslovakia, for example, J. Materna, "Results of the Research into Air Pollutant Impact on Forest in Czechoslovakia," *Symposium on the Effects of Air Pollution on Forest and Water Ecosystems*, 23–24 April 1985, Foundation for Research of Natural Resources, Helsinki, Finland, 1985.

11. Hordijk, "Use of the RAINS Model," 598.

12. Di Primio, "Monitoring and Verification," 2.

13. IIASA continues to improve and expand RAINS. Recent developments are not included in the description below, which presents RAINS as it was at the initial phases of the negotiations leading to the 1994 Second Sulfur Protocol under the UN/ECE Convention on Long-Range Transboundary Air Pollution. Currently RAINS is extensively used by the European Commission to arrive at its acidification strategy. Since 1992 a RAINS model for Asia has been developed in which the IIASA team has been extensively assisted by its large network of modelers in Europe and the United States.

14. Alcamo et al., *The RAINS Model of Acidification*, 82.

15. Ibid., 18.

16. The negotiations leading to the OSLO Protocol started with a set of scenarios calculated with a revised RAINS model. A major revision was the inclusion of critical load maps in RAINS, replacing the original forest soil, lake, groundwater, and forest submodels.

17. Jorgen Wettestad, "Acid Lessons? Assessing and Explaining LRTAP Implementation and Effectiveness," IIASA Working Paper WP-96-18, March 1996, 61.

18. Ibid., 35.

19. Marc Levy, "European Acid Rain: The Power of the 'Tote Board' Diplomacy," in P. M. Haas, R. Keohane, and M. Levy, eds., *Institutions for the Earth: Source of Environmental Protection* (Cambridge: MIT Press, 1993).

20. C. S. Holling, ed., *Adaptive Environmental Assessment and Management* (Chichester, England: Wiley, 1978); G. A. Norton and C. S. Holling, eds., *Pest Management: Proceedings of an International Conference, October 25–29, 1976* (Oxford: Pergamon Press, 1978).

21. M. L. Parry, T. R. Carter, and N. T. Konijn, eds., *The Impact of Climatic Variations on Agriculture*, vol. 1: *Assessments in Cool Temperate and Cold Regions*; vol. 2: *Assessments in Semi-Arid Regions* (Dordrecht: Kluwer Academic Publishers, 1988).

22. H. Peyton Young, *Equity in Theory and Practice* (Princeton: Princeton University Press, 1994).

14

RAND, IIASA, and the Conduct of Systems Analysis

Roger E. Levien

Introduction

During the fifty years from the beginning of World War II to the end of the Cold War, the two previously unrelated fields of science and government operations and policy were forced together in a series of ever more intimate liaisons. What began as convenient cohabitation under the pressure of global combat—analysis of military operations—eventually matured to a somewhat rocky marriage of common interests—analysis of national and international policy issues.

I have been privileged to participate in research and management positions over twenty-five of those years at two of the principal organizations at which the relationship developed—The RAND Corporation in Santa Monica, California, and Washington, D.C.; and the International Institute for Applied Systems Analysis (IIASA) in Laxenburg, Austria. In this chapter, I shall describe from that inside perspective the evolution of systems (and policy) analysis at these two research organizations and, in particular, indicate what I perceived to be the lessons of RAND's success and how they were used to help to shape the organization and operations of IIASA.

Before turning specifically to RAND and IIASA, I will describe the most general features of the evolution of the relationship between science and government operations and policy as represented by the development of operations analysis, systems analysis, and policy analysis, and—through cooperation among nations—of international systems and policy analysis. My intent is not to be comprehensive, but rather to provide the context within which RAND and IIASA developed and to which they significantly contributed.

After that background, I shall describe the development of RAND and the lessons of its experience during its first twenty-five years (1948–1973). Then, turning to IIASA, I shall cover its development, the application of RAND's lessons to IIASA, and the new lessons that were learned in meeting the unique conditions of the institute's international context.

OPERATIONS, SYSTEMS, AND POLICY ANALYSIS

Operations Analysis: 1940s and World War II

Operations analysis (or operations research, as it is now more commonly called in the United States) arose out of necessity and opportunity during World War II. The necessity was the need to design effective operational procedures for the use of the new technologies of detection and destruction that rapidly entered the military throughout the war, without the time for conventional testing and field exercises. The opportunity was the availability to the military of scientists who could apply their analytical skills and tools to the rapid and efficient design and testing of operational procedures. The scientists came from a number of disciplines: physics, engineering, mathematics and statistics, even biology. What they brought to the problems of operations analysis (and design) were both the general analytical and experimental approaches of the physical and biological sciences, and the specific tools of mathematics, experimental design, and statistics that enabled them to find the best or even optimal procedures without the necessity for costly and time-consuming field exercises.

They achieved significant successes, recognized by the military, in designing the practices for operating the new radar detection and defense systems during the Battle of Britain and in designing the anti-submarine search procedures used during the Battle of the Atlantic. These successes were followed by many other similarly effective applications to both combat and support operations.

For the most part, the operations analysts had to take the properties of the military systems they were working with as fixed. Their focus was on how best to operate that system—whether a radar or sonar system, an aircraft, or a ship—to achieve the best military effect.

Systems Analysis: 1950s and the Cold War

The success of scientists and engineers in improving military operations (and in the design of new classes of weapons) during World War II led military leaders to seek new ways to have continued access to scientific and technological contributions during the postwar era. As a result, a number of new institutions were established that enabled the military to employ or contract with teams of scientists and engineers to assist in the design and implementation of new weapons systems.

Operations research became a permanent part of all the services and, as many of its practitioners returned to civilian life, found application to business and civilian government operations.

In peacetime, however, new opportunities arose for the application of scientific and technological talents to the improvement of military systems. For whereas during wartime the analysts had to take most of the system—particularly its hardware—as fixed, in the postwar era, some were given the opportunity to apply their analytical tools to the specification of the system itself. For example, they were able to participate in the specification of the desirable performance for satellite reconnaissance systems, bomber basing systems, ballistic missile weapon and silo systems, air defense systems, and many others. As their responsibilities broadened, new disciplines had to be engaged, particularly the economic and political sciences and the behavioral and social sciences. For now an appreciation of the likely capabilities and responses of potential enemies and allies had to be incorporated in the design of offensive and defensive systems, as did an understanding of the capabilities of our own forces in their deployment and operation.

The new field that developed from these activities was named "systems analysis" because it extended the scope of analytical attention from the operations of fixed systems to the specification of the systems themselves. What remained "fixed" for the most part to systems analysts were the governmental policies that the systems supported. (The phrase "for the most part" is significant here, for while system analysts took the government's containment policy against the Soviet Union as given, their analyses played a critical role in designing the policies that supported it—deterrence through a survivable second strike capacity—and in designing arms control policies.)

Policy Analysis: 1960s and the Great Society

The perceived success of national defense systems analysis during the 1950s and 1960s led to a large number of efforts to apply it to civilian systems, especially in the United States. Two elements of the practice that had developed in applications to military systems were prominent in those efforts: the "systems approach," which was interpreted as the desire to incorporate in the design of a civilian project all those interrelated elements that affected its performance; and the use of analytical tools and mathematical models to evaluate the performance of a civilian system before its deployment.

Many of the initial efforts failed. The reasons were numerous, among them:

1. The performance of many critical civilian systems is more highly dependent on the uncontrollable behavior of human and social subsystems than are military systems;

2. The state of social and behavioral science was not adequate to support an analytical design approach;

3. The models developed were too highly dependent upon unverifiable assumptions; and

4. Experience in military systems did not equip systems analysts to deal realistically with civilian systems.

Perhaps the most important reason was that while there was a national policy consensus on the goals of the military and, for the most part, on the objectives of its systems, no such policy consensus was available for most civilian systems. Indeed, the desirability of and appropriate performance of civilian systems—whether for public safety, health care, education, transportation, or housing—are the very stuff of political dispute. Consequently, systems analyses often became weapons or victims of those disputes.

To bring a more realistic sense of the context and content of these analytical tools to their use in civilian settings, scientists and practitioners from the subject domains were recruited. Thus, lawyers, health practitioners and public health specialists, educational researchers, sociologists, political scientists, and other social scientists who had engaged with policy problems more pragmatically joined with systems specialists. The resultant set of approaches and interests began to be called "policy analysis," emphasizing its concern with the full range of policy choices—not simply physical system design—relevant to the design or improvement of a civilian system. The design of public policies became the focus of analytical attention, in conjunction with systems and operational design where relevant.

New tools, drawn from economics and social science in particular, were added to the repertoire of the policy analyst. Regression analysis of multivariable data sets, design of experiments, and survey research became the basic tools in the policy analyst's kit.

These tools were applied to policies in health care—the design and experimental testing of health insurance systems; housing—analysis of rent control and the design of a housing allowance system; education—design and experimental testing of voucher systems. They also provided the methods required for more sophisticated design of public systems for transportation, water supply, and communications. Operations analysis tools were extensively deployed to improve the operational design of public systems for emergency service—police, fire, and ambulance—and sanitation.

Policy analysts could and did address policies that systems and operations analysts would take as fixed. But their domain was restricted to the general political, social, and cultural assumptions of a single nation. In the United States, for example, they could take the commitment to democracy and a market economy as given.

International Systems (and Policy) Analysis: 1970s and Détente

Operations research had spread worldwide during the postwar years, gaining adoption in the military and business sectors of most developed nations, but nowhere else as broadly and successfully as in the United States. As the digital computer found wider use in defense and commerce, the methods of operations research became more mathematical and computational, leading to the development of large computer models for the simulation and, in many cases, solution of complex operational problems in military and business operations. Because of its wide lead in the adoption of the computer, the United States led in this arena as well. Furthermore, in the sixties, the McNamara years in the Defense Department were characterized—initially to great public acclaim—by the widespread adoption of quantitative analysis of national security problems throughout the Pentagon under the heading of "systems analysis." The apparent success of these analytical tools of management in relatively well specified and quantifiable areas caught the attention of modernizing government officials in many nations, leading them to believe that these "scientific" tools could be used to solve complex governmental problems.

In addition to the attempts to apply the "systems approach" to civilian problems noted above, this belief gave rise in the late sixties to efforts to apply these tools to problems of international scope. These efforts took a number of forms. The best publicized was the series of studies of the global future based on "global models" initiated by the international Club of Rome, but pursued as well by organizations in the United States, Germany, Japan, and Argentina. Another result of this belief was the proposal and eventual establishment of an international institute—the International Institute for Applied Systems Analysis—intended to apply the methods of systems analysis to "common problems of the developed nations." The initial impetus for the creation of IIASA was the desire to build a bridge over the Cold War divide between East and West. It was hoped that the quantitative and "objective" methods of systems and policy analysis would enable scientists from both ideological camps to work together to solve public policy problems that they had in common.

Although there have been specific systems or policy analyses sponsored or conducted by various intergovernmental or nongovernmental international bodies, only at IIASA did there develop the range of activities and continuity of effort required to establish a style of "international policy and systems analysis." To some extent the practices of policy and systems analysis as they had developed in the United States (especially) were transferred to IIASA. However, to the interdisciplinary nature of the teams required to carry out those analyses, IIASA added a requirement for internationality. Each analytical team represented not only the mixture of disciplines required to deal with the system or policy, but also scientists from an appropriate sample of the sponsoring nations. Although social scientists were prominent in their participation and leadership in policy and systems analyses in the United States, they were less well represented—except for economists and demographers—in the international studies at IIASA. For, whereas in any single nation analysts could assume a national consensus on the political and economic system, at IIASA—which was sponsored by members from both sides of the East-West divide—no such consensus cut across its participants. Nor were the standards of social science the same on both sides of that divide. However, ecologists and environmental scientists assumed a prominent role in many IIASA studies.

In the post–Cold War era, IIASA's role has changed and it has shifted its focus to the overall question of global change. According to its Agenda for the Third Decade:

> IIASA's primary goal will be to develop the means to assess the interactions between human development and the environment.... Strategies to mitigate or adapt to global environmental change must be formulated at both global and regional levels.... Global change includes both economic and environmental change.... The formerly centrally planned economies have begun complex transitions to political pluralism, market economies, and participation in the global trading system.... Analysts at IIASA will focus on several critical issues facing these countries. (Agenda for the Third Decade, pp. 5, 7, 8, 9)

The Future

As mankind enters the twenty-first century, the once youthful activities of operations, systems, and policy analysis have matured. Although each has had substantial successes, they have also encountered the limits of their applicability, and the ambition and promise of their early years has moderated to a more realistic appreciation of the role that they can play in the complex human and social interplay of forces that shape public policy decisions. The best way to insure that they achieve their full potential is to apply the same critical analysis to their own institutions and practices as analysts apply to government. In the remainder of this chapter, I shall try to draw the lessons from RAND and IIASA's experiences that bear on the design of future independent analysis organizations.

THE RAND CORPORATION

The Beginning: Project RAND at Douglas Aircraft, 1946

As World War II drew to a close, a group of scientists and engineers who had played advisory roles to the military, as well as senior military leaders who had gained respect for their contributions to the understanding and improvement of military operations, sought a way to maintain the relationship during peacetime. Both the civilian and the military participants agreed that in order to obtain the maximum benefit for the military (and the nation), the civilian scientists had to be outside the military and to be as free as possible from the normal procurement practices. As one of the civilians characterized their view:

> I believe we set a precedent in recognizing that we cannot do intelligent, long-range, strategic planning without taking into consideration our scientific and technological resources and their future development, nor can we give proper direction to research and technological development without its leadership having some concept of our strategic plans. (Dr. Edward L. Bowles in 1946 letter to General H. H. Arnold, quoted in Scott, 1966)

A direct result of their deliberations was the establishment of Project RAND[1] in March 1946 under the sponsorship of Gen. H. H. "Hap" Arnold, chief of the Army Air Force. To keep it away from the traditional procurement bureaucracies, he also established a special high-level office, Deputy Chief of Air Staff for Research and Development, whose first incumbent, General Curtis LeMay, used his authority to insure that the project had and maintained its independence. Under LeMay's guidance, Project RAND received a "broad and permissive" (Scott, 1966) statement of work to conduct a program of research on "intercontinental warfare, other than surface, with the object of advising the Army Air Forces on devices and techniques."

Perhaps the most foresighted, unusual, and valuable element of Project RAND's establishment was General Arnold's allocation to it of $10 million left over from his research budget as a result of the war's conclusion in 1945. These funds were to be expended over several years and had no specific objectives beyond the general statement of work. These conditions of "endowed independence" enabled Project RAND to experiment widely in its early years without suffering unduly for its failures.

A number of the civilians who participated in the discussions leading to Project RAND were engineers and executives of the Douglas Aircraft Company. With their enthusiastic support, the project was housed administratively and physically within Douglas's headquarters in Santa Monica, California. Not incidentally, the location 3,000 miles from the Pentagon—before the era of the jet airliner, fax, and e-mail—was another means of assuring Project RAND's ability to think broadly without being drawn in to the short-term and often narrow day-to-day interests of its client organization.

The initial study undertaken by the fledgling organization demonstrated its ability to think beyond the narrow and short term, although it was done at the request of the air force. It was documented in Project RAND's first report in May 1946, "Preliminary Design of an Experimental World-Circling Spaceship." Although this was primarily a hardware report, containing little of the political-strategic analysis or even economic-cost considerations that would typify later RAND studies, it did contain the following sentences, whose prescience was demonstrated upon the launch of Sputnik eleven years later:

> The achievement of the satellite craft by the United States would inflame the imagination of mankind, and would probably produce repercussions in the world comparable to the explosion of the Atomic Bomb. . . . To visualize the impact on the world, one can imagine the consternation and admiration that would be felt here if the U.S. were to discover suddenly that some other nation had already put up a successful satellite. (quoted by R. Cargill Hall, *Congressional Record*, vol. 109, 7 October 1963, A6279)

It was not long before the leadership of Project RAND, of Douglas Aircraft, and of the air force came to the realization that the location of an objective long-term research organization privy to air force plans within one of the airplane companies competing for air force contracts was not a good idea. In addition to the appearance of possible conflict of interest, the management style and culture of engineering organizations is in sharp contrast to the more academic style that RAND required to attract and retain first-class research staff.

The RAND Corporation, Not-For-Profit, 1948

The resolution of the problem was to establish in March 1948 a separate not-for-profit corporation, the RAND Corporation, chartered "to further and promote scientific, educational, and charitable purposes, all for the public welfare and security of the United States of America." In November of that year, the air force transferred its contract from Douglas to the new organization, which had moved to another site in Santa Monica.

Frank Collbohm, a Douglas executive who had been named director of Project RAND after a number of notable outside scientists had declined the position, was appointed president of the RAND Corporation.

An important air force policy statement was issued during the transition from Douglas to independent not-for-profit research organization. It provided the basic framework within which RAND–air force relations developed. It asserted that:

> RAND is a background research organization—not a development project.
>
> Project RAND will continue to have maximum freedom for planning its work schedules and research program.

RAND benefited from that policy and used it to maintain its independence. In particular, it retained the freedom to initiate its own studies and to make its own problem identification. It also held onto the right to refuse to do studies that it did not feel equipped to perform or that would be inconsistent with its mission. By establishing these principles early and vigorously defending them, it built strong momentum that deflected the inevitable efforts to curtail its independence (Scott, 1966, 79).

With its new independence, RAND entered a period of high creativity and productivity. Social scientists were added to the original staff of physical scientists, engineers, and mathematicians. The first systems analyses were carried out on the next-generation bomber aircraft[2] and on the design of an air defense system for the continental United States. In addition to those studies, targeted specifically at the requirements of the air force, there were research activities that made major contributions to a wide range of disciplines.

In 1967, Frank Collbohm retired as president of RAND and was succeeded by Harry Rowen. One of Rowen's first initiatives was to seek to diversify RAND's clients both within the defense community and into the civilian agencies of government. The result of his initiatives was a rapid growth in domestic business, including the establishment of a New York City RAND Institute that worked for the administration of Mayor John Lindsay.

It would go beyond the purpose of this chapter to describe the full range of RAND's accomplishments through the 1970s. However, a simple listing will give a sense of the diverse contributions that RAND made both to the scientific world and to its clients.

Disciplinary Successes

Among the fields to which RAND scientists contributed in a substantial way during its first twenty-five years are the following:

1. Mathematics: Nonlinear and dynamic programming, game theory, network theory
2. Economics: program budgeting, design of social experiments, research and development management
3. Computer science: artificial intelligence, user interface design, survivable networks, computer graphics
4. Social studies and international studies: Soviet and Chinese studies, deterrence, arms control
5. Management and system sciences: cost analysis for public programs, public systems analysis, design of logistics systems
6. Engineering: remotely piloted vehicles, spacecraft and shuttle design

System/Policy Analyses

Among the subjects on which RAND performed successful systems or policy analyses during its first years were the following:

1. National Security

- Design for an earth-circling spacecraft (1946)
- Deterrence and its refinements (early 1950s onward)
- Nuclear weapons security and design of weapons and warheads

Personal Note

My beliefs about RAND were developed during an eighteen-year association, from 1956 to 1974. In the summer of 1956, I was selected to initiate a new summer program for graduate students at RAND. After returning in a similar capacity in 1957 and 1958, I was appointed a full-time member of the research staff in October 1960. After participating for several years in National Security studies, I initiated work in computer science, designing and implementing the Relational Data File, an early relational database system. In 1967, the new president, Harry Rowen, named me head of a new department, System Sciences, which had as its responsibility the development of systems analysts and the recruitment of the professional specialists in fields such as medicine and law who would be needed as RAND expanded into the domestic arena. During that time, I oversaw a wide range of studies in the domestic arena and personally undertook a number of studies that concerned education. I was appointed, additionally, to co-leadership of the Education Program. After moving back to Washington to participate actively in design of the National Institute of Education, I was given responsibility for the Washington Office Domestic Program. I also became deputy vice president of domestic programs for RAND.

In 1974 I left RAND on a one-year sabbatical to IIASA; it extended to a seven-year leave of absence. When I left for IIASA, I took with me my understanding of the lessons of RAND's first twenty-five years and attempted to adapt them to that new international institute. What were those lessons?

- Role of remotely piloted vehicles in combat
- Air defense system design and implementation

2. Domestic Research

- Housing studies: abandonment and rent control in New York City, housing allowance experiment
- Health-care studies: hospital effectiveness, health insurance experiments
- Education: educational use of the computer, design of the National Institute of Education
- New York City: police, fire, ambulance, and housing studies

The Lessons of RAND's First Twenty-Five Years: 1948–1973

From the late 1940s through the early 1960s, a number of institutions in addition to RAND were established by agencies of the military to provide operational and systems analysis support. Among them were the Operations Research Office and, later, the Research Analysis Corporation that were set up by the army; the Center for Naval Analysis established by the navy; the Weapon Systems Evaluation Group and the Institute for Defense Analysis created to serve the office of the secretary of defense; and ANSER, formed by the air force to provide the short-term support that RAND could not. Although most of these organizations served their sponsors effectively, none achieved the prominence or widespread success that RAND did. Why? What factors in RAND's design and operations accounted for its accomplishments?

Recognition of Requirements for First-Class Systems Analysis

The leadership of RAND recognized that first-class systems analysis in support of government required four conditions:

1. *Interdisciplinary teams comprising excellent disciplinary specialists participating under the integrative leadership of a talented leader.* Three points are significant about this condition. *Interdisciplinary teams* are required by realistic systems and policy problems because of all the factors—physical, organizational, and human—that influence their design and performance. Those teams must comprise *first-rate disciplinary specialists* because they often confront issues that lie on the frontier of single disciplines or at the intersection of several. *Talented integrative leadership* is essential to both formulate and manage the analysis and to interrelate effectively with each of the disciplinary specialists. Individuals with such skills are extremely rare; RAND attracted or developed many, although never enough of them.

2. *A broad scope of work to enable the analysis to follow the problem where it led.* It was commonplace at RAND for a problem as perceived by the client to turn out to be the result of an entirely separate issue. Often the major success of a RAND study was identification of the nature of the true problem. The terms of the Project RAND contract with the air force enabled this kind of redefinition. In later years, when RAND was forced to compete for work in response to client-defined problem statements, it found its ability to identify the real problem severely restricted. I believe that the creativity characteristic of RAND's early successes was reduced as a consequence.

3. *Sufficient continuity of the relationship with the government client to permit the analysts to develop deep understanding of the client's organization and responsibilities and for mutual trust to develop.* The RAND–air force relationship, though occasionally strained, functioned extremely well by enabling RAND analysts over time to develop a deep knowledge of air force operations, often better than that possessed by the air force incumbent who had recently been rotated into his position. Reciprocally, through the

discretion of its activities and the value it delivered, RAND built a strong relationship of mutual trust with the air force.

4. *Enough independence from the client agency to be able to avoid succumbing to pressure to produce the answers desired by one or another faction within the client organization.* Because of the terms of its establishment and the tradition that they established, RAND was able to protect its independence. However, challenges always arose and it was a primary responsibility of the president, Frank Collbohm, to patrol RAND's independence boundaries and assure that they were never breached. It was a responsibility that he took extremely seriously and at which he rarely failed.

In contrast, none of the other institutions was able to achieve these four conditions. Most organized and conducted their studies according to problem or functional structures and with specialized teams lacking broad interdisciplinary representation. None of them was able to attract the large number of highly qualified disciplinary specialists with strong links to excellent university departments that characterized RAND. All were highly dependent upon one controlling client, and they often had narrow, short-term-oriented scopes of work.

Part of the reason that RAND was able to succeed lay in the nature of its charter, statement of work, funding arrangements, and initial relationship with its client. Another portion might be ascribed to good fortune in the talents and character of its leadership and professional staff. But a significant part was due to the way that RAND was organized and managed so as to enable high quality systems and policy analysis to be carried out.

Balancing Academic Research and Systems Analysis

RAND's "secret" to achieving high-quality interdisciplinary systems analysis lay in the way in which its organization and management balanced disciplinary excellence with systems analysis through a matrix organization in which one dimension was disciplinary departments and the other, problem-focused programs.

In the early years, departments were the only organizational unit, with problem-oriented research teams established on an ad hoc basis, drawing their members from several departments. In the late sixties, the matrix was formalized by establishing a program structure. The programs became an especially important means for obtaining and managing grants and contracts as RAND took on a wider range of clients.

The departments were for the most part defined by discipline—economics, physics, engineering, computer science, mathematics, and social sciences, although there were some exceptions—logistics, cost analysis, and systems operations—during the early sixties. By good fortune and design, most of the disciplinary departments established close relationships with the leading university departments in the corresponding discipline.

One means that was very effectively used was to establish consulting relationships with leading academics, who would spend the summer months at RAND. In the sixties, RAND's roster of consultants numbered over 500, among them Henry Kissinger, Kenneth Arrow, and Herbert Simon. Another mechanism was RAND's policy of encouraging publication of its work in respected peer-reviewed academic journals and books. Often, a young academic could publish more as a result of a stay at RAND than would have been possible at a university with its teaching and committee obligations.

As a result, RAND was recognized as an academically respectable place for a talented young researcher to go after obtaining his or her Ph.D. from a leading university. These researchers retained the ability to return to academia from RAND, often with benefit to their careers. Consequently, RAND was able to attract the first-class disciplinary talent that is a critical element of leading edge interdisciplinary research.

The projects were defined by specific topics of concern to the air force or, more generally, the national security of the United States, such as bomber basing, next-generation aircraft requirements, or air defense system design. Programs were clusters of projects addressing an area of concern, such as strategic or tactical systems, logistics, weapon systems research and development, or command and control systems. The most effective programs were those that established a deep understanding of an area of air force interest and were able to anticipate future issues before they became an operational concern.

Generally, the effective programs were linked closely to real problems in association with the responsible organizations and the decision makers who faced them. That often entailed considerable travel and time in the field for the analysts. Through that close association the program teams developed a realistic understanding of what recommendations would be implementable with beneficial effect in the client organization.

If the departments had their constituencies in the disciplines, whose standards of academic rigor were naturally applied within RAND to any work in which the department's members participated, the projects and programs had their constituencies in the client agencies, whose standards of practical benefit and implementability were directly applied to the results of the project and program work.

RAND's management's skill lay in balancing the demands of those two constituencies so as to achieve work that met the different, and important, standards of each.

Incentive Systems and Quality Control

To achieve that balance required the implementation of a set of standards and rewards—an incentive system—different from both the purely academic and the purely bureaucratic.

One element was RAND's essential job security, not tenure achieved through publications, but rather an implicit continuity based on continued good performance in any of the various activities that RAND required—disciplinary excellence, analytical excellence, project leadership, other forms of management, client relationship management, and so on. No one, even those from academic backgrounds, felt pressure at RAND to publish. Good performance in the task at hand was, however, critical.

It was possible for a first-class disciplinary specialist to establish a career distinct in its path and stages from the traditional academic path, but to retain linkages so that at some future time transition to a tenured academic position could occur easily. Similarly, many of the problem-oriented specialists established careers distinct from those of the traditional governmental employee, but through their demonstrated expertise were appointed to high governmental positions. A significant number of RAND alumni assumed high level positions in federal agencies, among them Charles Hitch and Alain Enthoven, who brought program budgeting and systems analysis to McNamara'a Pentagon, and James Schlesinger and Fred Ikle, who occupied high positions during Nixon's presidency.

Operationally, RAND instituted and implemented a system of rigorous internal and external reviews prior to the communication of any of its results. Even in areas of national security, where public peer review was not possible, RAND insisted on tough critiques by internal experts and consultants with appropriate clearance.

Criticality of President's Vision

Perhaps most important to the initial and continuing success of RAND was the clear vision of its president, Frank Collbohm. His sense of what was right or wrong for RAND was precise, enabling him to serve as the "inertial guidance" for the organization, keeping it focused on the tasks for which it was created, protecting its independence and high standards, and deflecting efforts to take it into areas for which it was not suited. He took as his responsibility maintaining the support and respect of its clients. But he appeared to place an even greater priority on providing the context for the conduct of excellent work by the staff. Though many staff members achieved national and international prominence for the work they did at RAND, Collbohm left the limelight to them. The world came to associate RAND with Herman Kahn, Albert Wohlstetter, Charles Hitch, Alain Enthoven, or George Dantzig, but few outside of RAND and its clients knew the name of its founding president.

Personal Note

Early in 1967, McGeorge Bundy had sent one of his associates to the RAND Corporation to gather ideas for the then-proposed International Center for the Systematic Analysis of the Common Problems of Advanced Societies. After participating in the meetings, I had co-authored with Sidney Winter a draft proposal for such an institute based on RAND's experience and the ideas developed during the discussions (Levien and Winter, 1967). That draft document was used in the early stages of the five-year negotiations. As a result, I maintained an interest in their progress and met several times with Howard Raiffa when he visited RAND in the course of the negotiations.

In 1974, Raiffa invited me to come to the institute on a sabbatical from RAND to initiate a project to create a Handbook of Applied Systems Analysis, which was of particular interest to several of the member countries where practical experience with operations and systems analysis was small or nonexistent. I arrived in August of 1974, a year after the institute began active research.

Because Raiffa had taken on the directorship during a two-year sabbatical from Harvard, he had to return to Harvard in time for the 1975 academic year. He proposed to the council that I be named the next director and I was selected. I served two three-year terms, from the fall of 1975 through November 1981.

Although I have remained actively engaged with IIASA through the U.S. Committee for the Institute, I shall report here only on the period during which I was its director, for the influence of my RAND background and my learning about what was required for IIASA to succeed was greatest during that period.

THE INTERNATIONAL INSTITUTE FOR APPLIED SYSTEMS ANALYSIS

For a discussion of the founding and early years of IIASA, see the previous chapter by Brooks and MacDonald.

RAND Experience Applied to IIASA

Howard Raiffa had established a project structure for IIASA, forming more than ten distinct projects, each staffed by teams of specialists who for one reason or another were available to come to IIASA on relatively short notice during the first year. In this he was responding to the practical imperative of getting the institute off to a quick start and to the early achievement of some significant results. Despite the short notice, he was able to attract some first-rate Western scientists—the mathematician George Dantzig, the future Nobel laureate in economics Tjalling Koopmans, the Canadian ecologist C. S. Holling, and the West German nuclear engineer Wolf Haefele among them. In some cases, researchers who had worked together came as a group. In others, specialists in a common subject area arrived from different countries and were clustered together in a project.

Inevitably, there was a wide range of talent and experience among the initial recruits. Those from the East had often been chosen more on the basis of their political reliability and personal contacts than because of their professional capabilities. Those from the West were often better scientists, but not many of them had had experience with systems or policy research, had worked on or led interdisciplinary teams, or had participated in international studies.

I saw my task as taking the successful rapid start that Raiffa had given the institute and building under it the solid foundation on which it could grow and prosper. In doing so, I naturally drew upon what I believed were the lessons of the twenty-plus years of RAND's experience, recognizing that IIASA's situation was in many key respects different from RAND's. Not least, despite the obvious intent of its founders, the scientific organizations that were given the responsibility for its formal management were often more interested in the purely scientific nature of the research than in its relevance to international public policy issues. This led to tension over the criteria of "excellence" in identifying first-class researchers for the institute and, eventually, in assessment of the quality of its results.

Matrix Structure

I had no doubt that the goal of IIASA was to accomplish on an international scale the kind of excellent systems and policy analysis that RAND had performed in the United States. To do so, I strongly believed that it had to achieve the same kind of balance between disciplinary excellence and policy effectiveness that characterized RAND at its best. Thus, I set out to emulate the matrix structure of RAND with disciplinary and program dimensions.

However, while RAND had a professional staff of over 500, IIASA's scientific staff never exceeded about 100 during my tenure. Nevertheless, it included a wide range of disciplines: engineers, physicists, ecologists, water resource specialists, demographers, economists, computer scientists, operations researchers, physicians and regional planners among them. Had I instituted a traditional academic disciplinary organization, the result would have been a large number of departments with a small number of researchers in each. So, in their stead, I established four *research areas*, each of which encompassed a group of related disciplines:

1. *Resources and Environment*, which housed the ecologists, environmental scientists, and water resource specialists
2. *Human Settlements and Services*, which comprised demographers, regional planners, and health-care specialists
3. *Management and Technology*, which included some engineers, physicists, management researchers, and science policy specialists
4. *System and Decision Sciences*, which contained both computer scientists and the methodologists from mathematics, operations research, and economics

Each of these research areas managed a small number of applied research projects staffed primarily with its own members, although a good number engaged staff from other areas part time. These projects were easier to design and staff because they were close to the topics that the researchers had dealt with at their home institutions, which were generally neither interdisciplinary nor very applied. The challenge to their leaders was to integrate the staff, who came from different nations with somewhat different research styles, and to identify and succeed in analyzing a sufficiently interesting practical problem.[3]

To accomplish the more difficult and potentially more significant systems and policy studies, I established cross-cutting *research programs*. The energy project that had been initiated under Raiffa's directorship became the first program. Its leader, Wolf Haefele, had led West Germany's breeder reactor program. He was thoroughly familiar with nuclear energy and had substantial research leadership experience. Furthermore, he had a deep interest in a problem of profound global importance: *How would the energy needs of a growing global population that was also progressing economically be met over the next half-century?* Fifty years was the relevant time frame for such a study because of the long time constant of change in the global energy supply and distribution infrastructure. My intent was that Haefele would have a small core staff, but that most of the energy program's staff would be drawn—as they were at RAND—from the research areas. Although a certain amount of that happened, Haefele's management style required complete commitment from his core team, which grew to be substantially larger than had been intended. Furthermore, his team, though interdisciplinary, did not always comprise disciplinary specialists who met the standards of the research areas. The work that resulted, *Energy in a Finite World*, was a substantial intellectual and managerial accomplishment that received a mixed reception despite its having been subject to intense prepublication review. On the one hand, there was high praise from many associated with the traditional energy sector in many nations. On the other hand, there was sharp criticism from critics of traditional energy policy and from methodologists who discerned flaws in its analytical practices. Nevertheless, the energy program stands as a major result of IIASA's early years; it established a focus on global energy issues that has continued at IIASA to the present time, building upon and refining the base that the energy program established. Not least, that base included an international network of collaborating institutions that shared in its common development.

A few years after the beginning of the energy program, the institute established a second program—on food and agriculture. It was led initially by a Hungarian economist, Ferenc Rabar, and then by an Indian economist, Kirit Parikh. It too focused on the ability of global resources to meet the expanding needs of a growing and developing population. Not surprisingly, given the disciplinary background of its leaders, the food and agriculture program used an economic framework, examining the consequences of achieving supply and demand balance through the linked systems of global production and trade. Key to its conduct was the recruitment of a group of collaborating institutes in each of the key nations who had responsibility for analyzing their national systems. IIASA assumed responsibility for analyzing their global linkages through the trading system. The results of the food and agriculture program appeared in a variety of forms. Perhaps its greatest success was the creation of an international community of institutions, cutting across political boundaries, who shared a common understanding of the issues and facts of global food supply.

Both of these programs had successes and weaknesses that reflected the personal interests and skills of their leaders. Aside from RAND, there were essentially no places at which one could learn how to be the leader of a large, interdisciplinary policy research study, and even at RAND the primary way to learn was through experience.[4] No courses were in existence, although there were several RAND books that comprised most of the extant literature.[5] Consequently, everyone had to learn on the job. At IIASA there were the additional difficulties of dealing with a staff that was not only interdisciplinary, but also international, and with having at least a dozen "clients," each with somewhat different interests and policy environments.[6] Consequently, it was remarkable that large policy studies were designed, managed, and completed at all, let alone that they succeeded in influencing the policies of a great many national and international bodies.

Importance of Quality Control and Communications

A second aspect of RAND's experience that I sought to transfer to IIASA was an appreciation of the importance of communication and the need to bring work to a high level, both in its content and in its presentation, before exposing it to outside audiences, whether scientific or policy.

The requirement assumed additional importance and difficulty at IIASA because, although the founders had wisely decreed that the institute would have only one working language and that would be English, most of the institute's staff were not native English speakers. (It should be noted that the requirement of all staff members that they be competent in English was a significant impediment to recruitment from many countries.) Many have commented that "broken English" is the international language of science; IIASA was a case in point. Furthermore, the standards of scientific quality were not uniform across all IIASA's member organizations nor across all the research institutions from which IIASA's staff was drawn.

Consequently, I introduced two features of the RAND system. The first was a high standard of external peer review of quality before any finished work of the institute could be published.[7] The publication of the approved work was supported by a staff of editors who insured they achieved a high standard of exposition and language. The goal was that it should not be possible to distinguish the reports written by non-native speakers of English from those of the native speakers.

The second was a tough internal review of every presentation that was to be presented outside of the institute before it could be released. These reviews covered both the content and the effectiveness of its presentation. There was a presentations coach who helped all scientists to perfect the organization and visual and oral communication of their results. The presentations coach also ran workshops for new scientists to help them polish their presentation techniques. That was especially important for scientists from Eastern Europe where, at that time, even overhead projectors were rare.

Role of the Director

I drew upon RAND as well for my model of the role of the director of the institute. I understood that my responsibilities were

- To establish and communicate the direction of the institute;
- To protect its independence, while preserving the support of the member organizations;
- To see to the provision of the necessary resources and appropriate context;
- To establish and assure the operation of effective quality controls; and
- To get out of the way of the research staff.

New Aspects of IIASA

Although the experience of RAND proved useful at IIASA, there were aspects of IIASA that were substantially different from RAND and required, therefore, substantially different approaches. Among them were the following:

International Problems

Although RAND dealt with problems that were international in nature, such as the relationship between the United States and the Soviet Union, it always did so from the perspective of a single national client—the United States. IIASA, however, dealt only with problems that were of interest to multiple national clients, sometimes with conflicting interests.

It proved useful to distinguish between two categories of such problems. The first category was "global problems" that spanned national boundaries and could be solved only by the actions of many nations acting together. The purpose of these studies was to establish a common basis of understanding of the nature of the problem and of approaches to its resolution that could be used in international negotiations and by international bodies. Both of the programs—energy, and food and agriculture—fell into this grouping.

The second category was "universal problems" that existed within national boundaries and that single nations could act alone to resolve, but that all nations shared. Hence the purpose of studies of these problems was to identify approaches that could be applied in many, if not always all, nations. Most of the activities of the research areas fell into this grouping. Among them were studies of water quality and water demand management, large-scale regional development projects, health-care delivery, and population migration.

For the most part, the major difference introduced by this internationality was the requirement to account for the different interests and perspectives of the sponsor nations and to try to insure that the highest standards of apolitical neutrality were met.

Multiple Sponsors with Diverse Expectations

RAND was blessed with a single sponsor, the United States Air Force, during its formative years, who understood well what it expected RAND to contribute. That sponsor also allowed RAND a few years in which to develop its capabilities. Neither of those conditions prevailed at IIASA. Instead, from its beginning IIASA needed to account for the varied interests of its multiple sponsors. Between 1972 and 1976, the number of national member organizations (NMOs) grew from twelve to seventeen.[8] In some nations, the NMO was an existing scientific body, such as the National Academy of Sciences in the United States or the Royal Society in the United Kingdom. In others, it was an organization especially established to be the member, such as the French Association for the Development of Systems Analysis and the National Committee for Applied Systems Analysis and Management in Bulgaria. The composition of these ad hoc bodies reflected what each nation thought "applied systems analysis" might be, as well as the local politics of science and technology policy. In several cases, particular disciplines—often control theory—gained dominant positions on the committees.

In most instances, the NMOs were able to insulate the institute from direct government intervention. Nevertheless, the eventual source of funding in all cases was from government agencies. In the United States, the National Science Foundation was tasked with providing support. In the United Kingdom, it was the Ministry of the Environment. In many nations, it was the Ministry of Finance. Inevitably, the sources of funds would exert their right to see what they were getting for their money.

The result of this diversity of sponsorship was a diversity of expectations going beyond even the anticipated diversity of interest among different nations. For example, the United Kingdom's member, the Royal Society, viewed IIASA as a scientific research institute that ought to be judged primarily by its ability to add to the store of peer-reviewed science published in prestigious disciplinary journals. The Ministry of Finance in Bulgaria, however, was much more interested in practical results of immediate use to the Bulgarian economy. Almost every expectation between these extremes was present among one or another of the NMOs, and generally several different expectations were held by each NMO.

Although there were sponsoring organizations with individuals who were knowledgeable about some aspect of operations or systems analysis, there was no organization among the sponsors that had wide experience in or understanding of the practice of systems or policy analysis. Thus, IIASA had not only to learn to do systems analysis in an international setting; it also had to educate its sponsors in what it was striving to accomplish and convince them to modify their expectations appropriately. And while it was doing that, it had to shape its research portfolio so as to satisfy within reason the wide range of expectations that its sponsors continued to hold.

The responsibility for crafting a research program to satisfy these diverse wants fell in the first instance to the director, with the assistance of the senior research leaders. As the most senior person associated with IIASA who had had significant experience in systems and policy analysis, I tried to move the institute step-by-step toward activities that would meet the high standards that RAND had established, while at the same time holding at bay those who looked for more purely academic or more immediately operational research projects.

During the six years of my directorship, the financial support of the NMOs was at the maximum permitted by the charter, and the research plans were always approved by the institute's council. Nevertheless, there was continual criticism of the institute for not achieving enough high-quality academic research or enough results of short-term relevance to solving practical problems.

Transient Staff

IIASA had no permanent research staff. Because most of its scientists came from countries other than Austria, most of them had to leave good positions in their home countries in order to spend time at IIASA. If IIASA had limited itself only to scientists who were willing to move themselves and their families permanently to Austria, it would have narrowed the candidate pool too much. Thus, a transient and constantly changing research staff became a way of life at IIASA. In that respect, it differed significantly from RAND, where long tenure was the norm, and from most high quality academic institutions.

The average length of stay of a researcher at IIASA stabilized during the 1970s at about two years, but the distribution was wide, ranging from a few months to, at the time I left, eight years. All scientists had fixed-term, renewable contracts. During an average year in that period, there would be over 150 different researchers in residence, representing more than twenty countries and an even larger number of disciplines, whose combined effort amounted to about 100 full-time equivalents.

The transience of the scientific staff was a cause of concern to many observers more familiar with traditional university and governmental research operations. However, from my perspective, for several reasons, it was a great strength of the institute:

1. Since the director had full authority to hire and remove all employees, the fixed and short-term nature of the researchers' contracts provided a graceful way to eliminate unproductive research staff. Especially in the early years, the political nature of some early appointments would have caused lasting difficulty, if it had not been easily possible to not renew them.
2. Good performers, however, could have their contracts renewed, if they were available to stay, as many were after they and their families had been established in Austria and they had grown productive at the institute.
3. As a matter of policy, the senior research leaders were all expected to stay for longer than two years, providing the continuity that was needed to ensure successful work.

4. The research leaders became adept at scheduling the flow of short- and long-term appointments to satisfy the changing needs of their projects without creating long-term commitments to specialists required only at a specific point.

5. First-class researchers, who would never have given up their home positions, could spend short periods at the institute and, as available, return for extended stays during sabbaticals or leaves.

6. The flow of alumni created a natural constituency and base of contributors around the world.

Collaborating Institutions

Although RAND did some subcontracting during its early years, for the most part RAND operated alone, performing all its activities with its own staff and consultants. In contrast, IIASA came to rely heavily on networks of collaborating institutions around the world. These collaborations took many different forms, some minimal and others involving considerable closely coordinated work.

An example of close collaboration was the work done on population migration in each of IIASA's seventeen countries. Using a common methodology and research protocol developed at IIASA, a research organization in each country analyzed interregional migration within its national boundaries. The results of these seventeen studies, which were unique, were published by the institute in a common format together with a crossnational comparison and summary. Similar collaboration underlay the models of food supply and demand that were a key component of the Food and Agriculture Program.

The network of collaborating institutions was essential and valuable for several reasons:

1. They enabled the work of the institute to be conducted inside each of the member organization countries by researchers familiar with the specific character of that country and its government. Reciprocally, their knowledge of national situations was available to the researchers at the institute.

2. They served as a reliable source of staff appointments to the institute who were familiar with it and its work when they arrived and whose skills and capabilities were known to the research leaders. This became the most valuable source of recruits for the institute.

3. They became a constituency for the work of the institute within their country and helped to build and retain its support.

4. They became the vehicles for disseminating and implementing results of the institute's research within their own country.

National Balance in the Research Staff

As an international institute created to build bridges of trust across the East-West divide, IIASA has a primary requirement to insure that each of its research projects comprises representatives from several nations. And because one of the benefits of membership is the opportunity to have scientists in residence at the institute, there is a requirement across all projects that there be a "reasonable" balance of representation from each of the member organization countries. "Reasonable" was interpreted to mean in rough proportion to the financial contributions from each nation, both through membership payments and, in several cases, supplemental contributions from other national organizations such as foundations, corporations, or government agencies. No remotely similar requirement was present at RAND.

In the early years, this requirement was a severe constraint, especially when combined with the need for English language capability and the tendency of the Eastern countries to use political reliability as a primary criterion. Remarkably, over time the institute developed enough familiarity with the relevant communities and institutions in its member organization countries to be able to select good people with a reasonable geographic balance. Over time, as well, the criteria were employed flexibly so that a country that was short of representation during a period might be overrepresented for several years subsequently.

Dearth of Experience in Systems/Policy Analysis

In contrast to the problem of establishing a scientific research organization in a well-established field, the creation of IIASA—like the creation of RAND—could not draw upon an established community of practitioners and supporters. Thus, IIASA had to create through its own activities the practice of "international systems and policy analysis." In part, that has been accomplished through the adoption of experience from RAND and other organizations; but in large measure it has been formed by the experience of IIASA itself.

The problems that this lack of experience created in recruitment and in building and sustaining support from the member organizations have already been mentioned. A further problem, not yet mentioned, has been the transience of the individuals whose support of IIASA has been critical in each of the member organization nations. The founders—McGeorge Bundy, Philip Handler, Jermen Gvishiani, Sir Solly Zuckerman, Pierre Aigrain—provided a climate of support that enabled the institute to learn and grow during its early years. They were followed by a generation of council members whose appreciation of the institute and what it could and could not do evolved with IIASA's development. But they too left and with them, often, the strong links to the funding bodies that assured the institute's funding. Their understanding and government links have been essential to the continued viability of IIASA. A critical issue for the future of IIASA is its ability to build and maintain a strong constituency of customers and sponsors for its unique product—international systems and policy analysis—in each of the countries with NMOs.

Balance in the Research Agenda

RAND was under no obligation to provide an annual research agenda for approval by its sponsor. Beyond a broad statement of the areas in which it intended to work, its research program was within its own responsibility, subject of course to continued contacts with the air force. Generally, in the 1960s, RAND aimed for about one-third work suggested by the air force, one-third initiated by RAND, and one-third of mutual concern.

IIASA, however, has to submit to its council, and through the council members, the NMOs that they represent, a highly specific research plan for the following year, which receives critical review and suggestions from many of the NMOs. In the fall of the year, the research plan for the following year, and the associated budget, is approved by the council.

For all of the reasons described above, that plan has to reach a critical balance, reflecting the diverse expectations of the sponsors, the potential availability of appropriate research staff members, the interests and capabilities of available leaders, the need for national balance in staffing, and the appropriate trade-off between short- and long-term benefits. As a result, the crafting of the research plan is one of the most difficult and important tasks of the research leadership of the institute.

The research plan, consequently, has served as the lightning rod for criticism of the institute and efforts to shape it to the interests or needs of one or another of its constituencies. The diplomatic skills of the senior research leaders, the respectful understanding of the council, and the tendency for comments to cancel each other, have enabled what could be an impossible task to be accomplished. And in its best form, the research plan has been a valuable—indeed essential—tool for recruiting and retaining both research staff and the research sponsors who fund the institute's work.

IIASA Now and in the Future

The IIASA described above is the one that I was privileged to serve during the latter half of the seventies. Obviously, in the subsequent years it has undergone many changes, as it will continue to do as it enters the new century. However, much of what I described remains true in general, if not in specifics. Most important, IIASA remains the only truly international research institution performing international and interdisciplinary systems and policy analyses of global and universal issues. What it has learned in the twenty-five years of its existence is a valuable international asset. It warrants strong international support to insure that that asset continues to grow and that it achieves its considerable potential to contribute to the informed resolution of international policy problems.

NOTES

1. RAND was intended as an acronym for "research and development," although in later years it was waggishly, and more accurately, interpreted as "research and no development."

2. Ed Paxson, who carried out this study, coined the term "systems analysis" in that context (Digby, 1988).

3. It turned out that, for the most part, it was easier for an American engineer and a Russian engineer to understand each other than for an American engineer and an American economist to communicate.

4. Since the early 1970s, RAND has run the RAND Graduate Institute, whose goal is to train policy and systems analysts to the Ph.D. level.

5. See McKean, 1958; Hitch and McKean, 1960; Novick, 1966; Fisher, 1971; and Quade, 1975.

6. The *Handbook of Applied Systems Analysis* was eventually published in three volumes under the independent editorship of Hugh Miser and Ed Quade, who had been associated with the project at IIASA until its termination in the early eighties. See Miser and Quade, 1985; Miser and Quade, 1988; and Miser, 1995.

7. Outside reviewers were selected by research management and paid an honorarium for that effort.

8. The new members were the Netherlands, Sweden, Finland, Hungary, and Austria.

REFERENCES

Digby, James. "Operations Research and Systems Analysis at RAND, 1948–1967." *OR/MS Today* (October 1988): 10–13.

Fisher, Gene H. *Cost Considerations in Systems Analysis.* New York: Elsevier, 1971.

Hitch, Charles J., and Roland N. McKean. *The Economics of Defense in the Nuclear Age.* Cambridge: Harvard University Press, 1960.

International Institute for Applied Systems Analysis. *Agenda for the Third Decade.* Laxenburg, Austria, 1995.

Levien, Roger E., and Sidney G. Winter Jr. "Draft Proposal for an International Research Center and International Studies Program for Systematic Analysis of the *Common Problems of Advanced Societies*" (internal paper). The RAND Corporation, 1967.

McKean, Roland N. *Efficiency in Government through Systems Analysis: With Emphasis on Water Resource Development.* New York: John Wiley, 1958.

Miser, Hugh J., ed. *Handbook of Systems Analysis: Cases.* New York: John Wiley, 1995.

Miser, Hugh J., and Edward S. Quade, eds. *Handbook of Systems Analysis: Overview of Uses, Procedures, Applications, and Practice.* New York: North Holland, 1985.

Miser, Hugh J., and Edward S. Quade, eds. *Handbook of Systems Analysis: Craft Issues and Procedural Choices.* New York: North Holland, 1988.

Novick, David. *Program Budgeting: Program Analysis and the Federal Budget.* Cambridge: Harvard University Press, 1966.

Quade, Edward S. *Analysis for Public Decision.* New York: Elsevier, 1975.

Scott, Bruce L. R. *The RAND Corporation: Case Study of a Nonprofit Advisory Corporation.* Combridge: Harvard University Press, 1967.

15

How a Genetic Code Became an Information System

Lily E. Kay

Introduction

In the 1950s molecular biology underwent a striking discursive shift: it began to represent itself as a communication science, allied to cybernetics, information theory, and computers. Through the introduction of such terms as *information*, *feedback*, *messages*, *codes*, *alphabet*, *words*, *instructions*, *texts*, and *programs* molecular biologists came to view organisms and molecules as information storage and retrieval systems. Heredity came to be conceptualized as electronic communication—akin to guidance and control systems—governed by a genetic code: four nucleotides specifying the assembly of twenty amino acids into myriads of proteins. This semiotic and linguistic repertoire was absent from molecular biology before the 1950s. Based on these scriptural representations, the genome could be read and edited unambiguously by those who know, laying claims to new levels of control over life: beyond control of matter there was now control of information and the power of the word.

In a series of lectures in 1969 that formed the basis for his notorious book *Chance and Necessity*, Nobel laureate Jacques Monod celebrated these new representations as a turning point in biology. The organism was nothing but "a cybernetic system governing and controlling the chemical activity at numerous points," he argued. Within the scientific imagination of the missile age, gene enzyme regulations were "systems, thus [are] compatible to those employed in electronic automation circuitry, where the very slight energy consumed by a relay can trigger a large-scale operation such as, for example, the firing of a ballistic missile." He credited the ideas of his colleague Leo Szilard and their elaboration by the noted information theorist Leon Brillouin for providing the key to these fundamental biological concepts.[1]

"Heredity is described today in terms of information, message, and code," proclaimed Monod's partner, Nobel laureate François

Jacob, in the introduction to his book *The Logic of Life.* Analogizing the genetic code to a computer program, Jacob argued that "heredity functions like the memory of a computer. Organs, cells, and molecules are thus united by a communication network." He credited Norbert Wiener's cybernetics and Szilard's and Brillouin's information theory in supplying the crucial ideas of message and feedback regulation, through which "[h]eredity becomes the transfer of a message repeated from one generation to the next."[2]

And in his little book *The Book of Life* (1967), molecular biologist Robert Sinsheimer, the progenitor of the Human Genome Project, described the information stored in the human chromosomes as the written word.

> [T]he book of life. In this book are instructions, in a curious and wonderful code, for making a human being. In one sense—on a subconscious level—every human being is born knowing how to read this book in every cell of his body. But on the level of conscious knowledge it is a major triumph of biology in the past two decades that we have begun to understand the content of these books and language in which they are written.[3]

How did molecular biologists come to view organisms and molecules as information storage and retrieval systems, and heredity as information transfer? By which historical process did the genome come to be conceptualized as a coded text written in a natural language? What has been the import—epistemic, operational, and social—of these new informational and scriptural representations of life?

I suggest that on a very general level, the answers to these interrelated questions can be located in the emergent historico-cultural episteme variously referred to as the postindustrial and postmodern epoch, driven, among other things, by postwar information technologies, notably television, computers, and guidance and control weapons systems. Though first expressed by Alain Touraine, it was Daniel Bell who popularized the totalizing term "postindustrial society" to denote a major shift from industrial production toward an information-based economy: the production, processing, and distribution of information goods and services.[4] In probing this postmodern condition, where knowledge and information serve as global currency, Jean François Lyotard has examined the crisis of representation engendered by the sciences and technologies of language, such as cybernetics, computers, and informatics. Mark Poster has surveyed and analyzed some of the historical nodal points in that transition to an information technoculture.[5]

More specifically, in her "Cyborg Manifesto," Donna Haraway has outlined the "informatics of domination" of bodies and populations —the simultaneously material and ideological dichotomies that express the transition from industrial capitalism/modernity to postindustrialism/ postmodernity. Representation becomes simulation; organism, a biotic component; physiology, communication engineering; reproduction, replication; sex, genetic engineering; and so on—a shift in representation that she traces back to the 1950s, to Cold War technosciences of guidance and control.[6]

Similarly, Thomas Hughes has located the shift from modern to postmodern engineering in the emergent computer networks and complex systems engineering practiced in the military-funded aerospace universe during the 1950s and 1960s. He too has outlined some polarities defining modern and postmodern engineering, such as hierarchical/vertical versus horizontal/layered; centralization versus decentralization; mechanical components versus information packets; order versus messy complexity; industrial cooperation versus government-university-industrial complex.[7] Both Haraway's and Hughes's analyses spotlight not only the transformation of a world-picture but also the concomitant reconfiguration of power relations and discursive practices (knowledge/power nexus) in the postwar era.

The representation of phenomena in terms of information systems was not confined to molecular biology. Nearly every discipline in the social sciences (sociology, psychology, anthropology, political science, and economics), as well as in the life sciences (immunology, endocrinology, embryology, physiology, neuroscience, evolutionary biology, ecology, and molecular genetics) flirted with the seductive ideas of cybernetics and information theory in the 1950s, with different degrees of productivity and commitment. Though most pronounced in the United States and the Soviet Union, these trends occurred also in England and France.[8]

Thus informational representations of heredity did not arise from the inner logic of DNA genetics, nor were they the outcome of the elucidation of the architecture of the double helix in 1953. Linguistic tropes and textual metaphors in the life sciences, including molecular biology, preceded that landmark event. These descriptions of heredity—constructed still within the protein paradigm of the gene—were transported into molecular biology from the new techno-sciences of cybernetics, information theory, and computer design in the late 1940s. And while the idea of a code-script analogous to a Morse code was proposed by Erwin Schrödinger in 1944, the idioms of information theory and cybernetics were absent from his acclaimed but old-fashioned book, *What Is Life?* It was in the 1950s that the genetic code was signified as an information system.[9]

Eminent physicists, biophysicists, chemists, mathematicians, communication engineers, and computer analysts—whose own projects situated them at the hub of weapons design, operations research, and computerized cryptology—joined in the effort to "crack the code of life." And they transported the tropes and icons of the new communication sciences into molecular biology, thus extending and amplifying the discursive space that emerged in the late 1940s. We shall examine first the emergence of the information discourse in biology at the end of the 1940s, and then follow its instantiation in the later analyses of genetic codes, where it resulted in a deconstruction. While it now seemed technically legitimate to speak of molecules, genes, and organisms as information and linguistic systems, there was a fatal flaw in this discursive reconfiguration, for in information theory language was purely syntactic and information has no semantic value.[10] Ironically, these new scriptural representations of heredity (having little to do with human communication) endured as a metaphor for the eternal word, or book of life, but in their strict technical sense led to the erasure of its meaning.

ORGANISMS, INFORMATION, TEXTS

The impact of digital control systems on representations of life processes was both anticipated and promoted by MIT mathematician Norbert Wiener and his circle in the early 1940s, while they were working on computational problems of ballistics in Warren Weaver's mathematical division within the Office of Scientific Research and Development (OSRD). These interests were first articulated in 1943 in a noted joint article, "Behavior, Purpose, and Teleology," by MIT-trained engineer Julian Bigelow, Harvard physiologist Arturo Rosenblueth, and Wiener—the first link between servomechanims, physiological homeostasis, and behavioral processes.[11] Negative feedback as a paradigm of thought and action became Wiener's disciplinary mission. Even before the war ended, Wiener, the Hungarian émigré mathematician John von Neumann, and their circle began campaigning vigorously for the as-yet-unnamed field of automated digital control.[12]

By 1945 von Neumann, already a key figure in strategic military planning, was becoming interested in biology in general, and genetics in particular, as part of his investigation of self-reproducing machines. He focused on viruses and phages as the simplest models of reproduction. In linking viruses to information processing, von Neumann, like most researchers (especially in the United States), operated within the dominant paradigm in life science. He conceptualized reproduction within the protein view of heredity, in which autocatalytic mechanisms of enzymes served as explanations of gene action and virus replication, and genetic specificity arose from the combinatorial properties of amino acid sequences. Von Neumann made contacts with the biomedical community, participating in meetings and communicating with life scientists such as Max Delbrück, Sol Spiegelman, Joshua Lederberg, and John Edsall. Through these exchanges he was encouraged to develop his self-duplicating machine as a possible heuristic model of gene action. Although von Neumann's project in biology did not go far (he died in 1957), its discursivities began to permeate genetic representations.[13]

The cybernetic view of life gained momentum after 1948. That year Wiener's book *Cybernetics: Or Control and Communication in the Animal and Machine* was published simultaneously in France and the United States. In this remarkably influential book Wiener expounded two central notions: that problems of control and communication engineering were inseparable—communication and control were two sides of the same coin—and that they centered on the fundamental notion of the message: a discrete or continuous sequence of measurable events distributed in time. He expounded on what Michel Foucault would call a new episteme: the coming of the information epoch: "If the seventeenth and early eighteenth centuries are the age of clocks, and the eighteenth and nineteenth centuries constitute the age of steam engines, the present time is the age of communication and control."[14]

His historicization applied also to animate phenomenology. The twentieth-century body was sharply demarcated from that of the nineteenth century, having moved from material and energetic representations to informational ones. "In the nineteenth century," Wiener observed (supplying the discursive framework of information for Francis Crick's Central Dogma a decade later),

> the automata which are humanly constructed and those other natural automata, the animals and plants of the materialist, are studied from a very different aspect. The conservation and degradation of energy are the ruling principles of the day. The living organism is above all a heat engine. . . . The engineering of the body is a branch of power engineering. . . . [T]he newer study of automata, whether in the metal or in the flesh, is a branch of communication engineering, and its cardinal notions are those of the message, amount of disturbance or "noise," . . . quantity of information, coding techniques, and so on.

He stressed that the same conclusions held for both inanimate and animate systems: for enzymes, hormones, neurons, and chromosomes. And through regular dialogues and visits with his old friend John B. S. Haldane, Wiener's cybernetic view of heredity, like Haldane's, became grounded in the primacy of proteins. Like von Neumann, Wiener forecasted in 1948 that the combinatorial mechanisms by which amino acids organized into protein chains, which in turn formed stable associations with their likes, could well be the mechanisms of reproduction of viruses and genes.[15]

Also in 1948 the MIT-trained engineer Claude E. Shannon at Bell Labs published a major article, "The Mathematical Theory of Communication," in which he developed the salient features of information theory. His work suggested that information can be thought of in a manner entirely divorced from content and subject matter—devoid of semantic value—and established the basic unit of information as the binary digit, or bit. His highly technical article gained wide exposure through a joint effort with Warren Weaver (director of the Rockefeller Foundation's Natural Science division and molecular biology program), who explained the mathematical concepts of information in his characteristically lucid and eloquent manner.[16]

While developing the mathematical theory of communication, Shannon was also working on secrecy systems (indeed, the two were intertwined). His work transformed theories of coding, cryptanalysis, and linguistics by introducing concepts such as (mathematical) redundancy and binary coding. But these linguistic analyses had little to do with human language: they were designed for machine communications. Shannon was mindful of these differences. Though he pondered the relation between information theory and human communication, he scrupulously circumvented semantic questions. As we shall see, however, the new cryptology, coupled with electronic computing, would be applied to analyses of genetic codes.[17]

The Wiener-Shannon theory gained even wider circulation following the publication of Wiener's book *The Human Use of Human Beings*, in 1950. Here he proposed that contemporary society could be understood only through a study of messages and communication facilities. The individual and the organism must be recast in terms of information. In a key chapter entitled "The Individual as the Word," Wiener elaborated this concept and its corollary—the technical possibility of writing the book of life.

> The earlier accounts of individuality were associated with some sort of identity of matter, whether of the material substance of the animal or the spiritual substance of the human soul. We are forced nowadays to recognize individuality as something which has to do with continuity of pattern, and consequently with something that shares the nature of communication.

It is not matter but the memory of the form that is perpetuated during cell division and genetic transmission, he insisted. He prophesized that in the future it would be possible to transmit the coded messages that comprise organisms and human beings.[18]

We witness here the opening of a new discursive space in which the word, or the message, became bound up with concepts of the gene as the locus of scriptural-technological control, and with notions of control over bodies, in ways that bypassed their physicality. Their three-dimensionality was flattened into a one-dimensional magnetic tape, their material density symbolically represented as a digital code.

The work of Wiener, Shannon, and von Neumann affected a vast range of academic fields and cultural sensibilities, especially in the United States, but also in England, France, and the Soviet Union. Even in those disciplines where the technical promises of information theory did not materialize, the information discourse with its icons and linguistic tropes—information transfer, messages, words, codes, texts—shaped the thinking, imagery, and representations of phenomena. Such was the case with the genetic code, where the discursive framework of information outlasted authentic technical attempts to apply the theory to genes and the problem of protein synthesis.

The mission of building a new discipline—an information-based biology—was taken up by the Viennese émigré radiologist Henry

Quastler, who in the mid-1950s moved from the University of Illinois's Laboratory of Control Systems to Brookhaven Laboratories. From 1949 until his death in 1963, Quastler, inspired by Wiener and Shannon and funded through military sources, channeled relentless energy into rewriting biology as an information science. His output—articles, reports, symposia, and books—was prolific. He achieved a measure of acclaim in the 1950s, with hundreds of references to his work in the *Science Citation Index*. In 1952 he organized the first major symposium at the Control Systems Laboratory on information theory in biology.[19] Two things are striking about the proceedings: the discursive shift to scriptural representations of life, and the conceptualization of information storage and transfer within the protein paradigm of heredity.

Working out the measure of biological specificity in terms of information content, Quastler provided a quantitative formulation of what five years later Crick would articulate qualitatively as the sequence hypothesis. Similarly, several symposium participants investigated the informational specificity inherent in the combinatorial properties of proteins. Proteins were especially attractive for information theory, they observed, because "[t]hey are constructed much as a message ... the protein molecule could be looked upon as the message and the amino acids residues as the alphabet," so that researchers could study the intersymbol correlations in the protein text. Just two years later such texual analyses would become the prime resource in attempts to break the genetic code.

In 1949 Quastler and physicist Sydney Dancoff published a paper in which they calculated the information content of a human organism to be 5×10^{25} bits. Since the information content of single printed page is about 10^4 bits, this meant that the description of a human being would be the equivalent of 5×10^{21} pages of information—on the order of an enormous library. Based on this logic, the information content of a germ cell was set at about 10^{11} bits; and what they called the "genome catalogue" at the order of about a million bits.[20] These scriptural representations of life and heredity—texts, alphabets, letters, words, messages—were a new form of biology. They were grounded in the discursive framework of information theory, and they appeared well before DNA replaced proteins as the source of hereditary information.

Information theory seemed to promise a great deal in the early 1950s. A week before Watson and Crick's report on the double-helical structure of DNA appeared in *Nature* in 1953,[21] Watson, with geneticist Boris Ephrussi and physicists Urs Leopold and J. J. Weigle, sent a note to *Nature* suggesting new terminology in bacterial genetics. Arguing for imposing rhetorical order on the proliferating semantic confusion in bacterial genetics (transformation, recombination, induction, transduction, etc.), they proposed to replace these uses with the term "inter-bacterial information." "It does not imply necessarily the transfer of material substances, and recognizes the possible future importance of cybernetics at the bacterial level."[22]

This expectation of the relevance of the information concept to the study of the combinatorial properties of DNA clearly guided the thinking of Watson and Crick in their *Nature* report.

> It follows that in a long molecule many different permutations are possible, and it therefore seems likely that the precise sequence of bases is the code which carries the genetic information.[23]

To emphasize: this was no longer Schrödinger's quaint code-script, which had drawn on a fifty-year-old tradition (deriving from people like Emil Fischer and Charles Sherrington) of embedding biological specificity in the combinatorial elements of protein sequences.[24] And it did not merely reflect a paradigmatic shift from protein to nucleic acid sequences as the origin of genetic specificity. The very idea of biological specificity was being displaced by the notion of information, as were representations of texts and codes. Watson and Crick's code bore its own historicity, that of the postwar era: it governed the transfer of hereditary information, represented in terms of the discourse of electronically programmed communication system.

The Rise and Fall of Overlapping Codes

George Gamow, the Russian émigré physicist who taught at George Washington University and was also a cartoonist, science popularizer, and military strategist, had been enthusiastic about biology for more than a decade. But Watson and Crick's 1953 publication in *Nature* excited him even beyond his normal exuberance.

> I remember very well this day, I was for some reason visiting Berkeley and I was walking through the corridor in Radiation Lab, and there was Luis Alvarez going with *Nature* in his hand (Luis Alvarez was interested at this time in biology) and he said, "Look, what a wonderful article Watson and Crick have written." This was the first time I saw it. And then I returned to Washington and started thinking about it. . . .[25]

Gamow immediately perceived the relationship between DNA structure and protein synthesis as a cryptanalytic problem: how can a long sequence of four nucleotides determine the assignment of long protein sequences comprised of twenty amino acids? "And the question was to find, is it possible?"[26]

> And at that time I was consultant in the Navy and I knew some people in this top secret business in the Navy basement who were deciphering and broke the Japanese code and so on. So I talked to the admiral, the head of the Bureau of Ordnance. . . . So I told them the problem, gave them the protein things [list of amino acids], and they put it in the machine [computer] and after two weeks they informed me there is no solution. Ha!

Cryptanalysis too was being transformed by information theory and computing. Although it had utilized letter frequency analysis and their neighbor relations ever since the Renaissance, these methods had become mechanized and mathematized during World War II, when cryptology became a prime source of intelligence. But by the time Gamow consulted with the navy cryptologists, even those recent practices were already being transformed by Shannon's statistical theory of communication, which had set up a measure of redundancy in codes (redundancy would equal one minus the relative entropy, which in turn is a measure of information). Shannon argued that redundancy arose from an excess of linguistic rules, and that—backed by letter frequency counts—it could furnish the ground for cryptanalysis. Information theory showed how to raise the difficulty of cryptanalysis and how much ciphertext was needed to reach a valid solution. These approaches were now coupled with electronic computers. As such they did not merely raise the sophistication and efficiency of cryptanalysis but reconstituted it within the new sciences of electronic communication, symbolic logic, linguistics as logical syntax, and mechanical translation; cryptanalysis was resignified within the compelling logic of guidance and control.[27]

Gamow's initial solution to the cryptanalytic problem (how sequences of four nucleotide bases specified sequences comprised of twenty amino acids) became known as the "diamond code."[28] It was an overlapping triplet code (but not called so until a year later), based on a combinatorial scheme in which four nucleotides, arranged three at the time ($4 \times 4 \times 4 = 64$), were more than sufficient to specify twenty amino acids (whereas nucleotide doublets, $4 \times 4 = 16$, clearly fell short). The scheme was based on DNA-protein "translation": the amino acids would fit into the rhomb-shaped "holes" (diamonds) formed by the nucleotides in the DNA chain. Drawing on Quastler's work, Gamow argued that there was a mathematical correspondence between the "expected intersymbol correlation of the overlapping nucleic acids of the diamonds" and the "observed intersymbol correlation" between the amino acids in polypeptide chains. It so happened that there were exactly twenty such diamond-shaped configurations, corresponding to the twenty amino acids in the alphabet—"magic twenty," as it soon came to be called (see figure 15.1).

Several researchers—Linus Pauling, Erwin Chargaff, Francis Crick, and Martynas Ycas—raised objections to Gamow's model as soon as it was published. Both Crick and Ycas pointed out that his diamond code contradicted the small amount of sequence data then available (Fred Sanger had recently sequenced insulin). Ycas, a Russian émigré and Caltech-trained biologist working at the Pioneering Research Division, Quartermaster Research and Development Command, U.S. Army, became Gamow's decoding partner. As Gamow explained, "Dr. Ycas takes care of the biological part of the problems we are trying to attack, while I handle the more mathematical part of the picture."[29]

In the fall of 1954, while he was a visiting professor at Berkeley, Gamow engaged Caltech scientists Max Delbrück, molecular biophysicist Alexander Rich, and physicist Richard Feynman in the coding problem. They too were skeptical of his diamond code and proposed a coding scheme based on RNA; the role of RNA as an intermediary in protein synthesis was then increasingly accepted. But decoding along these new lines would require a computer, Gamow argued.

> As in the breaking of enemy messages during the war (hot, or cold!), the success depends on the available *length* of the coded text. As every intelligence officer will tell you, the work is very hard, and the success depends mostly on luck. There are $20! = 10^{17}$ possible assignments of the aa's [amino acids] to base triplets! . . . I am afraid that the problem cannot be solved without the help of electronic computer.

Gamow planned to put the problem on Los Alamos's Maniac in July.[30] With growing enthusiasm for the coding problem, Gamow formalized the coding network by establishing the RNA Tie Club, a group of twenty members (selected by Gamow) corresponding to the twenty

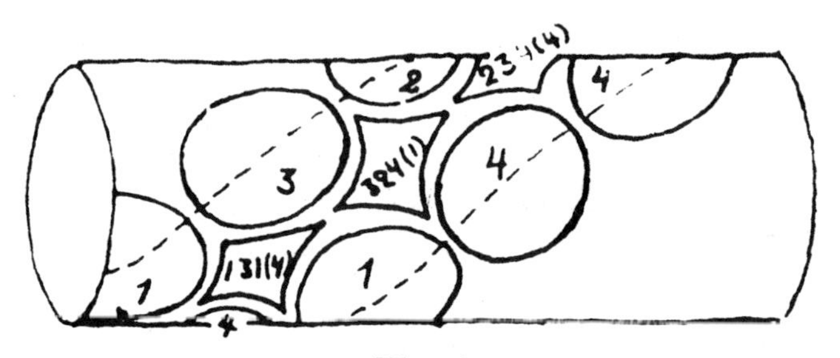

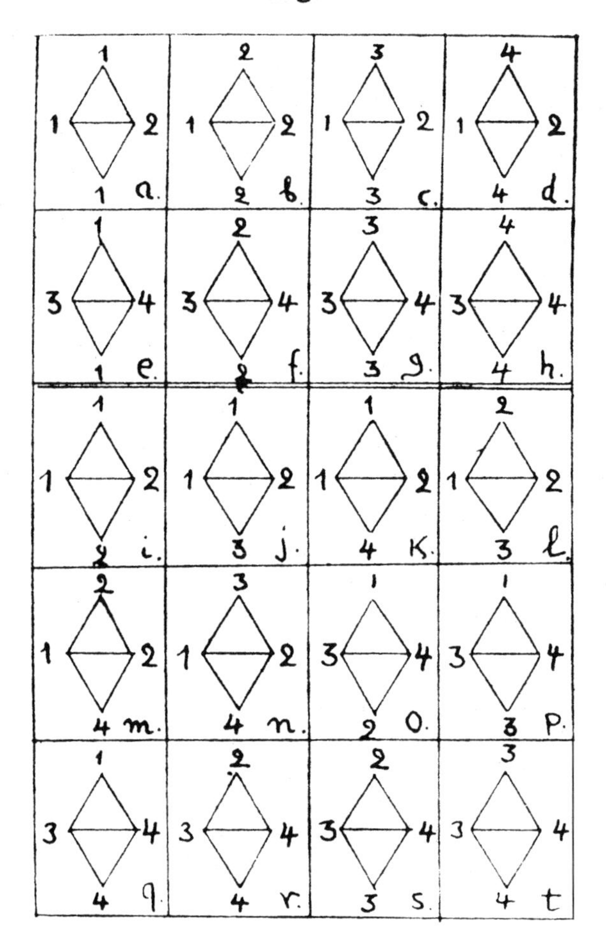

Figure 15.1
Top: Diamonds arranged along a schematic representation of Watson and Crick's double helix. Bottom: The twenty different types of diamonds. Source: *Nature,* 173 (1954), p. 318.

amino acids (another "magic twenty"?). Each member was to have a diagrammed tie, made to Gamow's design, and a tie pin bearing the abbreviation of the member's amino acid. Thirteen members were physical scientists—physicists, chemists, mathematicians, and computer scientists. The club's aim was to foster communications and camaraderie by circulating notes, manuscripts, and articles pertaining to the coding problem; its hope was to locate funding for regular meetings. (For a while it seemed certain that the army's Quartermaster Research and Development Command would sponsor the RNA Tie Club, but the scheme fell through.) Delbrück's motto, "Do or die, or don't try," graced the club's stationery.[31] Though it fell short of Gamow's high expectations, the club served to promote the coding problem, with all its discursive and material resources, among geneticists and later biochemists.

Several coding schemes were soon proposed by members of the RNA Tie Club: major-minor code, compact and loose triangular codes, and Edward Teller's sequential code. All were quite restrictive—either fully overlapping (sharing two out of three nucleotides in a triplet) or partially overlapping (sharing just one). They were all based on the principle of matching polypeptide fragments with overlapping nucleotide triplets along a single-stranded nucleic acid; and all were extraordinarily complicated, necessitating use of computerized cryptanalysis.[32]

In the summer of 1954, Gamow and Nicholas Metropolis—a theoretical physicist and expert in electronic computing and logic design—tested the various overlapping (restrictive) codes on the Maniac (using the so-called Monte Carlo method). Applying statistical frequency analysis, they compared artificial amino acid sequences based on the various proposed codes with naturally occurring sequences; the stronger the restrictions on the association between amino acids in the sequence, the smaller the number of different neighbors as compared to random distribution.

The results were negative. In spite of the strong intersymbol restrictions in the proposed codes, both the artificial sequences constructed based on them and the naturally occurring sequences produced a random amino acid distribution. Rather than question their guiding premise that the code was overlapping, or more fundamentally that the scheme was, in fact, a language-like code, the team inferred that the method employed was not sensitive enough.[33]

Another member of the RNA Tie Club offered a different angle on coding: symbolic logic. Mathematical biophysicist Robert Ledley from Johns Hopkins's Office of Operations Research applied his expertise in digital computational methods of symbolic logic. He showed how this method could work for the limited case of $3! = 6$ (three amino acids); this took only a hundred hours on the Maniac. But to solve the coding problem for the case of $20! = 2.3 \times 10^{17}$ (twenty amino acids), "a computer put to work in the days of the Roman Empire, at a rate of one million solutions per second, 24 hours a day, all year round, would not yet be close to finishing the job," he estimated.[34]

Two other decoding systems were devised in 1955: Gamow and Ycas's statistical analysis of the correlation between nucleic acid and protein composition in two viruses (TMV, tobacco mosaic virus, and TYV, turnip yellow virus), which led to contradictory results; and Rich and Gamow's neighbor distribution plots based on a table of numerous dipeptides meticulously compiled by the South African molecular biologist Sydney Brenner. The table showed that amino acid distribution in peptides followed a Poisson (random) distribution, and thus negated the restrictiveness (nonrandomness) of overlapping codes. Two years later, much too late, Brenner would provide an elegant and decisive proof that fully overlapping codes were logically impossible.[35]

By the end of 1955, when Gamow, Rich, and Ycas gathered all their findings in a major review, "The Problem of Information Transfer from the Nucleic Acids to Proteins," it was clear that the code was nonoverlapping. Yet no one confronted the broader epistemic implications of these findings: the problematic linguistic attributes of nucleic acids and proteins, and the questionable validity of the code idiom. The idea of information veiled these problems.

Indeed, Gamow, Rich, and Ycas did not deploy the notion of information unconsciously or casually, as did many life scientists in subsequent years. They used information—with all its valencies to mathematics, logic, cryptanalysis, linguistics, computers, operations research, and weapons systems—as a means of framing the coding problem. The trope of information served to integrate mechanisms of molecular specificity, structural considerations, mathematical relations, linguistic attributes, and coding within a single explanatory framework.

Even if one grants the problematic analogy between combinatorial elements in a molecule and the alphabetical elements of a language, the "translation" between the four nucleic acid bases and the twenty amino acids would obey the rules of cipher, not code. Codes operate on linguistic entities, dividing their raw material into meaningful elements like words and syllables, while ciphers will split the "t" from the "h" in "the," for example. More fundamentally, whether code or cipher, the tacit operating assumption behind all the so-called deciphering or decoding attempts was that the code operated on language-like entities. And once the analogy between combinatorial elements in nucleic acids or proteins and the letters of the alphabet took root, the comparison with language took on a life of its own, until the metaphor of language, with all its ambiguities, aporias, and tautologies, began to be taken literally. "The Code"—that logocentric icon preceding its experimental warrant—demanded the existence of language as its predicate. No language, no code.

The commitment to "The Code" was unwavering, even in the face of empirical obstacles. The authors even compared cryptoanalytically the neighbor frequency in the English language (using Milton's *Paradise Lost*) with distributions of artificially constructed amino acid sequences based on the various codes. The English-language sample (being highly redundant, or restrictive) deviated markedly from Poisson (randomness), while known amino acids followed random distribution. They even used Shannon's cryptanalytic innovations to determine the information content of these sequences, only to learn that, once again, their theoretical sequences deviated markedly from Poisson and from their experimental sample.[36] Herein lies another aporia. Clearly the amino acid "text" did not obey the rules of any known language (as Quastler's colleagues had already demonstrated); either the code overlaps and thus contradicts the experimental evidence, or it does not overlap and contradicts the meaning of language. But precisely because language and code necessitated each other in the authors' representations, the tautology survived intact.

Syntax to Semantics? Comma-Free and Other Codes

Once the empirical evidence persuaded the code workers that the code was nonoverlapping—that is, it had no intersymbol restrictions—the analogies with human communication lost their validity. Unlike human languages, possessing semantic attributes, the code's language, like machine languages, would have to be purely syntactic. By the time their article went to press, Gamow and Ycas had already constructed a nonoverlapping code, the so-called combination code, in which only the combination of the bases in a triplet and not their order mattered. The number of these combinations added up exactly to what became known as the "magic twenty"—the twenty amino acids.[37]

Nicholas Metropolis and Stanislaw Ulam at Los Alamos were already testing that code to address the problem of randomness of amino acids. And soon after Gamow called on his friend von Neumann to supply an analytical solution to the thorny problem of explaining the nonrandom distribution of amino acids predicted by the combination code in relation to the arrangements of nucleotide triplets in RNA. "What do *you* think about this 'new' trouble?" Gamow prodded Ycas. But after weighing the pros and cons of randomness, Ycas concluded that "it would be best to say that when such eminent gray matter as von Neumann and Ulam are working at high speed it is best for myself to remain blank."[38] By the time they submitted their paper for publication (1955) they managed to make some sense of the contradictions and were quite satisfied with their results.

But their code had drawbacks, as Ycas admitted to Crick.

> Chemically, of course, this [combination code] makes no *obvious* sense. Since the triplets are non-overlapping, we have a "punctuation mark" problem. Also the "degeneracy" to get magic 20 raises stereochemical problems.[39]

With the problem of "punctuation marks" Ycas articulated what would soon become a key preoccupation for analysts of nonoverlapping codes: how to distinguish between consecutive triplets along the nucleic acid sequence.

Crick concurred. Having just vetoed all extant codes, he pointed out that in the DNA double helix one chain runs up while the other runs down, as Delbrück had already emphasized, implying that "a base sequence read one way makes sense, and read the other way makes nonsense."[40] We see that molecular directionality was no longer merely analogized to the directionality of language, it was conflated with it through the act of reading and sense-making. Once again we witness how the tropes of language—reading and sense—became constitutive of the decoder's modes of reasoning, and how the information discourse (erroneously) bestowed semantic attributes on the (syntactic) arrangement of molecular symbols. We see the process by which Nature was being (re)textualized.

There was much discussion about the "punctuation mark problem," namely how the "reading mechanism" recognized the start and end of a triplet, Ycas recalled. Without a marker ("punctuation"), a sequence of bases could be read ambiguously, for example, ABC, DCC, BDA, . . . or, A, BCD, CCB, DA. . . . There were three ways out

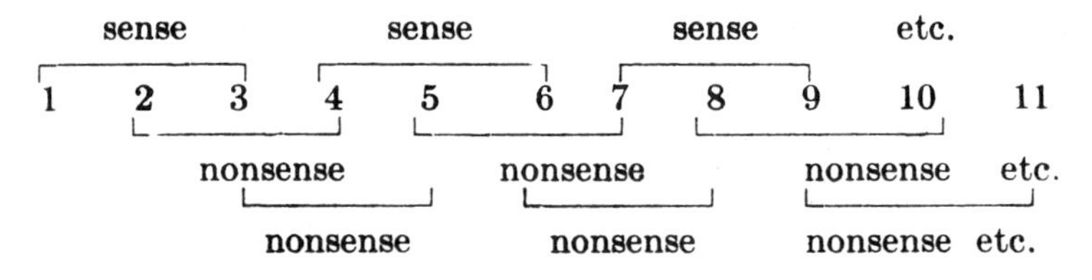

Figure 15.2
The numbers represent the positions occupied by the four letters A, B, C, and D. The diagram shows which triplets make sense and which nonsense. From F. H. C. Crick, J. S. Griffith, and L. E. Orgel, "Codes without Commas," *Proceedings of the National Academy of Science*, 43 (1957).

of this conundrum: there could be an initial configuration, marking the beginning of the count (as was later shown); the "coding" triplets could be separated by some kind of "noncoding" (nonspecific) arrangement of bases; or the nature of what soon became known as the "dictionary" might itself prevent ambiguities—namely, only those particular configurations of triplets that specified amino acids would have "meaning" and count as "words." It was the third option that Crick, with chemist Leslie Orgel and biophysicist John Griffith, chose to elaborate in their ingenious comma-free code in 1956. (See figure 15.2.) And as Ycas observed, this type of code was already well established in communication engineering.[41]

Almost miraculously it turned out to be possible to select twenty "sense" triplets in such a way that juxtaposition of any two produced overlapping "nonsense" triplets, thus preserving the "magic twenty." The solution was not unique, though: there were 288 solutions. The authors also offered a mechanism for the assembly of polypeptides. Although they were cautious about their scheme, the comma-free code generated much excitement, "It seemed so pretty, almost elegant," Crick recalled. Parenthetically, physicist Leo Szilard, then deeply immersed in molecular biology, came up independently with a similar model.[42]

Comma-free codes attracted considerable attention. While their novelty, pristine logic, and seductive elegance intrigued molecular biologists, this class of codes was already being used in advanced communications systems. It seemed natural, therefore, to draw on the expertise of mathematicians and communication engineers to analyze and elaborate the general properties of comma-free codes. Such expertise was readily available at Caltech's Jet Propulsion Laboratory (JPL)—

a stone's throw from Delbrück's group—which since the end of World War II had been sponsored by the U.S. Army and was a major center of the ICBM (intercontinental ballistic missile) program and other military aerospace technologies.[43]

Three JPL scientists responded to challenge of comma-free codes: Solomon W. Golomb, a young mathematician and communication engineer; mathematician Basil Gordon, a specialist in combinatorial analysis; and mathematician and electrical engineer Lloyd R. Welch, who two years later joined the Institute for Defense Analysis. In the summer of 1957 they worked on providing a mathematical generalization of the coding problem: what was the maximum size of a comma-free dictionary in the case of an arbitrary number of symbols and arbitrary length of words? They defined the set D of k-letter words ("sense") as a "comma-free dictionary" in such a way as to exclude all overlapping triplets ("nonsense").Their solution, though partial, demonstrated that it was possible to construct such codes.[44]

The term "dictionary" added yet another linguistic trope to the discursive and semiotic space configured by information theory, electronic computing, and linguistics, within which the genetic code was being constituted as a scriptural technology. Their usage of that idiom seemed reasonable from the standpoint of cryptanalysis based on information theory, where a dictionary simply signified a specific set of symbols. But technically, these symbols could never transcend their syntactic bonds to attain the semantic value—or "sense"—of human language.

Rather unexpectedly, comma-free codes could even contribute to the mathematical theory of coding, since they were a subset of a larger class of related codes, the authors argued. In their search for optimum coding techniques, Claude Shannon and Brockway McMillan had studied theoretical codes in which the entire message was available. But in actual communication application—as with the problem of the genetic code—only disjointed portions of a message were likely to be received. This is where comma-free codes could be useful, they concluded. Genetic codes were now becoming "boundary objects," migrating along the two-way traffic between molecular genetics and the militarized world of mathematics and communication engineering.[45]

The treatment of Golomb, Gordon, and Welch was purely mathematical. Soon, however, Delbrück teamed up with Golomb and Welch to address comma-free codes as both a mathematical and a genetic problem. They showed that there were five classes of such codes (a total of 408 codes). They also expanded the range of possible coding errors—"misprints" as they called them—to include what they termed "missense."

> Certain misprints in the coded message will produce nonsense (the resulting triplet does not code for any amino acid), other misprints will produce missense (the resulting triplet codes for a different amino acid). The codes were studied with respect to missense/nonsense ratios produced by various classes of misprints.[46]

But the most thoughtful feature of their model—from a biological standpoint—underscored that such genetic codes had to meet another requirement: transposability (directionality relative to a DNA strand). As Delbrück had stressed for years, DNA had no intrinsic sense of direction, since its two helices run in opposite directions. And since there were no known cases of one genetic locus coding for two different proteins, it seemed likely that only one DNA strand was read. Thus, Delbrück argued, the dictionary had to be not just comma-free but transposable as well. Namely its reversed complement had to be all nonsense.

> We wish to emphasize that we consider the postulate of comma-freedom and the postulate of transposability to be almost on the same footing. Indeed the principal virtue of comma-freedom is that any message can be read unambiguously starting at any point, with the proviso, however, that one must know in advance *in which direction to proceed*. Since the equivalence of the two opposite directions in a structural sense seems to be one of the more firmly established features regarding the DNA molecule the advance knowledge as to the direction in which to read cannot come from the basic structure. Comma-freedom would therefore seem to be worthless virtue unless it is coupled with transposability.[47]

As he had done for several years, Delbrück framed his arguments within the discourse of information. Given the mounting evidence that DNA functioned through the intermediary RNA, he felt that "it would be unwise not to give some currency to 'information transfer' as possible replication mechanism."[48]

Delbrück did not invoke the information idiom lightly. Like most molecular biologists, he tended to use the term generically (rather than in its mathematical sense), but unlike them he was well versed in information theory. "I am teaching information theory this term," he wrote in the fall of 1955 to his colleague Robert Sinsheimer, then at Iowa State College, "and we have been using your DNA data to illustrate the notions of intersymbol influence and statistical properties of information sources."[49] Clearly, his close interactions with the mathematicians and communication engineers at JPL shaped his thinking, thus reinforcing the growing trend toward representing heredity as a communication system of guidance and control; even if by the conflation of technical and genetic usages of information he actually undermined the validity of DNA, RNA, and proteins as cryptanalytic texts possessing sense.

Robert Sinsheimer also tried his hand at decoding in 1958, after he had moved to Caltech. He took a bold leap: a binary code. Sinsheimer had become familiar with feedback controls, cybernetics, and communication theory as a graduate student, when he worked at MIT's Radiation Laboratory during the war. "It therefore seemed very natural to apply the concepts of information content, stability to thermal and other interference (noise), etc. to the issue of genetic inheritance, mutation, et al.," he recalled. At Iowa State University he gave a series of lectures on these subjects, in John W. Gowen's genetics seminar, based on Wiener's *Cybernetics*, Shannon and Weaver's *The Mathematical Theory of Communication*, and Quastler's *Essays on the Use of Information Theory in Biology*.[50]

Given the recent finding that base composition in nucleic acids varied widely in different organisms and that the pattern of variation followed the equivalence of bases A to C (Adenine and Cytosine, with amino groups) and T to G (Thymine and Guanine, with keto groups), only two symbols mattered, he reasoned.

> There is an elementary theorem in information theory (Shannon & Weaver, 1949) that if a message is to be written in a code of T symbols, the message can be written most efficiently (i.e., make use of the least quantity of symbols) if each symbol is used to an equal extent. In our case the message would be the protein content of a cell; this is to be expressed in a two-symbol (*N*, *K*) RNA code.[51]

In his informationally efficient two-letter code the quantity of the letters (*N*, *K*) was always equal—thus satisfying Shannon's theorem—and the effective composition of DNA was invariant.

With such a proliferation of codes—several hundreds of them including combination codes, comma-free codes, transposable comma-free codes, binary codes, quadruplet codes, and sextuplet codes—the optimism surrounding the coding problem had dwindled markedly by the end of the 1950s. The sedentary life of the RNA Tie club came to an end; its president, Gamow—preoccupied with his new life at the University of Colorado and beleaguered by alcohol-related ills—had become primarily a spectator and commentator. And it was not much of a spectacle. Assessing the "Present Position of the Coding Problem" at the 1959 summer symposium at Brookhaven National Laboratory, Crick was glum. He now was ready to reassess some of the fundamental premises of the coding problem.[52]

There was little to hold on to but the cardinal belief in the existence of "The Code"—that object prefixed in the mind and preceding empirical evidence. For five years now—despite the contradictory data and dubious results, and with some of the most "eminent gray matter working at high speed," as Ycas had put it—there were intriguing theories but few concrete returns. Yet the champions of the genetic code clung tenaciously to their logos. And in the next two years there were several other attempts to "crack the code of life," relying primarily on the recently sequenced TMV as a "Rosetta stone."

As late as the spring of 1961 Delbrück's colleague, Golomb, was still enthusiastically refining his promising comma-less code. "New Way to Read Life's Code Found," announced the *New York Times*. "A 'Dictionary' of 24 Words Appears Able to Describe Inheritance Mechanism."[53]

> The scientist, 29-year-old Dr. Solomon W. Golomb of the Jet Propulsion Laboratory of Pasadena Calif., told a meeting of American Mathematical Society at the New Yorker Hotel of building what might be called a "dictionary of life".... Biologists who have examined the new theory say it fits all of the facts now known about the way nucleic acids and proteins work in transmitting hereditary information.... Scientists from several fields have been trying to crack the code of life since the middle Nineteen Fifties.... Dr. Golomb decided that this [nature's alphabetic] extravagance probably produced the redundancy in the coded messages that would be necessary to minimize mistakes and assure a high degree of reliability in transmitting genetic information ... he is satisfied that his code is a good and workable one and could be found in nature—if only on another inhabited planet.

But within a few weeks of this announcement something no one expected happened. The genetic code was broken by two obscure young biochemists, Marshall Nirenberg and Heinrich Matthaei at

NIH, who broached the problem from a radically different point. And within a few years the problem of correlating amino acids with nucleotide triplets was solved biochemically.

CONCLUDING REMARKS

Scientists have generally tended to regard the theoretical work on the genetic code in the 1950s as naively optimistic at best, or erroneous and unfruitful at worst: a defeat of the Pythagorean ideal by the material world. But as one of the code breakers, Carl E. Woese (physicist, microbiologist, and author of the important text *The Genetic Code*), reflected,

> What has not been generally appreciated is that the subsequent spectacular advances in the field, occurring in the second period [1961–1967], were interpreted and assimilated with ease, their value appreciated and new experiments readily designed, precisely because of the conceptual framework that had already been laid.[54]

But it is not merely the conceptual framework that had been laid down. Rather a knowledge-power nexus had formed within which molecular biology reconfigured itself as a (pseudo)information science and represented its objects in terms of communication systems (including linguistics).

Several scholars have noted the cognitive and disciplinary impact of the physical sciences on molecular biology, as well as the impact of Cold War technosciences on the life and social sciences. As Evelyn Fox Keller has observed, physics was a social resource for biology; by borrowing physics-like agendas, language, and attitudes, biologists eventually reframed the character and goals of their field.[55] But physics was not the only source of influence. By examining the work on the genetic code in the 1950s—cognitive strategies, discursive practices, semiotic tools, and institutional resources—we have seen how the old problem of genetic specificity was reframed. It was recast through scriptural technologies poised at the interface of several interlocking postwar discourses: physics, mathematics, information systems, cryptanalysis, electronic computing, and linguistics.

Thus to extract the levels of significance of the genetic code as an information system and its various tropes—language, dictionary, message, instructions—we must appreciate the scientific legitimation it received through information theory, and through the regimes of signification of Cold War military technoculture. Ironically, these modes of signification led to a deconstruction: since the language of information theory is purely syntactic, the book of life conveys no semantic meaning.

But such epistemological pedantries have hardly retarded the enormous technological and cultural momentum of genomic representations as advanced information systems. For example, for Leroy Hood, czar of the human genome project, mapping the human genome, or creating an encyclopedia of life, is essentially a project both driven by and generating technology, notably computer techniques for inputting, storing, and accessing the three billion base pairs, and fast microchips for pattern recognition.[56] "Three billion bases of sequence can be put on a single compact disk (CD)," promises molecular biologist and Nobel laureate Walter Gilbert. "One will be able to pull a CD out of one's pocket and say, 'Here is a human being; it's me!'... To recognize that we are determined, in a certain sense, by a finite collection of information that is knowable will change our view of ourselves. It is a closing of an intellectual frontier, with which we will have to come to terms."[57] Though probably unaware of its sources, this manifestation of technoscientific imagination and its discourses was shaped in the 1950s by emergent representations of nature, life, and society as information systems.

Notes

1. Jacques Monod, *Chance and Necessity: An Essay on the Natural Philosophy of Modern Biology* (New York: Vintage Books, 1972), especially chaps. 3 and 4; quotes on 45 and 68, respectively.

2. François Jacob, *The Logic of Life: A History of Heredity* (New York: Pantheon Books, 1974), introduction and chap. 5; quotes on 1 and 254, respectively. Originally published in France in 1970.

3. Robert L. Sinsheimer, *The Book of Life* (Reading, Mass.: Addison-Wesley, 1967), 5–6.

4. Alain Touraine, *Le Société Post-industrielle: Naissance d'une Société* (Paris: Denöel, 1969); and Daniel Bell, *The Coming of the Post-Industrial Society* (New York: Basic Books, 1973). For an interesting overview of theories of postindustrial or information society see David Lyon, *The Information Society: Issues and Illusions* (Cambridge: Polity Press, 1988), and Boris Frankel, *The Post-Industrial Utopians* (Madison: University of Wisconsin Press, 1987).

5. Jean François Lyotard, *The Postmodern Condition: A Report on Knowledge* (Minneapolis: University of Minnesota Press, 1984); David Harvey, *The Condition of Postmodernity* (Cambridge, Mass.: Basil Blackwell, 1989); and Mark Poster, *The Mode of Information: Poststructuralism and Social Context* (Chicago: University of Chicago Press, 1990).

6. Donna Haraway, "A Cyborg Manifesto: Science, Technology, and Socialist-Feminism in the Late Twentieth Century," in her *Simians, Cyborgs, and Women: The Reinvention of Nature* (New York: Routledge, 1991), chap. 8.

7. Thomas Hughes, "Modern and Postmodern Engineering," Seventh Annual Arthur Miller Lecture on Science and Ethics, MIT, April 1993.

8. On the influence of cybernetics and information theory on the social sciences see Steve J. Heims, *The Cybernetics Group* (Cambridge: MIT Press, 1991), and Paul Edwards, *The Closed World: Computers and the Politics of Discourse in Cold War America* (Cambridge: MIT Press, 1996). For a contemporary overview on information in biology see Walter M. Elsasser, *The Physical Foundation of Biology: An Analytical Study* (New York: Pergamon Press, 1958). On the cybernetic influence in evolutionary biology see Donna J. Haraway, "The High Cost of Information in Post–World War II Evolutionary Biology: Ergonomics, Semiotics, and the Sociobiology of Communication Systems," *The Philosophical Forum* 13 (Winter–Spring 1981–1982): 244–278, and *idem*, "Signs of Dominance: From Physiology to a Cybernetics of a Primate Society," *Studies in the History of Biology* 6 (1983): 129–219. On cybernetics' influence in ecology see Gregg Mirman, *The State of Nature: Ecology, Community, and American Social Thought, 1900–1950* (Chicago: University of Chicago Press, 1992), 5 and passim. An extensive literature search conducted in journals of endocrinology and immunology for the period 1950–1970 has revealed numerous articles that employed concepts from cybernetics and information; probably the best known is F. Macfarlane Burnet, *Enzyme, Antigen, and Virus: A Study of Macromolecular Pattern in Action* (Cambridge: Cambridge University Press, 1956), in which he devotes considerable space to discussion of the information concept in immunology. Embryology was particularly responsive to the information discourse; see, for example, P. Raven, *Oogenesis: The Storage of Developmental Information* (New York: Pergamon Press, 1961); C. H. Waddington, "Form and Information," in Waddington, ed., *Toward a Theoretical Biology*, 14 (Edinburgh: Edinburgh University Press, 1972); and Evelyn Fox Keller, "The Body of a New Machine: Situating the Organism between Telegraphs and Computers," in her *Refiguring Life: Changing Metaphors of Twentieth-Century Biology* (New York: Columbia University Press, 1995), 79–91. On Soviet cybernetics, see Slava Gerovitch, "Beyond the Rhetoric: The Construction of Soviet Cybernetics," second-year paper, Program in Science, Technology, and Society, MIT, 1994. Mark B. Adams, in "Molecular Answers in Soviet Genetics," a paper presented at the second Mellon Workshop, "Building Molecular Biology: Comparative Studies of Ideas, Institutions, and Practices," MIT, April 1992, showed that Soviet molecular biology derived its institutional legitimation from cybernetics. My search through the journal *Problems of Cybernetics* for the years 1960–1965 reveals that, as in the United States, the widespread cybernetic influence in biology was mainly

rhetorical. For an examination of the early reception of cybernetics see Lily E. Kay, "Cybernetics, Information, Life: The Emergence of Scriptural Representations of Heredity," *Historical Studies in the Physical and Biological Sciences* (Berkeley, Calif.: University of California Press, 1996).

9. Erwin Schrödinger, *What Is Life?* (Cambridge: Cambridge University Press, 1944). Schrödinger's proposal was incorporated into the historiography of molecular biology by Robert C. Olby, *The Path to the Double Helix* (London: Macmillan, 1974); Gunther S. Stent and Richard Calendar, *Molecular Genetics: An Introductory Narrative* (San Francisco: W. H. Freeman, 1978); Horace F. Judson, *The Eighth Day of Creation: The Makers of the Revolution in Biology* (New York: Simon & Schuster, 1979); and recently by Richard Doyle, "On Beyond Living: Rhetorics of Vitality and Post-Vitality in Molecular Biology" (Ph.D. diss., University of California, Berkeley, 1993). For a reassessment of Schrödinger's code, see Lily E. Kay, *Who Wrote the Book of Life? A History of the Genetic Code* (Stanford: Stanford University Press, forthcoming), chap. 2.

10. For a detailed analysis, see Lily E. Kay, "Who Wrote the Book of Life? Information and the Transformation of Molecular Biology, 1945–1955," *Science in Context* 8, no. 4 (1996): 24–48.

11. Arturo Rosenblueth, Norbert Wiener, and Julian Bigelow, "Behavior, Purpose, and Teleology," *Philosophy of Science* 10 (1943): 18–24. Kay, "Who Wrote the Book of Life?"; and Peter Galison, "The Ontology of the Enemy: Norbert Wiener and the Cybernetic Vision," *Critical Inquiry* 21 (1994): 228–266. See also Steve J. Heims, *John von Neumann and Norbert Wiener: From Mathematics to the Technologies of Life and Death* (Cambridge: MIT Press, 1980).

12. Massachusetts Institute of Technology Archives (MIT), Wiener Papers, MC 22, box 2.66; Wiener to von Neumann, 17 October 1944; and von Neumann to Wiener, 16 December 1944.

13. Ibid., box 2.72, von Neumann to Wiener, 29 November 1946; Library of Congress (LC), von Neumann Papers, box 7.1, Spiegelman to von Neumann, 3 December 1949, and von Neumann to Speigelman, 10 December 1946. John von Neumann, "The General and Logical Theory of Automata," in *Cerebral Mechanism in Behavior* (New York: Hafner Publishing Co., 1951), 1–32; and William Aspray, *John von Neumann and the Origins of Modern Computing* (Cambridge: MIT Press, 1990).

14. Norbert Wiener, *Cybernetics: Or Control and Communication in the Animal and the Machine* (New York: MIT Press and John Wiley, 1961, 2nd ed.), introduction and 39. Michel Foucault, *On the Order of Things* (New York: Vintage Books, 1970).

15. Wiener, *Cybernetics*, 41–42 and 93–94.

16. Claude Shannon, "The Mathematical Theory of Communication," *Bell System Technical Journal* 27, no. 3 and no. 4 (1948): 379–423 and 623–656; Claude Shannon and Warren Weaver, *The Mathematical Theory of Communication* (Urbana: University of Illinois Press, 1949).

17. Claude Shannon, "Communication Theory of Secrecy Systems," *Bell System Technical Journal* 28, no. 4 (1949): 656–715. David Kahn, *The Codebreakers: A Story of Secret Writing* (New York: Macmillan, 1967), 743–756. See also Paul Edwards, *The Closed World: Computers and the Politics of Discourse in Cold War America* (Cambridge: MIT Press, 1995).

18. Norbert Wiener, *The Human Use of Human Beings: Cybernetics and Society* (Boston: Houghton Mifflin, 1950), 103–109.

19. Henry Quastler, *Essays on the Use of Information Theory in Biology* (Urbana: University of Illinois Press, 1953). For a detailed account see Kay, *Who Wrote the Book of Life?* chap. 2.

20. Henry Quastler, "The Measure of Specificity," 41–74; Herman R. Branson, "Information Theory and the Structure of Proteins," 84–104, quotation on 84–85; Leroy Augustine, Herman Branson, and Eleanore B. Carver, "A Search for Intersymbol Influence in Protein Structure," 105–118; Sydney Dancoff and Henry Quastler, "Information Content and Error Rate in Living Things," 263–273; all in Quastler's *Information Theory in Biology.*

21. James D. Watson and Francis H. C. Crick, "Molecular Structure of Nucleic Acids: A Structure for Deoxyribose Nucleic Acid," *Nature* 171 (1953): 737–738.

22. B. Ephrussi, U. Leopold, J. D. Watson, and J. J. Weigle, "Terminology in Bacterial Genetics," *Nature* 171 (1953): 701.

23. James D. Watson and Francis H. C. Crick, "Genetical Implications of the Structure of Deoxyribonucleic Acid," *Nature* 171 (1953): 964–967, quote on 966.

24. Edward Yoxen, "Where Does Schrödinger's 'What Is Life?' Belong in the History of Molecular Biology?" *History of Science 17* (1979): 17–52; and Kay, *Who Wrote the Book of Life?* chap. 2.

25. See, for example, California Institute of Technology Archives (CIT), Delbrück Papers, box 8.21, Gamow to Delbrück, 13–22 April 1941. American Institute of Physics (AIP), George Gamow, Oral History; interview by Charles Wiener, 4 April 1968, 76.

26. Ibid., 80.

27. Kahn, *The Codebreakers*; Fletcher Pratt, *Secret and Urgent: The Story of Codes and Ciphers* (New York: Blue Ribbon Books, 1942); Gordon Welchman, *The Hut Six Story: Breaking of the Enigma Codes* (New York: McGraw-Hill, 1982); F. H. Hinsley and Alan Stripp, eds., *Codebreakers: The Inside Story of Bletchley Park* (New York: Oxford University Press, 1993); James Bamford, *The Puzzle Palace: A Report on America's Most Secret Agency* (Boston: Houghton Mifflin, 1982); Jeffrey Richelson, *The U.S. Intelligence Community* (Cambridge, Mass.: Ballinger, 1989, 2nd ed.). See also G. J. A. O'Toole, *Honorable Treachery: Intelligence, Espionage, and Covert Action from the American Revolution to the CIA* (New York: Atlantic Monthly Press, 1991).

28. George Gamow, "Possible Relation between Deoxyribonucleic Acid and Protein Structure," *Nature* 173 (1954): 318; and Gamow, "Possible Mathematical Relation between Deoxyribonucleic Acid and Proteins," *Det Konelige Danske Videnskabernes Selkab, Biologiske Meddelesker* 22, no. 3 (1954): 1–13.

29. Oregon State University Archives (OSU), Ava Helen and Linus Pauling Papers, box 263.39; Gamow to Pauling, 22 October 1953; American Philosophical Society (APS), Chargaff Papers, Gamow folder, Chargaff to Gamow, 2 March 1954. Judson, *Eighth Day of Creation*, 250–253; Francis Crick, *What Mad Pursuit: A Personal View of Scientific Discovery* (New York: Basic Books, 1988), 94; Library of Congress (LC), Gamow Papers, box 7, Ycas folder (1954); Ycas to Gamow, 16 March 1954; Gamow to W. G. Parks (director of Gordon Research Conferences, AAAS), 5 April 1956.

30. LC, Gamow Papers, box 7, Ycas folder (1954), Gamow to Chargaff, 6 May 1994, 2.

31. Ibid., box 7.30, RNA Tie Club; Judson, *The Eighth Day of Creation*, 264–265; Crick, *What Mad Pursuit*, 95.

32. LC, Gamow Papers, box 7, Ycas folder (1954) passim. For a detailed review, see George Gamow, Alexander Rich, and Martynas Ycas, "The Problem of Information Transfer from Nucleic Acids to Proteins," *Advances in Biological and Medical Physics*, 4 (New York: Academic Press, 1956), 41–51. For a technical analysis of these codes, see Sahotra Sarkar, "Reductionism in Molecular Biology" (Ph.D. diss., University of Chicago, 1989).

33. George Gamow and Nicholas Metropolis, "Numerolog of Polypeptide Chains," *Science* 120 (1954): 779–780.

34. Robert S. Ladley, "Digital Computational Methods in Symbolic Logic, with Examples in Biochemistry," *Proceedings of the National Academy of Sciences* 41 (1955), 498–511; quote on 498.

35. Gamow, Rich, and Ycas, "The Problem of Information Transfer." CIT, Delbrück Papers, box 30.5, Sydney Brenner, "On the Impossibility of All Overlapping Codes," note to the RNA Tie Club, September 1956; and Brenner, "On the Impossibility of All Overlapping Triplet Codes in Information Transfer from Nucleic Acid to Proteins," *Proceedings of the National Academy of Sciences* 43 (1957): 687–694.

36. Gamow, Rich, and Ycas, "The Problem of Information Transfer," 64–66.

37. George Gamow and Martynas Ycas, "Statistical Correlation of Protein and Ribonucleic Acid Composition," *Proceedings of the National Academy of Sciences* 41 (1955), 1011–1019.

38. On the nearly daily communications about their study see LC, Gamow Papers, box 7, Ycas folder (1955), passim; Gamow to Ycas, 1 August 1955; Ycas to Gamow, 5 August 1955.

39. Ibid., Ycas to Crick, 15 February 1955.

40. Francis Crick, "On Degenerate Templates and the Adaptor Hypothesis," a note to the RNA Tie Club, undated but mid-January 1955. I am grateful to Crick for sending me a copy of this paper.

41. Martynas Ycas, *The Biological Code* (New York: John Wiley, 1969), 31.

42. Francis Crick, Leslie Orgel, and John Griffith, "Codes without Commas," *Proceedings of the National Academy of Sciences* 43 (1957): 416–421; Crick, *What Mad Pursuit*, 99–100. University of California, San Diego (UCSD), Szilard Papers, MSS.32, box 26.3; Leo Szilard, "How Amino Acids Read the Nucleotide Code?" 1–14 plus appendix, 7 June 1957. For a detailed account see Kay, *Who Wrote the Book of Life?* chap. 4.

43. Stuart W. Leslie, *The Cold War and American Science: The Military-Industrial Complex* (New York: Columbia University Press, 1993), passim.

44. Solomon W. Golomb, Basil Gordon, and Lloyd D. Welch, "Comma-Free Codes," *Canadian Journal of Mathematics* 10 (1958): 202–209.

45. Ibid., 209; and Brockway McMillan, "Two Inequalities Implied by Unique Decipherability," *IRE Transactions on Information Theory* 2 (1956): 115–116. On this two-way traffic see Solomon W. Golomb, "Efficient Coding for the Desoxyribonucleic Channel," *Proceedings of Symposia in Applied Mathematics* 14 (1962): 87–100, where he refers to E. T. Jayne, "Note on Unique Decipherability," *IRE Transactions on Information Theory* IT-5 (1959): 98–102, and to "Recent Results in Comma-Free Codes," Research Summary, Jet Propulsion Laboratory, California Institute of Technology, 15 February 1961. On the meaning and significance of boundary objects see Joan Fujimura, "The Molecular Bandwagon in Cancer Research: Where Social Worlds Meet," *Social Problems* 35 (1988): 261–283; and Susan L. Star and James R. Griesemer, "Institutional Ecology, 'Translations,' and Boundary Objects: Amateurs and Professionals in Berkeley's Museum of Vertebrate Zoology," *Social Studies of Science* 19 (1989): 387–420.

46. Solomon W. Golomb, Lloyd R. Welch, and Max Delbrück, "Construction and Properties of Comma-Free Codes," *Biologiske Meddelelsket Det Kongelige Danske Videnskabernes Selskab* 23 (1958): 1–34; quote on unpaginated abstract.

47. Ibid., 11. See also, Ycas, *The Biological Code*, 33.

48. Max Delbrück and Gunther Stent, "On Mechanisms of DNA Replication," in William D. McElroy and Bentley Glass, eds., *The Chemical Basis of Heredity* (Baltimore: Johns Hopkins University Press, 1957), 699–736; quote on 730.

49. CIT, Delbrück Papers, box 20.3, Delbrück to Sinsheimer, 21 November 1955.

50. Sinsheimer to Kay, 19 July 1994, personal communication (letter). I am grateful to Sinsheimer for providing me with this information. See also Sinsheimer, *The Strands of Life*, 56.

51. Robert Sinsheimer, "Is the Nucleic Acid Message in a Two-Symbol Code?" *Journal of Molecular Biology* 1 (1959): 218–220.

52. Francis H. C. Crick, "The Present Position of the Coding Problem," *Brookhaven National Laboratory Symposia*, June (1959), 35–39.

53. John A. Osmundsen, "New Way to Read Life's Code Found," *New York Times*, 7 April 1961. The report refers to Golomb's paper, "Efficient Coding for the Desoxyribonucleic Channel" (note 45).

54. Carl E. Woese, *The Genetic Code: The Molecular Basis for Genetic Expression* (New York: Harper & Row, 1967), 17.

55. For example, Donald Fleming, "Emigre Physicists and the Biological Revolution," *Perspectives in American History* 2 (1968): 152–189; Olby, *The Path to the Double Helix*, section 4; Robert Kohler, "Management of Science: The Experience of Warren Weaver and the Rockefeller Foundation Programme in Molecular Biology," *Minerva* 14 (1976): 279–306; Pnina Abir-Am, "The Discourse of Physical Power and Biological Knowledge in the 1930's: A Reappraisal of the Rockefeller Foundation's 'Policy' in Molecular Biology," *Social Studies of Science* 12 (1982): 341–382; Lily E. Kay, "Conceptual Models and Analytical Tools: The Biology of Physicist Max Delbrück," *Journal of the History of Biology* 18 (1985): 207–246; Donna Haraway, "The Biological Enterprize: Sex, Mind, and Profit from Human Engineering to Sociobiology," *Radical History Review* 29 (1979): 206–237; Evelyn Fox Keller, "Physics and the Emergence of Molecular Biology: A History of Cognitive and Political Synergy," *Journal of the History of Biology* 23 (1990): 389–409.

56. Leroy Hood, "Biology and Medicine in the Twenty-First Century," in Daniel J. Kevles and Leroy Hood, eds., *The Code of Codes* (Cambridge: Harvard University Press, 1992), 136–163.

57. Walter Gilbert, "A Vision of the Grail," in *The Code of Codes*, 96.

Notes on Contributors

Atsushi Akera is a lecturer in the Department of Science and Technology Studies at Rensselaer Polytechnic Institute. His completed his dissertation, "Calculating the Natural World: Scientists, Engineers, and Computers in the United States, 1937–1968," at the University of Pennsylvania in 1998.

Harvey Brooks is Benjamin Peirce Professor of Technology and Public Policy Emeritus and Professor of Applied Physics (Emeritus) at Harvard University. He is currently Senior Research Associate in the Science, Technology, and Public Policy Program of the Belfer Center for Science and International Affairs in the John F. Kennedy School of Government at Harvard.

Glenn E. Bugos is president of The Prologue Group, a business history consultancy in Redwood City, California. His academic work focuses on engineering management in modern aerospace, construction, and biotechnology. He is the author of *Engineering the F-4 Phantom II: Parts into Systems* (1996).

Davis Dyer is a historian with The Winthrop Group. He is the author of *TRW: Pioneering Technology and Innovation since 1900* (1998) and, most recently, coeditor (with Alan Brinkley) of *The Reader's Companion to the American Presidency* (2000).

Paul N. Edwards is associate professor in the School of Information at the University of Michigan and chair of the Science Program in the university's Residential College. He is the author of *The Closed World: Computers and the Politics of Discourse in Cold War America* (1996). His latest project is tentatively titled *The World in a Machine: Computer Models, Data Networks, and Global Atmospheric Politics.*

Gabrielle Hecht is associate professor in the Department of History and in the Residential College at the University of Michigan. She is the author of *The Radiance of France: Nuclear Power and National Identity after*

World War II (1998), winner of the Herbert Baxter Adams Prize of the American Historical Association.

David A. Hounshell is Henry R. Luce Professor of Technology and Social Change at Carnegie Mellon University. His publications include *From the American System to Mass Production, 1800–1932* (1984), winner of the Dexter Prize of the Society for the History of Technology, and (with John Kenly Smith, Jr.) *Science and Corporate Strategy: Du Pont R&D, 1902–1980* (1988), winner of the Newcomen Prize of the Business History Conference.

Agatha C. Hughes (1924–1997) was a dedicated editor, ceramicist, and gardener. She coedited (with Thomas P. Hughes) *Lewis Mumford: Public Intellectual* (1990) and published essays in *Technology and Culture* and *American Heritage*. She taught at the Chestnut Hill Academy.

Thomas P. Hughes is Mellon Professor at the University of Pennsylvania (Emeritus) and Visiting Professor at the Massachusetts Institute of Technology, Stanford University, and the Royal Institute of Technology (Stockholm). His most recent book is *Rescuing Prometheus* (1998).

David Jardini is director of Hatch Beddows, a global metals industry strategy group. His first book will be *Thinking through the Cold War: RAND, National Security, and Domestic Policy, 1945–1973* (2001).

Stephen B. Johnson is an assistant professor in the Department of Space Studies at the University of North Dakota and the editor of *Quest, the History of Spaceflight Quarterly*. He has two books soon to be published on the development of systems management methods: *The United States Air Force and the Culture of Innovation, 1945–1965* and *The Secret of Apollo: R&D Management in American and European Space Programs, 1945–1979.*

Arne Kaijser is professor in the Department of History of Science and Technology, Royal Institute of Technology, Stockholm. His main research interest is comparative studies of the historical development of large technical systems. His books include *Stadens ljus: Etableringen av de första svenska gasverken* (1986) and *I fädrens spår: Den svenska infrastrukturens historiska utveckling och framtida utmaningar* (1994); he is the coeditor (with Pär Blomkvist) of *Den konstruerade världen: Stora tekniska system i historiskt perspektiv* (1998).

Lily Kay was formerly associate professor at MIT and is currently affiliated with Harvard University. She is the author of *The Molecular Vision of Life: Caltech, the Rockefeller Foundation, and the Rise of the New Biology* (1993) and *Who Wrote the Book of Life? A History of the Genetic Code* (2000). As a Guggenheim Fellow, she is currently working on a book on the history of neuroscience.

Roger Levien is the founder of Strategy & Innovation Consulting, which supports senior executives in developing longer-term strategic direction. He has worked at The RAND Corporation, the International Institute for Applied Systems Analysis (where he served as general director from 1975 through 1981), and the Xerox Corporation. He is the author of *Taking Technology to Market* (1997) and coauthor of *The Emerging Technology: Instructional Uses of the Computer in Higher Education* (1972) and *R&D Management: Methods Used by Federal Agencies* (1975).

Donald MacKenzie holds a Personal Chair in Sociology at the University of Edinburgh. He is the author of *Inventing Accuracy: A Historical Sociology of Nuclear Missile Guidance* (1992), which won the 1993 Ludwig Fleck Prize of the Society for the Social Studies of Science and was co-winner of the 1993 Robert K. Merton Award of the American Sociological Association. He is currently working on a book on computing, risk, and the sociology of proof.

Alan McDonald is the program officer of the Environmentally Compatible Energy Strategies Project at the International Institute for Applied Systems Analysis. He has worked on refueling system design and reactor safety policy for General Electric's Fast Breeder Reactor Department, on energy facility siting procedures for the California Energy Resources Conservation and Development Commission, and as an independent consultant for automotive manufacturers on automobile safety standards. From 1982 to 1997 he was Executive Directive of the U.S. Committee for IIASA at the American Academy of Arts and Sciences.

David A. Mindell is Dibner Assistant Professor of the History of Engineering and Manufacturing in the Program in Science, Technology, and Society at MIT and a visiting investigator in the Deep Submergence Laboratory of the Woods Hole Oceanographic Institution. His is the author of *War, Technology, and Experience aboard the USS*

Monitor (1999), and he is now working on a history of feedback control and computing from 1916 to 1945.

Erik Rau is visiting assistant professor of history at Drexel University. He earned his Ph.D. in the history of technology and engineering at the University of Pennsylvania in 1999 with a dissertation on operational research in the United States and Great Britain during World War II.

Joar Tiberg is a journalist currently working for *Ordfront Magasin* in Stockholm. While a student in the Department of History of Science and Technology of the Royal Institute of Technology he wrote on the history of futures studies and completed a masters dissertation on the debate over nuclear waste disposal in Sweden.

Index